ENDOCRINOLOGY AND METABOLISM CLINICS OF NORTH AMERICA

Growth Hormone

GUEST EDITOR
Ken K.Y. Ho, FRACP, MD

CONSULTING EDITOR
Derek LeRoith, MD, PhD

March 2007 • Volume 36 • Number 1

SAUNDERS

An Imprint of Elsevier, Inc.
PHILADELPHIA   LONDON   TORONTO   MONTREAL   SYDNEY   TOKYO

W.B. SAUNDERS COMPANY
A Division of Elsevier Inc.

1600 John F. Kennedy Boulevard • Suite 1800 • Philadelphia, Pennsylvania 19103-2899

http://www.theclinics.com

ENDOCRINOLOGY AND METABOLISM
CLINICS OF NORTH AMERICA
March 2007
Editor: Rachel Glover

Volume 36, Number 1
ISSN 0889-8529
ISBN 1-4160-3871-X
978-1-4160-3871-9

Copyright © 2007 by Elsevier Inc. All rights reserved. No part of this publication may be reproduced or transmitted in any form or by any means, electronic or mechanical, including photocopy, recording, or any information retrieval system, without written permission from the Publisher.

Single photocopies of single articles may be made for personal use as allowed by national copyright laws. Permission of the publisher and payment of a fee is required for all other photocopying, including multiple or systematic copying, copying for advertising or promotional purposes, resale, and all forms of document delivery. Special rates are available for educational institutions that wish to make photocopies for non-profit educational classroom use. Permissions may be sought directly from Elsevier's Rights Departments in Philadelphia, PA, USA. Phone: (+1) 215-239-3804; fax: (+1) 215-239-3805; e-mail: healthpermissions @elsevier.com. Requests may also be completed on-line via the Elsevier homepage (http://www.elsevier. com/locate/permissions). In the USA, users may clear permissions and make payments through the Copyright Clearance Center, Inc., 222 Rosewood Drive, Danvers, MA 01923, USA; phone: (+1) 978-750-8400, fax: (+1) 978-750-4744, and in the UK through the Copyright Licensing Agency Rapid Clearance Service (CLARCS), 90 Tottenham Court Road, London WIP 0LP, UK; phone: (+44) 171-436-5931; fax: (+44) 171-436-3986. Other countries may have a local reprographic rights agency for payments.

The ideas and opinions expressed in *Endocrinology and Metabolism Clinics of North America* do not necessarily reflect those of the Publisher. The Publisher does not assume any responsibility for any injury and/or damage to persons or property arising out of or related to any use of the material contained in this periodical. The reader is advised to check the appropriate medical literature and the product information currently provided by the manufacturer of each drug to be administered to verify the dosage, the method and duration of administration, or contraindications. It is the responsibility of the treating physician or other health care professional, relying on independent experience and knowledge of the patient, to determine drug dosages and the best treatment for the patient. Mention of any product in this issue should not be construed as endorsement by the contributors, editors, or the Publisher of the product or manufacturers' claims.

Endocrinology and Metabolism Clinics of North America (ISSN 0889-8529) is published quarterly by Elsevier Inc., 360 Park Avenue South, New York, NY 10010-1710. Months of publication are March, June, September, and December. Business and editorial offices: 1600 John F. Kennedy Boulevard, Suite 1800, Philadelphia, PA 19103-2899. Customer Service Office: 6277 Sea Harbor Drive, Orlando, FL 32887-4800. Periodicals postage paid at New York, NY and additional mailing offices. Subscription prices are USD 193 per year for US individuals, USD 319 per year for US institutions, USD 99 per year for US students and residents, USD 242 per year for Canadian individuals, USD 383 per year for Canadian institutions, USD 264 per year for international individuals, USD 383 per year for international institutions and USD 138 per year for Canadian and foreign students/residents. To receive student/resident rate, orders must be accompanied by name of affiliated institution, date of term, and the *signature* of program/residency coordinator on institution letterhead. Orders will be billed at individual rate until proof of status is received. Foreign air speed delivery is included in all *Clinics* subscription prices. All prices are subject to change without notice. POSTMASTER: Send address changes to *Endocrinology and Metabolism Clinics of North America*, Elsevier Periodicals Customer Service, 6277 Sea Harbor Drive, Orlando, FL 32887-4800. **Customer Service: (+1) 800-654-2452 (US). From outside of the US, call (+1) 407-345-4000; e-mail: hhspcs@harcourt.com.**

Reprints. For copies of 100 or more, of articles in this publication, please contact the Commercial Rights Department, Elsevier Inc., 360 Park Avenue South, New York, NY 10010-1710; phone: (+1) 212-633-3813; fax: (+1) 212-462-1935; e-mail: reprints@elsevier.com.

Endocrinology and Metabolism Clinics of North America is covered in *Index Medicus, EMBASE/Excerpta Medica, Current Contents/Clinical Medicine, Current Contents/Life Sciences, Science Citation Index, ISI/BIOMED, BIOSIS, and Chemical Abstracts.*

Printed in the United States of America.

GROWTH HORMONE

CONSULTING EDITOR

DEREK LeROITH, MD, PhD, Chief, Division of Endocrinology, Metabolism, and Bone Diseases, Mount Sinai School of Medicine, New York, New York

GUEST EDITOR

KEN K.Y. HO, FRACP, MD, Director, Pituitary Research Unit, Garvan Institute of Medical Research; Professor of Medicine, University of New South Wales; and Chairman, Department of Endocrinology, St. Vincent's Hospital, Darlinghurst, Sydney, Australia

CONTRIBUTORS

ANDREA F. ATTANASIO, MD, Cascina del Rosone, Agliano Terme, Italy

XAVIER BADIA, MD, PhD, Principle, Health Economics and Outcome Research, IMS Health, Barcelona, Spain

INDRANEEL BANERJEE, MBBS, MD, MRCPCH, Department of Pediatric Endocrinology, Royal Manchester Children's Hospital, Manchester, United Kingdom

ARIEL BARKAN, MD, Professor of Medicine and Neurosurgery, Division of Metabolism, Endocrinology and Diabetes, University of Michigan Medical Center, Ann Arbor, Michigan

MARTIN BIDLINGMAIER, MD, Head, Endocrine Research Laboratories, Medizinische Klinik-Innenstadt, Ludwig-Maximilians University, Munich, Germany

NILS BILLESTRUP, PhD, Steno Diabetes Center, Copenhagen, Denmark

PIERRE BOUGNÈRES, MD, PhD, Department of Pediatric Endocrinology, Hopital St. Vincent de Paul-Cochin; and Université Paris V-Réne Descartes, Paris, France

JENS S. CHRISTIANSEN, MD, DMSc, Medical Department M (Endocrinology and Diabetes) and Institute of Experimental Clinical Research, Aarhus University Hospital, Aarhus, Denmark

PETER E. CLAYTON, MBChB, BSc, MD, MRCP, FRCPCH, Professor of Child Health and Paediatric Endocrinology, Endocrine Science Research Group, Division of Human Development, School of Medicine, University of Manchester, Manchester, United Kingdom

DAVID R. CLEMMONS, MD, Professor of Medicine, Division of Endocrinology, University of North Carolina School of Medicine, University of North Carolina, Chapel Hill, North Carolina

KENNETH C. COPELAND, MD, Jonas Professor, Department of Pediatrics, University of Oklahoma College of Medicine, Oklahoma City, Oklahoma

VINCENT GOFFIN, PhD, Faculté de Médecine Necker; and Université Paris V-Réne Descartes, Paris, France

NAILA GOLDENBERG, MD, Lecturer/Research Fellow, Division of Metabolism, Endocrinology and Diabetes, University of Michigan Medical Center, Ann Arbor, Michigan

KEN K.Y. HO, FRACP, MD, Director, Pituitary Research Unit, Garvan Institute of Medical Research; Professor of Medicine, University of New South Wales; and Chairman, Department of Endocrinology, St. Vincent's Hospital, Sydney, Australia

GUDMUNDUR JOHANNSSON, MD, PhD, Department of Endocrinology, Sahlgrenska University Hospital and the Sahlgrenska Academy, Gothenburg University, Gothenburg, Sweden

JENS O.L. JØRGENSEN, MD, DMSc, Medical Department M (Endocrinology and Diabetes) and Institute of Experimental Clinical Research, Aarhus University Hospital, Aarhus, Denmark

MORTEN KRAG, MD, Medical Department M (Endocrinology and Diabetes) and Institute of Experimental Clinical Research, Aarhus University Hospital, Aarhus, Denmark

UDO J. MEINHARDT, MD, Pituitary Research Unit, Garvan Institute of Medical Research, Sydney, Australia

LOUISE MØLLER, MD, Medical Department M (Endocrinology and Diabetes) and Institute of Experimental Clinical Research, Aarhus University Hospital, Aarhus, Denmark

NIELS MØLLER, MD, Professor and Consultant Physician, Medical Research Laboratories, Medical Dep M., Århus University Hospital, Århus C, Denmark

PRIMUS E. MULLIS, MD, Professor of Paediatrics and Paediatric Endocrinology and Diabetology, Paediatric Endocrinology and Diabetology and Metabolism, University Children's Hospital, University of Berne, Bern, Switzerland

K. SREEKUMARAN NAIR, MD, PhD, Professor of Medicine, Division of Endocrinology, Mayo Clinic, Rochester, Minnesota

RALF NASS, MD, Research Assistant Professor, Division of Endocrinology and Metabolism, Department of Medicine, University of Virginia, Charlottesville, Virginia

JENNIFER PARK, MD, Resident, Department of Internal Medicine, University of Virginia, Charlottesville, Virginia

CHARMIAN A. QUIGLEY, MBBS, FRACP, Senior Clinical Research Physician, Lilly Research Laboratories, Indianapolis, Indiana

STEPHEN M. SHALET, MD, Department of Endocrinology, Christie Hospital, Manchester, United Kingdom

CHRISTIAN J. STRASBURGER, MD, Chief, Division of Clinical Endocrinology, Charité Universitätsmedizin Berlin, Germany

MICHAEL O. THORNER, MB, BS, DSc, FRCP, MACP, David C. Harrison Medical Teaching Professor of Internal Medicine, Division of Endocrinology and Metabolism, Department of Medicine, University of Virginia, Charlottesville, Virginia

SUSAN M. WEBB, MD, PhD, Associate Professor of Medicine, Department of Endocrinology, Hospital de Sant Pau, Autonomous University of Barcelona, Barcelona, Spain

GROWTH HORMONE

CONTENTS

Foreword xiii
Derek LeRoith

Preface xvii
Ken K.Y. Ho

The Growth Hormone Receptor in Growth 1
Pierre Bougnères and Vincent Goffin

> The growth hormone receptor (GHR) is a major effector of human growth. Functional variants of the GHR include very rare loss-of-function mutations (pathology) and very common polymorphisms (physiology). Recent experimental data have clarified the mechanisms through which mutations of the GHR or Stat5 lead to growth hormone insensitivity and major monogenic growth defects. Recent pharmacogenetic studies support that the response to growth-promoting administration of growth hormone is influenced by exon 3 polymorphism of the GHR.

Genetics of Growth Hormone Deficiency 17
Primus E. Mullis

> When a child is not following the normal, predicted growth curve, an evaluation for underlying illnesses and central nervous system abnormalities is required and appropriate consideration should be given to genetic defects causing growth hormone (GH) deficiency. This article focuses on the *GH* gene, the various gene alterations, and their possible impact on the pituitary gland. Transcription factors regulating pituitary gland development may cause multiple pituitary hormone deficiency but may present initially as GH deficiency. The role of two most important transcription factors, POU1F1 (Pit-1) and PROP 1, is discussed.

Factors Regulating Growth Hormone Secretion in Humans 37
Naila Goldenberg and Ariel Barkan

> Growth hormone (GH) secretion is pulsatile in nature in all species. The periodic pattern of GH release plays an important role in transmitting the GH message in a tissue-specific manner. The question of what regulates the pulsatile GH secretion pattern is an issue of not only theoretical interest but of considerable practical importance for designing different GH therapies for a variety of human diseases. This article provides a brief introductory overview of the different regulators of GH secretion and concentrates primarily on human studies.

Regulation of Growth Hormone Action by Gonadal Steroids 57
Udo J. Meinhardt and Ken K.Y. Ho

> Sex steroids modulate growth hormone (GH) secretion and action. Estrogen attenuates GH action in a dose- and route-dependent manner by inhibiting GH-regulated endocrine function of the liver. Testosterone amplifies the metabolic action of GH while exhibiting similar but independent effects of its own. The strong modulatory effect of gonadal steroids on GH responsiveness provides insights into the biologic basis of sexual dimorphism in growth, development, and body composition and practical information for the clinical endocrinologist in the treatment of hypopituitary patients.

Effects of Growth Hormone on Glucose and Fat Metabolism in Human Subjects 75
Jens O.L. Jørgensen, Louise Møller, Morten Krag, Nils Billestrup, and Jens S. Christiansen

> This article focuses on in vivo data from tests performed in normal subjects and in patients who had abnormal growth hormone (GH) status. Experimental data in human subjects demonstrate that GH acutely inhibits glucose disposal in skeletal muscle. At the same time GH stimulates the turnover and oxidation of free fatty acid (FFA), and experimental evidence suggests a causal link between elevated FFA levels and insulin resistance in skeletal muscle. Observational data in GH-deficient adults do not indicate that GH replacement is associated with significant impairment of glucose tolerance, but it is recommended that overdosing be avoided and glycemic control be monitored.

Growth Hormone Effects on Protein Metabolism 89
Niels Møller, Kenneth C. Copeland, and
K. Sreekumaran Nair

> Growth hormone (GH) has a pivotal role in regulating in vivo protein metabolism. GH enhances protein anabolism at the whole-body level, mainly by stimulating protein synthesis. It remains incompletely understood whether this important GH effect on

protein synthesis occurs in all tissues. This effect of GH may be different with acute versus chronic administration. These differences in the GH exposure may have different effects based not only on direct GH stimulation of protein synthesis but also the variable effects at the level of gene transcription that ultimately affect protein metabolism. Other GH effects are likely to be mediated by changes in various metabolites and hormones that also likely differ based on the duration of GH administration.

What Endocrinologists Should Know About Growth Hormone Measurements   101
Martin Bidlingmaier and Christian J. Strasburger

Determination of human growth hormone (GH) concentration in serum plays a key role in the diagnosis of GH deficiency and GH excess (acromegaly). Methods of measuring GH still lack standardization and show considerable between-method variability. Therefore, correct interpretation of GH test results requires knowledge of measurement techniques and awareness of potential problems in applying recommendations for cut-off values given in the literature. This article focuses on the molecular, structural, and methodologic background of the heterogeneity of assay results and on possible next steps toward standardization.

Value of Insulin-like Growth Factor System Markers in the Assessment of Growth Hormone Status   109
David R. Clemmons

Insulin-like growth factor-I (IGF-I) has been measured extensively in a variety of clinical settings. Total IGF-I frequently is used to assess the clinical impact of disorders of GH secretion and to monitor patients' response to therapy. It does not have sufficient precision to be used as a stand-alone test in the diagnosis of GH deficiency. Free IGF-I, IGF binding protein-3, or acid-labile subunit may provide useful information regarding GH secretion in specific conditions but are not superior to IGF-I for making the diagnosis of GH deficiency or acromegaly.

Growth Hormone Treatment of Non–Growth Hormone-Deficient Growth Disorders   131
Charmian A. Quigley

Although a large body of data on efficacy and safety of growth hormone (GH) treatment for various non–growth hormone-deficient (GHD) growth disorders has accumulated from a combination of clinical trial and postmarketing sources in the last 20 years or more, there remain limitations. Clinical trial data have the advantage of direct comparison of well-matched, randomized patient groups receiving treatment (or not) under comparable conditions and, as such, provide the highest quality evidence of efficacy. Clinical trials, however, are typically too small for any statistically valid assessment for safety, which is more comprehensively addressed using postmarketing data. Consequently, while the efficacy of GH treatment in children with non-GHD growth disorders has been solidly established and, based on the combination of the rigor of the clinical trial data and numerical power of the postmarketing data, no major concerns exist regarding safety, additional long-term data are required.

Growth Hormone and the Transition from Puberty into Adulthood 187
Andrea F. Attanasio and Stephen M. Shalet

With modern growth hormone (GH) replacement algorithms, children with a diagnosis of growth hormone deficiency achieve at the end of pediatric GH treatment an adult height that is on the average in the normal range. Recent experience with GH replacement in young adults with childhood-onset growth hormone deficiency, however, has shown that these patients present with variable degrees of somatic immaturity. As childhood GH treatment is discontinued when final height is attained, attention moves to the phase of somatic development that follows the end of longitudinal growth, called "transition," which had been excluded previously from consideration for either pediatric or adult GH replacement. This article reviews the changes taking place during this phase of development and their relevance for the attainment of adult body maturation. The critical role of GH in this process is described.

Management of Adult Growth Hormone Deficiency 203
Gudmundur Johannsson

The management of adults who have severe growth hormone deficiency (GHD) includes its recognition in susceptible patients, including all subjects who have any form of hypothalamic pituitary disorder. This article also focuses on the overall management of adults who have hypopituitarism and severe GHD. This includes aspects of diagnosis and management of growth hormone replacement therapy and the current status of its long-term safety. Some special circumstances, including the pregnant woman who has hypopituitarism, are discussed.

Quality of Life in Growth Hormone Deficiency and Acromegaly 221
Susan M. Webb and Xavier Badia

Quality of life (QoL) has emerged as an end point in the evaluation of adults with growth hormone deficiency and acromegaly. QoL is measured with questionnaires designed to be used in general population or any kind of disease (generic) or aimed at the specific dimensions affected in a determined condition; these latter ones are more likely to identify the impairments caused by the underlying disease and the benefits of treatment. QoL, which is severely impaired in adults with growth hormone deficiency, improves and normalizes after growth hormone replacement therapy and this effect is maintained over several years. Acromegalic patients also exhibit severe impairment of QoL, which despite improvement after successful therapy still remains below the reference values of normal population. QoL in these chronic endocrine diseases can be used as an outcome measure for clinical and therapeutic evaluation.

Growth Hormone Supplementation in the Elderly 233
Ralf Nass, Jennifer Park, and Michael O. Thorner

The imminent rapid increase in numbers of aging adults will significantly impact society and the health system in the coming years. The imperative is to allow older adults to maintain their independence and health for as long as possible and to prevent the ravages of chronic diseases and disabilities or the development of frailty. Significant hormonal changes are observed with aging (eg, decline in growth hormone [GH]), and strikingly similar changes in body composition are seen in GH-deficient adults and during aging. This article reviews some of the studies describing the effects of GH and GH secretagogues on body composition and functionality in the elderly. In addition, safety concerns and the need for further controlled studies of these therapies in the elderly are discussed.

Growth Hormone Treatment and Cancer Risk 247
Indraneel Banerjee and Peter E. Clayton

Increasing numbers of children receive growth hormone (GH) to treat a range of growth disorders, including those rendered GH deficient (GHD) by tumors or their treatment. Young persons with persistent growth hormone deficiency (GHD) and adults with severe GHD are also eligible to receive GH treatment. As in vitro and in vivo studies and epidemiologic observations provide some evidence that the GH—insulin like growth factor-I (IGF-I) axis is associated with tumorigenesis, it is important to assess, in practice, the incidence of tumors related to GH treatment. Reassuringly, surveillance studies in large cohorts of children and in smaller cohorts of adults indicate that GH is not associated with an increased incidence of tumor occurrence or recurrence. Nevertheless, all children who have received GH, in particular cancer survivors and those receiving GH in adulthood, should be in surveillance programs to assess whether an increased rate of late-onset and rare tumours may occur.

Index 265

FORTHCOMING ISSUES

June 2007
Andrology
Ronald Tamler, MD, Guest Editor

September 2007
Thyroid Disorders
Kenneth Burman, MD, Guest Editor

December 2007
Endocrinopathies of Transplant Medicine
Tracy L. Breen, MD, Guest Editor

RECENT ISSUES

December 2006
Acute Endocrinology
Greet Van den Berghe, MD, PhD, Guest Editor

September 2006
Impaired Glucose Tolerance and Cardiovascular Disease
Willa A. Hseuh, MD, Preethi Srikanthan, MD, and Christopher J. Lyon, PhD, Guest Editors

June 2006
Molecular Basis of Inherited Pancreatic Disorders
Markus M. Lerch, MD, FRCP,
Thomas Griesbacher, MD,
David C. Whitcomb, MD, PhD, Guest Editors

THE CLINICS ARE NOW AVAILABLE ONLINE!

Access your subscription at:
http://www.theclinics.com

Foreword

Derek LeRoith, MD, PhD
Consulting Editor

As a leading endocrinologist in Australia, with expertise in pituitary gland physiology and a special interest in growth hormone (GH) physiology, Dr. Ken Ho was recruited to edit this issue on GH. As the reader will rapidly see, he has performed this task remarkably well!

The issue begins with an article by Pierre Bougnères and Vincent Goffin on the role of the GH receptor (GHR) in growth. They begin their article with a description of the GHR signaling pathways and then describe mutations of the receptor that affect growth. These include failure of expression, failure of activation of the receptor, and mutations that result in a failure to activate signaling pathways immediately distal to the receptor, namely the Jak/STAT-5 activation. More recently, STAT-5 mutations that cause growth retardation have also been described. GHR polymorphisms have also been pursued, and one, a polymorphism with a deletion of exon 3, appears to be common in Caucasians, although the exact function consequence is still not well defined.

The genetics of GH deficiency is discussed by Primus E. Mullis. The development of the pituitary gland is complex, because there must be differentiation that leads to multiple cell types producing and secreting the array of pituitary hormones. POU1F1 (PIT 1) and PROP 1 are two very important transcription factors. Mutational defects in these transcription factors affect GH-, prolactin (Prl)-, and, to some degree, thyroid-stimulating hormone (TSH)-producing cells.

Isolated GH deficiency maybe sporadic or familial. There are two forms that have autosomal recessive inheritance (IGHD type IA, IB), as well as autosomal dominant (IGHD type II), and X-linked (IGHD III) forms. The variation in clinical presentation depends on the genetic defect, which is discussed in detail in this article.

The major known regulators of GH secretion are GH releasing hormone, somatostatin, and ghrelin. However, the pulsatile secretion and the sexual dimorphism are also produced by other factors, including sex steroids. Sleep has powerful effects on the pattern of secretion, and negative feedback systems, both ultra-short feedback loops and short-feedback loops, are functionally important. Insulin-like growth factors (IGFs) are powerful negative regulators, although their mode of action is still unclear. Nutritional influences such as serum levels of glucose and free fatty acids (FFAs) also feed back to regulate GH secretion. Naila Goldenberg and Ariel Barkan, who have contributed to many of the human studies in this area, discuss these aspects.

Udo Meinhardt and Ken Ho discuss the effect of gonadal steroids on GH secretion and GH action. They describe in more detail the effect of androgen and estradiol on GH. Androgens must be aromatized to estrogen, which then stimulates the secretion of GH, an effect that may explain the increased GH secretion seen during puberty, and perhaps explain some of the decline of GH levels after menopause. On the other hand, although gonadal steroids may enhance GH secretion, when taken in pharmacological doses, they interfere with the ability of GH to induce IGF-1 synthesis in the liver and result in lower circulating IGF-1 levels. This not the case with testosterone, which may enhance the action of GH.

Jørgensen, Møller, Krag, and Christiansen address the issue of GH's effect on carbohydrate and fat metabolism. GH is lipolytic on fat cells and inhibits insulin's effect on other tissues. This often results in abnormalities in carbohydrate and fat metabolism in states of altered GH secretion. Thus, acromegalics often have a worsening of diabetes, and untreated GH-deficient children show a tendency to fasting hypoglycemia. On the other hand, GH-deficient children develop abdominal obesity, and GH replacement reduces fat stores. However, the mechanisms whereby these effects are brought about are largely poorly understood. One possible mechanism that is fairly widely accepted is the lipolytic effect of GH, which leads to increased FFAs; this is known to inhibit insulin action in cells and tissues and may explain the anti-insulin effect of GH.

Møller, Copeland, and Nair, on the other hand, discuss the effects of GH on protein metabolism. GH is undoubtedly an anabolic hormone and has marked effects on tissues such as bone and muscle. Its effects are mediated by both IGF-1 and insulin, and as they point out, the sensitivity of various tissues to these factors plays a role in the anabolic response to GH. Most importantly, GH levels and action are affected by aging, and many of the changes related to aging maybe secondary to a lack of GH or tissue insensitivity.

Bidlingmaier and Strasburger discuss the issues involved in GH measurements. There is hetereogeneity in the GH molecules, effects of GH-binding protein present in serum, and inconsistencies in GH measurements between laboratories. Although this makes comparisons between clinical studies very difficult, as they point out, each individual laboratory has its own cutoff level, which should suffice for the endocrinologist to ensure adequate care for individual patients. David Clemmons, on the other hand, approaches the question of assessing GH status by the use of markers from the IGF system. In GH deficiency in children and young adults, total serum IGF-1 is low; however a GH stimulation test is still a necessity, because a normal total IGF-1 test does exclude GH deficiency. In acromegaly, an elevated total IGF-1 is consistently found, and a normal level excludes the diagnosis. Although IGF-1 is widely used, GH measurements—random and, particularly, stimulated levels—remain the "gold standard," because of more accumulated data with the GH levels in various conditions.

Management of various conditions using recombinant human GH (rhGH) is discussed in a number of articles. In the first of these, Charmian Quigley discusses management of non–GH-deficient children with rhGH. There are quite a number of conditions that respond to rhGH with increased height. These include chronic renal insufficiency, Turner's and SHOX deficiency, idiopathic short stature, and children born small for gestational age. Although rhGH therapy is not approved in all cases, when it is used by pediatric endocrinologists, very careful monitoring is critical. Andrea Attanasio and Steve Shalet discuss the transition period between adolescence and adulthood. Although the effect of rhGH on height may end after puberty and young adulthood, the importance of GH in body composition and other features does not end, and, as they point out, one should allow adult GH deficiency to occur in these cases. Re-examination of the GH/IGF-1 status should be made to help in therapeutic decisions. This article is followed naturally by the article by Gudmundur Johannsson on adult GH deficiency and rhGH therapy. Patients should be evaluated by measuring GH and IGF-1 status, and serum IGF-1 levels are used to monitor therapy. Nass, Park, and Thorner address a more taxing issue, namely GH replacement in the elderly. There is compelling evidence that GH levels decline with aging, and there is also muscle wasting at this age, which leads to musculo-skeletal problems. rhGH for GH secretagogues could restore muscle strength by inhibiting the loss of muscle. However, as noted, there are significant side effects to rhGH therapy, and more studies are necessary before making recommendations for its use in the aging population.

Critical issues in GH therapies are raised when Webb and Badia present information on quality of life (QOL) in GH-deficiency and acromegaly patients. As discussed, there are studies on QOL for both situations that demonstrate improvements upon therapy. Physicians should be cognizant of these important issues when making decisions that guide therapy.

Finally, in an article by Banerjee and Clayton, the oft-debated issue of whether GH therapy aggravates cancer is discussed. Their conclusions relate to GH therapy in children with previously treated tumors that led to the presence of GH deficiency. Here, treatment is indicated. However, after treatment of malignant tumors, a year's grace should be instituted before treatment of GH deficiency that maybe present. In addition, in adults receiving GH replacement, although the relationship between GH therapy and tumor growth is still under intensive study, careful monitoring is mandated, and total IGF-1 serum levels should be used to ensure the right dose of GH.

As stated above, Dr. Ho has created a very scholarly issue by recruiting world experts to address these important concerns related to GH.

Derek LeRoith, MD, PhD
Division of Endocrinology, Metabolism, and Bone Diseases
Mount Sinai School of Medicine
One Gustave L. Levy Place, Box 1055
Atran 4-36
New York, NY 10029-6574, USA

E-mail address: derek.leroith@mssm.edu

Preface

Ken K.Y. Ho, FRACP, MD
Guest Editor

This issue of the *Endocrinology and Metabolism Clinics of North America* reviews the mechanistic, physiologic, diagnostic, and therapeutic aspects of growth hormone (GH) that are relevant to clinical practice. The past few decades have seen major strides in our understanding of the molecular mechanisms of GH action and its physiologic role in adult life, the latter having been brought about by the abundant availability of recombinant hormone allowing critical evaluation of the benefits of replacement and supplementary therapy in children and adults.

The first two articles in this issue deal with the molecular and genetic mechanisms underlying GH-dependent growth failure. Bougnères and Goffin review the structure and function of the human GH receptor and polymorphisms that display variable responsiveness to GH. Mullis reviews the genetics of GH deficiency, covering diverse causes from mutation of transcription factors that regulate pituitary gland development to gene deletions and splice mutations, causing major disruption to the secretory process and affecting somatotroph viability.

Four articles are dedicated to GH physiology. Goldenberg and Barkan provide an excellent review of factors regulating GH secretion in humans, while Meinhardt and Ho provide an update on factors regulating the action of GH, with emphasis on the modulatory roles of gonadal steroids. With the recognition that GH plays a major role in the metabolic process throughout life, the contribution from Jørgensen and colleagues covers regulation of lipid and carbohydrate metabolism. The article by Nair and colleagues is a complementary perspective on protein metabolism, with data drawn from elegant human studies.

Two articles are devoted to the diagnosis of GH disorders. Bidlingmaier and Strasburger present a concise rendition on what endocrinologists should know about GH measurements, reviewing important information on standards, assay design, sensitivity, and specificity. Clemmons gives an authoritative perspective on the value of insulin-like growth factor-1 system markers in the assessment of GH status.

The last section, the largest in this issue, contains six articles devoted to management, three of which address GH deficiency. Attanasio and Shalet underscore the important role of GH in somatic maturation after epiphysial closure. Johannsson provides a comprehensive update on the long-term safety and efficacy of GH replacement in adults. Webb's timely contribution on quality of life assessment is a salutary reminder to the practicing endocrinologist of the importance of assessing this aspect of health in patients with GH deficiency or acromegaly. Clayton provides a balanced review on the safety of GH treatment in relation to cancer risk. Finally, this issue contains excellent reviews on two controversial areas of GH use: the elderly (Thorner and Nass) and non–GH-deficient short stature (Quigley). Quigley has extensively reviewed the global experience on five causes, the indications for which have been approved by the US Food and Drug Adminisration. These are idiopathic short stature, Turner syndrome, chronic renal failure, Short Stature Homeobox-containing gene deficiency, and children born small for gestational age. Her excellent and comprehensive review includes consensus statements from key professional societies on these conditions.

I wish to thank the authors of these articles for their time and effort in making this issue worthy of a place in the library of the practicing endocrinologist. I also wish to thank Rachel Glover and her staff at Elsevier for their superb support.

Ken K.Y. Ho, FRACP, MD
Department of Endocrinology
St. Vincent's Hospital
Sydney, NSW 2010, Australia

Pituitary Research Unit
Garvan Institute of Medical Research
384 Victoria Street, Darlinghurst
Sydney, NSW 2010, Australia

E-mail address: k.ho@garvan.org.au

The Growth Hormone Receptor in Growth

Pierre Bougnères, MD, PhD[a,c,*], Vincent Goffin, PhD[b,c]

[a]*Inserm, Unit 561, Department of Pediatric Endocrinology, Hôpital St Vincent de Paul-Cochin, 82 Avenue Denfert Rochereau, 75014 Paris, France*
[b]*Inserm, Unit 808, Faculté de Médecine Necker, 75015 Paris, France*
[c]*Université Paris V-René Descartes, Paris, France*

The growth hormone receptor (GHR) is a major effector of growth, as demonstrated by the consequences of loss-of-function mutations of this receptor. Without a functional GHR, the final stature of patients does not exceed 70% to 80% of normal. In the normal population, people can inherit GHRs of different molecular structures depending on assortments of coding polymorphisms of the *GHR* gene. The GHR mediates the effects of growth hormone (GH) on target tissues. The GHR also produces the circulating GH binding protein (GHBP), a truncated monomer of the receptor that corresponds to the ectodomain that carries the GH binding domain. The functional consequences of rare mutations (nature mistakes) and of frequent polymorphisms (issued from human genomic evolution) of the *GHR* gene are the subject of this short review.

Overview of growth hormone receptor signaling

The GHR is a member of the class I hematopoietic cytokine receptor superfamily [1], which currently includes more than 40 members. In humans, the GHR protein consists of an extracellular domain of 246 amino acids (aa), a single transmembrane domain (24 aa), and a cytoplasmic domain of 350 aa, for a total length of 620 aa [2]. It is established that the active form of the receptor is a receptor homodimer [3,4]. Like all cytokine receptors, the GHR is devoid of enzymatic activity and mediates GH-induced actions by way of the activation of associated kinases, which have been

---

* Corresponding author. Inserm, Unit 561, Department of Pediatric Endocrinology, Hôpital St Vincent de Paul-Cochin, 82 Avenue Denfert Rochereau, 75014 Paris, France.
  E-mail address: pierre.bougneres@wanadoo.fr (P. Bougnères).

identified in the past decade. Soon after its discovery, the Janus tyrosine kinase Jak2 was identified as a major component of GHR signaling [5]. Together with MAP kinase and phosphatidylinositol-3 (PI-3) kinase, the pathways involving these three kinases are now assumed to account for most signaling events triggered by the GHR [6]. Soon after these pathways were identified, mutagenesis and signaling studies were performed to identify cytoplasmic subdomains of the GHR individually required for their activation (Fig. 1). Like for all cytokine receptors, the membrane proximal sequence rapidly appeared to be an absolute requirement to get a functional receptor. This region encompasses "box 1," a proline-rich domain (aa 276–287) that is involved in the interaction of the GHR with the associated Janus kinase. It is of interest that this membrane proximal domain is also required for the activation of MAP and PI-3 kinase pathways. The rest of the receptor, especially the distal (C-terminal) domain, also plays a critical role in GHR signaling. It is tyrosine phosphorylated at many sites by Jak2 and

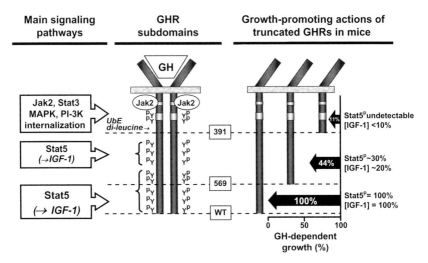

Fig. 1. Structure-function representation of the GHR related to growth-promoting actions. The GHR is schematically represented as a homodimer bound to its ligand, GH. The proximal domain contains box 1 (interacting with Jak2) and box 2 (both shown in light gray), and internalization motifs (UbE and di-leucine). Tyrosine residues phosphorylated by Jak2 are symbolized as Y^P. Numbers 391 and 569 refer to the truncated GHR expressed in knock-in mice. Mapping of GHR cytoplasmic domains required for the activation of the main signaling pathways is shown on the left. The C-terminal domain (including the five distal phosphotyrosines) is essential for full activation of the Jak2/Stat5/IGF-1 cascade. The relationships between these domains and features related to growth as revealed by the analysis of mice expressing GHR truncated after residues 569 or 391 are shown on the right. These parameters include body growth (expressed in percentage of GH-dependent growth), Stat5 activation (monitored by Stat5 tyrosine phosphorylation, $Stat5^P$), and circulating IGF-1 levels. See text for additional details. IGF, insulin-like growth factor; WT, wild-type. (*Data from* Rowland JE, Lichanska AM, Kerr LM, et al. In vivo analysis of growth hormone receptor signaling domains and their associated transcripts. Mol Cell Biol 2005;25:66–77.)

has been shown to recruit various partners by way of interactions involving Stat SH2 domains. These partners include signal transducers and activators of transcription (Stat), SHC, or negative regulators of GHR signaling such as tyrosine phosphatases (SHP1, SHP2) and suppressors of cytokine signaling (SOCS, CIS), which down-regulate the cascades activated by the receptor or serve as adapters for other signaling molecules. Older studies and recent studies converge to the evidence that Jak2 is the central molecule in GHR signaling. In addition to initiating downstream signaling, Jak2 also impacts on multiple features of the GHR that ultimately modulate its activity, such as receptor stability [7] or proteolysis leading to the generation of GHBP, a process also known as receptor shedding [8]. Receptor ubiquitination and degradation by way of the proteasome appear to be one of the rare JAK2-independent properties of the GHR [9]. Detailed description of the classic features of this receptor is beyond the scope of this article, and the authors invite the reader to refer to the numerous reviews previously published in the field (eg, [6,9–14]).

Growth hormone receptor signaling and growth

In mammals, GH is the main regulator of postnatal growth. It has traditionally been accepted that most of these actions are indirect, meaning they are mediated by insulin-like growth factor (IGF)-1, which is produced by the liver under GH stimulus and, in turn, exerts growth-promoting actions by way of its own receptor expressed in many target cells [15]. This original concept of the "somatomedin hypothesis" has considerably evolved during the past decade and is discussed later in the article.

Given the multiplicity of signaling cascades activated by the GHR, the question of which pathways are involved in GH regulation of growth has been addressed. In vitro studies involving engineered mutations of GHR cytoplasmic domain identified the membrane-proximal box 1 region as necessary and sufficient to activate Jak2 and MAP kinase pathways and for mediating GH-induced cell proliferation [16,17]. It is only very recently that GH-induced production of IGF-1 was shown to be regulated by way of the Jak2-Stat5 pathway, which involves Stat5 binding elements in the IGF-1 promoter [18–20]. Because activation of this pathway requires box 1 integrity (for Jak2 activation) and the presence of C-terminal tyrosines (for Stat5 recruitment) [21], this suggests that a full-length (fl) cytoplasmic domain is necessary for indirect growth-promoting actions of the GHR (ie, mediated by IGF-1; see Fig. 1).

Although these and other studies achieved relatively precise structure–function mapping of the GHR, cell models were intrinsically limited to elucidate the mechanisms by which GH controls body size in vivo. Two approaches involving genetically modified mouse models addressed this question. The first involved mice deficient for components involved in GHR-activated cascades (knock-out models), whereas the second involved mice expressing a truncated GHR (knock-in models). As expected, knockout mice for the GHR exhibited a dwarf phenotype and are referred to as Laron mice because they provide a model of GH resistance similar to Laron syndrome (see later text) [22]. These mice exhibit normal size at birth and start diverging from wild-type littermates from the age of 2 weeks, emphasizing the major role of GHR in postnatal growth. Knockout of Stat5B [23] and of Stat5A and Stat5B [24] but not of Stat5A alone abrogated sexual dimorphism in mice because body size and IGF-1 levels of STAT5B−/− male mice were equivalent to female mice, in which these parameters were not affected. Knockout of SOCS2 [25] (but not of other SOCS proteins [26]) produced giant mice, underscoring the role of this protein in the negative regulation of GH-induced postnatal growth [27]. Together, these observations strongly suggest that the Jak2/Stat5B/IGF-1 pathway plays a critical role in the control of postnatal body growth, which of course does not discard the involvement of other mechanisms and is evidenced by the fact that growth retardation of Stat5B knock-out mice is not as dramatic as in GHR knock-outs [28]. Very elegant knock-in studies were recently performed by Rowland and colleagues [29] to address this issue in more detail and to map the cytoplasmic regions of the GHR involved in its growth-promoting actions in vivo (see Fig. 1). Two types of truncated receptors were expressed in these animals using homologous recombination strategy to maintain physiologic regulation and expression levels. One mutation involved removal of the C-terminal region of the receptor (mutant 569), including five tyrosine residues identified as Stat5 and SOCS-2 recruitment sites, whereas the other involved truncation downstream to the membrane proximal region containing box 1 (mutant 391). Body size, circulating IGF-1 levels, and activation of Jak2/Stat5 and MAP kinase pathways were among the parameters analyzed in homozygous mice expressing one of these truncated GHRs [29]. It is of note that these in vivo studies largely confirmed the earlier in vitro observation regarding receptor regions required for activation of these pathways [6]. More important, a strong correlation was found among Stat5 activation, IGF-1 levels, and body size. In mice expressing mutant 569, loss of 70% of Stat5 signaling resulted in reduction of IGF-1 levels by around 80% and reduction of GH-dependent growth by 56%. Growth impairment and IGF-1 levels were even more affected when Stat5 activity was completely abolished in mice expressing mutant 391 GHR [29]; however, the latter retained 11% body growth compared with wild-type littermates, again suggesting that other pathways can, in part, support GHR-mediated growth. Because MAP kinase (Erk 1 and 2) activation remained intact in all these animals, this pathway is a good candidate. It is of note that retention of the ubiquitin-dependent endocytosis (UbE) [9] motif in these mutated GHRs (see Fig. 1) prevented their accumulation at the plasma membrane, which probably explains why they did not exert dominant-negative activity in heterozygous mice harboring one intact and one mutated allele for the GHR [30].

Although the somatomedin hypothesis has long assumed that most, if not all, growth-promoting actions of GH are mediated by circulating IGF-1 originating in the liver, the possibility that postnatal growth is also controlled by direct effects of GH has been specifically addressed within the last few years by comparing various genetically modified models involving components of GH and IGF axes [15,31,32]. Knockout of any component of the IGF axis (ligands, receptors) affects growth, confirming the importance of indirect effects of GH for controlling body size. Conditional knock-out of IGF-1 in the liver, however, had surprisingly little impact on postnatal growth despite circulating IGF-1 levels reduced by 75% [33]. Many recent studies have revisited the original somatomedin hypothesis [15], not only by conferring a major role to locally produced IGF-1 (acting by way of the autocrien/paracrine mechanism) but also by suggesting that circulating IGF-1 may not be so important for mediating GH-induced postnatal growth, at least in the mouse. In the current state of the art, it has become obvious that some actions of GH are IGF-1 independent [34], although GH-independent functions also clearly exist for IGF-1 [15], including the regulation of its production, which can involve factors other than GH, such as gamma interferon or steroids [18].

Growth disorders related to mutations of the growth hormone receptor

In a small number of humans, growth defects result from rare mutations of the GHR. Cell and animal models allow to us better understand how these mutations impair GHR functions, partly or totally. Mutations have been found to alter various intrinsic features of the GHR that are required for proper activation of the receptor itself or some components of GHR signaling cascades.

Growth hormone receptor mutations

GH insensitivity syndrome was first described in 1966 by Laron and colleagues [35]. Laron syndrome is an autosomal, fully penetrant recessive disease resulting from GH resistance. A few hundred patients have been identified worldwide who share, among other characteristics, severe growth retardation, acromicria, small gonads and genitalia, truncal adiposity, and high GH and very low IGF-1 circulating levels. The pathogenesis of this syndrome is due to various molecular defects mainly affecting the GHR, including exon deletion or mutations (nonsense, frameshift, missense) [36]. Most of the mutations affect the extracellular domain of the GHR, which is classically diagnosed by the absence of circulating GHBP. In contrast, normal (or high) GHBP denotes that the mutation resides within transmembrane or cytoplasmic domains of the GHR or affects postreceptor components [37]. Three types of recently identified mutations are described in the following text as representative examples of mutations affecting GHR expression, activation, or signaling.

Expression failure

A novel type of mutation leading to an atypical form of GH insensitivity was recently described by Maamra and colleagues [38]. This point mutation activates an intronic pseudoexon leading to the insertion of 36 residues after residue 207 (ie, within the dimerization domain of the GHR). This elongated (656-aa) receptor remains trapped around the nucleus and is therefore poorly expressed at the cell membrane, reflecting a trafficking defect possibly due to folding troubles. Despite this defect, mutated GHR displays normal binding affinity for GH in membrane preparations; its low abundance at the cell membrane results in poor activation by GH, leading to reduced (but not nil) downstream signaling (Jak2-Stat5). These properties may explain the relatively mild phenotype of these patients.

Activation failure

Substitution of histidine for aspartate 152 (D152H mutation) has been described in a context of familial GH resistance. This residue is located at the interface of the two extracellular domains of the receptor homodimer [3]. This mutation was initially suggested to interfere with receptor homodimerization because it was found to abolish homodimerization of GHBPs harboring this mutation [39]. Surface plasmon resonance studies recently performed by Bernat and colleagues [40] confirmed this observation and further demonstrated that it is the histidine of the second receptor (ie, the one interacting with the second binding site of GH) that is crucial for GHBP homodimerization because any substitution at position 152 has a deleterious effect. The unexpected finding came from the recent observation that D152H mutation failed to affect homodimerization of membrane-bound GHR [41]. The dichotomy between GHR dimerization and activation was partly elucidated in 2005 by Waters and colleagues [4,42]. Using various elegant and complementary approaches, these investigators showed that the GHR is constitutively homodimerized at the cell membrane, which presumably involves contacts between transmembrane or juxtamembrane domains of each receptor chain. The activation process mediated by the ligand is assumed to involve a conformational change such as relative rotation of upper and lower domains of the receptor [4]. Based on this model, one can speculate that D152H mutation observed in some Laron patients interferes with this conformational change, although this remains to be demonstrated.

Signaling failure

Ten years ago, a familial history of short stature revealed heterozygous expression of a severely truncated GHR mutant resulting from mutation of a splice acceptor site of exon 9. This truncated receptor virtually lacks the entire cytoplasmic domain because only seven residues remain [43]. It is easily understandable that in the absence of cytoplasmic domain, this receptor is devoid of any signaling capacity. Because internalization is also impaired in the absence UbE motif (see Fig. 1), such truncated receptors accumulate at the membrane and act as dominant-negative toward fl receptors [44], explaining severe growth retardation observed in this family despite the heterozygosity of the mutation. Less dramatically truncated GHRs were recently reported in patients who had severe GH insensitivity due to frameshift and to premature stop codon. The first resulted in the truncation of GHR after residue 449 (and nonsense sequence of residues 424–449) [45], and the other was truncated after residues 581 (nonsense sequence of residues 560–581) [46]. In both cases, receptor expression; cellular distribution; ligand binding; and activation of Jak2, Stat3, or MAP kinases were not affected by GHR deletion. In contrast, the ability to activate Stat5 was markedly impaired in both cases because Stat5 phosphorylation or Stat5-mediated reporter gene transactivation were close to background levels. These reports are in good agreement with the observations performed in mice [29,30] and emphasize the critical role of the Jak2-Stat5 cascade in the growth-promoting actions of GH in humans.

Post–growth hormone receptor defects

Increasing evidence for the role of Stat5 in human growth recently led Kofoed and colleagues [47] to identify the first mutation in Stat5B linked to GH insensitivity. The combination of severe growth retardation and immunodeficiency in this 16-year-old female patient was suggestive of a genetic defect shared by cytokines. The homozygous mutation that was found in STAT5B involved a single substitution of a proline for alanine 630. This mutation has several consequences on Stat5B properties. First, Stat5B^{A630P} was shown to be partly unfolded, which induces its accumulation as stable aggregates in cells, including those from the patient [48]. Protein aggregation was assumed to protect this Stat5B mutant from degradation by the proteasome, as usually observed for unfolded proteins. Second, because this residues lies in the src-homology (SH)-2 domain by which all Stat proteins dock to the phosphotyrosines of activated receptors, it is presumed that misfolding of the C-terminal SH2 domain prevents proper interaction with the GHR, which normally leads to Stat phosphorylation by Janus kinases. Third, even when phosphorylated by a constitutively active form of Src (another tyrosine kinase), Stat5B^{A630P} was found to be unable to bind DNA and to act as a transcription factor [49]. As a consequence, this mutated Stat5 is inactive and blocks all cellular events downstream in the Jak2-Stat5 cascade, including IGF-1 deficiency. Although this Stat5b "knockout" further confirms the importance of this pathway in body growth, this human female case is somehow puzzling regarding the regulation of sexual dimorphism in mammals because only male growth was reported to be affected in Stat5B knock-out mice [23,50]. Another mutation was recently found in STAT5B that led to the total absence of detectable mature protein due to early termination, suggesting that as for the GHR, various mutations of Stat5B probably underline this new syndrome that combines severe GH insensitivity and immunodeficiency [51,52]. Mutations of IGF-1 [53] or of IGF-1 receptor [54] leading to intrauterine and postnatal growth retardation have also been documented, the discussion of which is not the subject of this article [55].

Polymorphisms of the growth hormone receptor

Like most proteins of the body, the GHR does not have the same structure in all individuals because the *GHR* gene bears coding polymorphisms that alter the receptor protein. Most of these polymorphisms are single nucleotide polymorphisms or microdeletions that can be retrieved from public databases or from the Celera database. To the authors' knowledge, there has not yet been a systematic study or report of the GHR haplotypes generated by these single nucleotide polymorphisms. A fortiori, the functional role of these polymorphisms at the molecular level and the way they can modify GHR function at the physiologic level remain unknown for most of them.

One of these polymorphisms, the deletion of exon 3, has been studied more extensively because of its frequency. When the existence of two GHR transcripts of different length in certain tissues was first recognized, several groups of investigators [56–58] initially thought that the short form of the GHR (that with a deletion of exon 3) was a splice variant of the receptor, generated by an exon-skipping process. They hypothesized that this process was tissue dependent and obeyed regulations that were unclear. This error came from the fact that different tissues from different persons were studied. The fact that the GHR exon 3 difference was not tissue dependent but individual dependent became apparent a few years later [59]. The deletion of exon 3 was soon after shown to be a genomic event that, according to Pantel and colleagues [60], occurred in several steps during the evolutionary lineage from Old World monkeys genome to Pongina and Hominina genomes. Following the long-lived integration of proviral sequences in the vicinity of exon 3 of the *GHR* gene, a homologous recombination seems to have occurred in humans, resulting in the excision of exon 3 in one of our ancestors. Why this polymorphism has become frequent in up to approximately 50% of contemporary Caucasian people remains unknown. Because it is now frequent, it is clear that the homologous recombination was a very ancient event. It is also possible to hypothesize that exon 3 deletion (d3) has played a favorable role, fitting with the Darwinian evolution of our species and favoring its propagation in given human subgroups, but no functional role has been identified that substantiates this speculation. To the authors' knowledge, there has not been any systematic study of the distribution of the d3 in various human subgroups. The only population data that the authors have indicate that the d3 GHR polymorphism seems less frequent in West African Cameroonese and Burkinabe people (Pierre Bougnères, MD, PhD, unpublished data, 2006) and in Japanese people [61].

The effect of the d3 on the function of the receptor protein has been investigated by several groups. As usual for receptor molecules, initial experiments have focused on the ligand-binding properties of the variant GHR. When expressed in different cell models of variable relevance, d3 and fl GHR molecules appeared to bind 22K and 20K GHs with comparable affinity [62,63]; which led to the idea that d3 had no functional impact on GHR functions. Accordingly, a single allele of fl or d3 GHR was shown to be sufficient for normal growth [64]. This view was challenged, however, by the authors' later demonstration that d3 and fl GHR molecules transduce the GH binding signal with different intensity (Fig. 2) [65]. This functional difference has started to prompt studies for GH-related endocrinology and for potentially GH-related oncogenic processes [66]. The structural and functional mechanisms through which the d3 variant could have an increased transducer activity compared with the fl GHR have not been investigated.

Pharmacogenomics of human growth

Although skeletal growth is a complex phenomenon, the regulation of growth induced by GH therapy is restricted to the pathways encompassing GH transport and action. Because the GHR is a central player in transport and action of GH, it is not surprising that genomic variants attracted the attention of investigators interested in the efficacy of GH therapy. The first study observed that the growth response to administration of GH to 76 simply short children (45 idiopathic short stature [ISS] or 31 small for gestational age [SGA]) was distributed normally and that the short

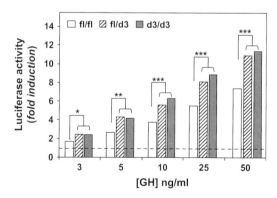

Fig. 2. In vitro activity of the d3 GHR is greater than fl GHR. HEK 293 transiently expressing fl GHR, d3 GHR, or both were stimulated by increasing concentrations of GH for 8 hours; 22 to 38 experiments were used for each condition. Fold-induction of the GH-responsive LHRE-luciferase reporter gene is expressed with respect to unstimulated cells (value = 1, *horizontal dotted line*); $*P < .005$, $**P < .0005$, $***P < .0001$. (*From* Dos Santos C, Essioux L, Teinturier C, et al. A common polymorphism of the growth hormone receptor is associated with increased responsiveness to growth hormone. Nat Genet 2004;36:720–4; with permission.)

children bearing one or two copies of the d3 variant grew faster than fl/fl homozygotes after correction for varying GH doses (Fig. 3) [65]. Children were carefully selected for being of European or Mediterranean ancestry and for being strictly prepubertal at the end of the study. Growth gain was calculated for each individual as the difference between spontaneous growth rate before and during GH administration. Multivariate regression analysis of age, sex, GH dose, and GHR genotype allowed to approximate an equation describing the dependence of growth gain on GH dose and genotype in this population: Gain (in centimeters per year) = 2.8(GH dose) + 2.9(GHR genotype) + ε, where *GH dose* is units per kilogram per week and *GHR genotype* is 1 for d3 GHR carriers and 0 for non-d3 carriers [65]. The observation was replicated in another French cohort of short children treated during 2 years [65]. Genotypic association of the d3 variant with accelerated growth was also reported by Binder and colleagues [67] in girls who had Turner syndrome treated with GH and in 60 prepubertal SGA children, although with borderline significance ($P = .067$) and a lesser size effect than observed for the mixed ISS and SGA children studied by Dos Santos and colleagues [65].

A positive effect of the d3 variant was reported by Jorge and colleagues [68] in a Brazilian population of 80 carefully studied patients who had GH deficiency (GHD). GHD was defined using stringent criteria and a cutoff of 3 µg/dL for the response of GH to stimulation tests, safeguarding studies

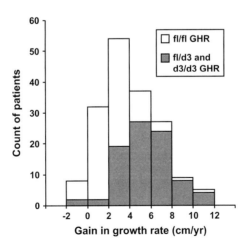

Fig. 3. The d3 GHR variant is more frequent in "good responders." Frequency histogram showing the distribution of individual growth rate increments (centimeters per year) during the first year of GH therapy. The distribution of values was normal and had a large variance. Both cohorts of GH-treated short children were pooled for analysis. Children who had fl/d3 or d3/d3 GHR genotypes are shown as gray boxes. (*From* Dos Santos C, Essioux L, Teinturier C, et al. A common polymorphism of the growth hormone receptor is associated with increased responsiveness to growth hormone. Nat Genet 2004;36:720–4.)

from incorporating "false GHD" cases [68]. The difference between the variants was observed not only for the growth rates recorded during GH therapy but also for the final stature reached by the patients [68]. Other studies in Italian [69] or Japanese [61] patients who had GHD did not find the differences reported by Jorge and colleagues [68] (Table 1). In a multiethnic cohort from India, Asia, and Europe, another study found no difference between d3/fl GHR genotypes [70], whereas a small study in French children who had GHD suggested that d3 carriers had a better response to GH [71]. The fact that genotypic effects are not consistent in all reports is an almost constant hallmark of association studies [72]. There is no general explanation for this observation, but part of the discordances often comes from positive bias from initial observations; negative bias linked to type II errors in following studies; statistically underpowered samples; phenotypic heterogeneity; or genetic, demographic, or clinical stratification of samples [73]. Regarding pharmacogenetics, one should note that the use of different doses of the active agent can modify the conditions for analyzing phenotype–genotype relationships. For example, if sensitivity to GH is modified by the GHR genotype, then it is likely that increasing the dose of GH to a certain posology may abolish the observation of differences across d3 and fl GHR genotypes (see the equation mentioned earlier). It is also likely that other gene variants at other loci, implicated in resistance/sensitivity to GH, alter the importance of coexisting d3 GHR effects. It is to be remembered that common polymorphisms should not be studied with the same functional hypotheses and experimental approaches simply because rare mutations impair protein functions to a degree that is generally easy to document in clinical medicine. For example, the groups who initially studied the d3 GHR variant tested whether it could influence normal growth and normal stature in samples of 150 healthy adult subjects [60] and found no effect. Genotype–phenotype associations require much larger samples because most complex, multifactorial quantitative traits like normal growth or stature, even if they are oligogenic, are genetically determined by variants whose contribution is not expected to exceed 10% of global variance (the "exponential model") [74]. Detecting the d3/fl GHR polymorphism to be a modifier of human height would require the study of thousands of human adults in a sample of population that would not be stratified by ethnic effects on stature and on d3/fl relative prevalence.

In conclusion, the d3/fl GHR is an example of genetic variation that could affect certain aspects of hormone physiology and pharmacology. The implications of this observation are not intended to create a revolution for the endocrinology of growth because the simple increase of GH dose can be expected to bring "poor genetic responders" to the level of good responders, which many practitioners would have anticipated based on simple pragmatic experience and follow-up of individual cases. It will be interesting, nonetheless, to test the effect of the GHR polymorphism in more samples, not only for growth purposes but also for potential prediction of

Table 1
Presence or absence of association between exon 3 deletion/full-length growth hormone receptor genotypes and growth response to growth hormone in different groups of growth hormone–deficient children varying in age, ethnicity, diagnostic criteria, sample size, and severity of phenotype

	Jorge et al, 2006 [68]	Pilotta et al, 2006 [69]	Ito et al, 2005 [61]	Blum et al, 2005 [70]	Thomas-Teinturier et al, 2005 [71]	Comment
d3 carriers N (%)	28 (50)	25 (49)	8 (20)	48 (48)	7 (49)	Genotype variability, two studies underpowered
IGHD/CPHD	19/39	53/1	unknown	107/0	15/0	Phenotypic heterogeneity
Criteria for GHD[a]	Stringent	Mild	Mild	Mild	Stringent	Diagnostic heterogeneity
Mean age (y)	8.9	7.8	6.8	7.1	5.8	Age variability
Ht SDS baseline	−4.3	−1.9	−2.6	−3.3	−3.1	Variable severity
Geography	Brazil	Europe	Japan	Europe	Europe	Ethnic differences
GH dose (μg.k.d)	31	29	25	29	30	Comparable dose
d3 effect	Present	Absent	Absent	Absent	Present	Apparent discordances

Abbreviations: CPHD, combined pituitary hormone deficiency; Ht, height; IGHD, isolated GHD; SDS, standard deviation score.
[a] Stringent = GH peaks <4 ng/mL; Mild = GH peaks <10 ng/ML.

metabolic or oncogenic effects of GH. The prediction of adverse effects may become useful in this field of pharmacogenomics.

References

[1] Kelly PA, Djiane J, Postel-Vinay MC, et al. The prolactin/growth hormone receptor family. Endocr Rev 1991;12:235–51.
[2] Leung DW, Spencer SA, Cachianes G, et al. Growth hormone receptor and serum binding protein: purification, cloning, and expression. Nature 1987;330:537–43.
[3] De Vos AM, Ultsch M, Kossiakoff AA. Human growth hormone and extracellular domain of its receptor: crystal structure of the complex. Science 1992;255:306–12.
[4] Brown RJ, Adams JJ, Pelekanos RA, et al. Model for growth hormone receptor activation based on subunit rotation within a receptor dimer. Nat Struct Mol Biol 2005;12:814–21.
[5] Argetsinger LS, Campbell GS, Yang X, et al. Identification of Jak2 as a growth hormone receptor-associated tyrosine kinase. Cell 1993;74:237–44.
[6] Herrington J, Carter-Su C. Signaling pathways activated by the growth hormone receptor. Trends Endocrinol Metab 2001;12:252–7.
[7] He K, Loesch K, Cowan JW, et al. Janus kinase 2 enhances the stability of the mature growth hormone receptor. Endocrinology 2005;146:4755–65.
[8] Loesch K, Deng L, Cowan JW, et al. Janus kinase 2 influences growth hormone receptor metalloproteolysis. Endocrinology 2006;147:2839–49.
[9] Strous GJ, dos Santos CA, Gent J, et al. Ubiquitin system-dependent regulation of growth hormone receptor signal transduction. Curr Top Microbiol Immunol 2004;286:81–118.
[10] Finidori J. Regulators of growth hormone signaling. Vitam Horm 2000;59:71–97.
[11] Moutoussamy S, Kelly PA, Finidori J. Growth hormone receptor and cytokine-receptor family signaling. Eur J Biochem 1998;255:1–11.
[12] Kelly PA, Finidori J, Edery M, et al. The prolactin/growth hormone/cytokine receptor superfamily. In: Lee AG, editor. Biomembranes. Greenwich (UK): JAI Press; 1996. p. 129–45.
[13] Carter-Su C, Smit LS. Signaling via JAK tyrosine kinases: growth hormone receptor as a model system. Recent Prog Horm Res 1998;53:61–82.
[14] Argetsinger LS, Carter-Su C. Mechanism of signaling by growth hormone receptor. Physiol Rev 1996;76:1089–107.
[15] Butler AA, Le Roith D. Control of growth by the somatotropic axis: growth hormone and the insulin-like growth factors have related and independent roles. Annu Rev Physiol 2001;63:141–64.
[16] Colosi P, Wong K, Leong SR, et al. Mutational analysis of the intracellular domain of the human growth hormone receptor. J Biol Chem 1993;268:12617–23.
[17] Postel-Vinay MC, Finidori J. Growth hormone receptor: structure and signal transduction. Eur J Endocrinol 1995;133:654–9.
[18] Hwa V, Little B, Kofoed EM, et al. Transcriptional regulation of insulin-like growth factor-I by interferon-gamma requires STAT-5b. J Biol Chem 2004;279:2728–36.
[19] Chia DJ, Ono M, Woelfle J, et al. Characterization of distinct Stat5b binding sites that mediate growth hormone-stimulated IGF-I gene transcription. J Biol Chem 2006;281:3190–7.
[20] Woelfle J, Chia DJ, Rotwein P. Mechanisms of growth hormone (GH) action. Identification of conserved Stat5 binding sites that mediate GH-induced insulin-like growth factor-I gene activation. J Biol Chem 2003;278:51261–6.
[21] Finidori J, Kelly PA. Cytokine receptor signalling through two novel families of transducer molecules: Janus kinases, and signal transducers and activators of transcription. J Endocrinol 1995;147:11–23.
[22] Zhou Y, Xu BC, Maheshwari HG, et al. A mammalian model for Laron syndrome produced by targeted disruption of the mouse growth hormone receptor/binding protein gene (the Laron mouse). Proc Natl Acad Sci U S A 1997;94:13215–20.

[23] Udy GB, Towers RP, Snell RG, et al. Requirement of Stat5b for sexual dimorphism of body growth rates and liver gene expression. Proc Natl Acad Sci U S A 1997;94: 7239–44.
[24] Teglund S, McKay C, Schuetz E, et al. Stat5a and Stat5b proteins have essential and nonessential, or redundant, roles in cytokine responses. Cell 1998;93:841–50.
[25] Metcalf D, Greenhalgh CJ, Viney E, et al. Gigantism in mice lacking suppressor of cytokine signalling-2. Nature 2000;405:1069–73.
[26] Greenhalgh CJ, Alexander WS. Suppressors of cytokine signalling and regulation of growth hormone action. Growth Horm IGF Res 2004;14:200–6.
[27] Greenhalgh CJ, Rico-Bautista E, Lorentzon M, et al. SOCS2 negatively regulates growth hormone action in vitro and in vivo. J Clin Invest 2005;115:397–406.
[28] Coschigano KT, Clemmons D, Bellush LL, et al. Assessment of growth parameters and life span of GHR/BP gene-disrupted mice. Endocrinology 2000;141:2608–13.
[29] Rowland JE, Lichanska AM, Kerr LM, et al. In vivo analysis of growth hormone receptor signaling domains and their associated transcripts. Mol Cell Biol 2005;25:66–77.
[30] Rowland JE, Kerr LM, White M, et al. Heterozygote effects in mice with partial truncations in the growth hormone receptor cytoplasmic domain: assessment of growth parameters and phenotype. Endocrinology 2005;146:5278–86.
[31] Lupu F, Terwilliger JD, Lee K, et al. Roles of growth hormone and insulin-like growth factor 1 in mouse postnatal growth. Dev Biol 2001;229:141–62.
[32] Dupont J, Holzenberger M. Biology of insulin-like growth factors in development. Birth Defects Res C Embryo Today 2003;69:257–71.
[33] Yakar S, Liu JL, Stannard B, et al. Normal growth and development in the absence of hepatic insulin-like growth factor I. Proc Natl Acad Sci U S A 1999;96:7324–9.
[34] Sotiropoulos A, Ohanna M, Kedzia C, et al. Growth hormone promotes skeletal muscle cell fusion independent of insulin-like growth factor 1 up-regulation. Proc Natl Acad Sci U S A 2006;103:7315–20.
[35] Laron Z, Pertzelan A, Mannheimer S. Genetic pituitary dwarfism with high serum concentration of growth hormone—a new inborn error of metabolism? Isr J Med Sci 1966;2: 152–5.
[36] Laron Z. Natural history of the classical form of primary growth hormone (GH) resistance (Laron syndrome). J Pediatr Endocrinol Metab 1999;12(Suppl 1):231–49.
[37] Laron Z. Growth hormone insensitivity (Laron syndrome). Rev Endocr Metab Disord 2002; 3:347–55.
[38] Maamra M, Milward A, Esfahani HZ, et al. A 36 residues insertion in the dimerization domain of the growth hormone receptor results in defective trafficking rather than impaired signaling. J Endocrinol 2006;188:251–61.
[39] Duquesnoy P, Sobrier ML, Duriez B, et al. A single amino acid substitution in the exoplasmic domain of human growth hormone (GH) receptor confers familial GH resistance (Laron syndrome) with positive GH-binding activity by abolishing receptor homodimerization. EMBO J 1994;13:1386–95.
[40] Bernat B, Pal G, Sun M, et al. Determination of the energetics governing the regulatory step in growth hormone-induced receptor homodimerization. Proc Natl Acad Sci U S A 2003; 100:952–7.
[41] Gent J, Van Den EM, van Kerkhof P, et al. Dimerization and signal transduction of the growth hormone receptor. Mol Endocrinol 2003;17:967–75.
[42] Waters MJ, Hoang HN, Fairlie DP, et al. New insights into growth hormone action. J Mol Endocrinol 2006;36:1–7.
[43] Ayling RM, Ross R, Towner P, et al. A dominant-negative mutation of the growth hormone receptor causes familial short stature. Nat Genet 1997;16:13–4.
[44] Ross RJM, Esposito N, Shen XY, et al. A short isoform of the human growth hormone receptor functions as a dominant negative inhibitor of the full-length receptor and generates large amounts of binding protein. Mol Endocrinol 1997;11:265–73.

[45] Milward A, Metherell L, Maamra M, et al. Growth hormone (GH) insensitivity syndrome due to a GH receptor truncated after Box1, resulting in isolated failure of STAT 5 signal transduction. J Clin Endocrinol Metab 2004;89:1259–66.
[46] Tiulpakov A, Rubtsov P, Dedov I, et al. A novel C-terminal growth hormone receptor (GHR) mutation results in impaired GHR-STAT5 but normal STAT-3 signaling. J Clin Endocrinol Metab 2005;90:542–7.
[47] Kofoed EM, Hwa V, Little B, et al. Growth hormone insensitivity associated with a STAT5b mutation. N Engl J Med 2003;349:1139–47.
[48] Chia DJ, Subbian E, Buck TM, et al. Aberrant folding of a mutant Stat5b causes growth hormone insensitivity and proteasomal dysfunction. J Biol Chem 2006;281: 6552–8.
[49] Fang P, Kofoed EM, Little BM, et al. A mutant signal transducer and activator of transcription 5b, associated with growth hormone insensitivity and insulin-like growth factor-I deficiency, cannot function as a signal transducer or transcription factor. J Clin Endocrinol Metab 2006;91:1526–34.
[50] Rosenfeld RG. Gender differences in height: an evolutionary perspective. J Pediatr Endocrinol Metab 2004;17(Suppl 4):1267–71.
[51] Hwa V, Little B, Adiyaman P, et al. Severe growth hormone insensitivity resulting from total absence of signal transducer and activator of transcription 5b. J Clin Endocrinol Metab 2005;90:4260–6.
[52] Rosenfeld RG, Kofoed E, Buckway C, et al. Identification of the first patient with a confirmed mutation of the JAK-STAT system. Pediatr Nephrol 2005;20:303–5.
[53] Woods KA, Camacho-Hubner C, Savage MO, et al. Intrauterine growth retardation and postnatal growth failure associated with deletion of the insulin-like growth factor I gene. N Engl J Med 1996;335:1363–7.
[54] Abuzzahab MJ, Schneider A, Goddard A, et al. IGF-I receptor mutations resulting in intrauterine and postnatal growth retardation. N Engl J Med 2003;349:2211–22.
[55] Rosenfeld RG, Hwa V. New molecular mechanisms of GH resistance. Eur J Endocrinol 2004;151(Suppl 1):S11–5.
[56] Urbanek M, MacLeod JN, Cooke NE, et al. Expression of a human growth hormone (hGH) receptor isoform is predicted by tissue-specific alternative splicing of exon 3 of the hGH receptor gene transcript. Mol Endocrinol 1992;6:279–87.
[57] Esposito N, Paterlini P, Kelly PA, et al. Expression of two isoforms of the human growth hormone receptor in normal liver and hepatocarcinoma. Mol Cell Endocrinol 1994;103: 13–20.
[58] Mercado M, Davila N, McLeod JF, et al. Distribution of growth hormone receptor messenger ribonucleic acid containing and lacking exon 3 in human tissues. J Clin Endocrinol Metab 1994;78:731–5.
[59] Wickelgren RB, Landin KL, Ohlsson C, et al. Expression of exon 3-retaining and exon 3-excluding isoforms of the human growth hormone-receptor is regulated in an interindividual, rather than a tissue-specific, manner. J Clin Endocrinol Metab 1995;80:2154–7.
[60] Pantel J, Machinis K, Sobrier ML, et al. Species-specific alternative splice mimicry at the growth hormone receptor locus revealed by the lineage of retroelements during primate evolution. J Biol Chem 2000;275:18664–9.
[61] Ito Y, Makita Y, Matsuo K, et al. Influence of the exon 3 deleted isoform of GH receptor gene on growth response to GH in Japanese children. Horm Res 2005;64(Suppl 1):1–397 [Abstracts of the ESPE/LWPES 7th Joint Meeting Paediatric Endocrinology. Lyon, France, September 21–24, 2005. p. 1–150].
[62] Urbanek M, Russell JE, Cooke NE, et al. Functional characterization of the alternatively spliced, placental human growth hormone receptor. J Biol Chem 1993;268: 19025–32.
[63] Sobrier ML, Duquesnoy P, Duriez B, et al. Expression and binding properties of two isoforms of the human growth hormone receptor. FEBS Lett 1993;319:16–20.

[64] Pantel J, Grulich-Henn J, Bettendorf M, et al. Heterozygous nonsense mutation in exon 3 of the growth hormone receptor (GHR) in severe GH insensitivity (Laron syndrome) and the issue of the origin and function of the GHRd3 isoform. J Clin Endocrinol Metab 2003;88: 1705–10.
[65] Dos Santos C, Essioux L, Teinturier C, et al. A common polymorphism of the growth hormone receptor is associated with increased responsiveness to growth hormone. Nat Genet 2004;36:720–4.
[66] Wagner K, Hemminki K, Grzybowska E, et al. Polymorphisms in the growth hormone receptor: a case-control study in breast cancer. Int J Cancer 2006;118:2903–6.
[67] Binder G, Baur F, Schweizer R, et al. The d3-growth hormone (GH) receptor polymorphism is associated with increased responsiveness to GH in Turner syndrome and short small-for-gestational-age children. J Clin Endocrinol Metab 2006;91:659–64.
[68] Jorge AA, Marchisotti FG, Montenegro LR, et al. Growth hormone (GH) pharmacogenetics: influence of GH receptor exon 3 retention or deletion on first-year growth response and final height in patients with severe GH deficiency. J Clin Endocrinol Metab 2006;91: 1076–80.
[69] Pilotta A, Mella P, Filisetti M, et al. Common polymorphisms of the growth hormone (GH) receptor do not correlate with the growth response to exogenous recombinant human GH in GH-deficient children. J Clin Endocrinol Metab 2006;91:1178–80.
[70] Blum WF, Machinis K, Shavrikova EP, et al. The growth response to growth hormone (GH) treatment in children with isolated GH deficiency is independent of the presence of the exon 3-minus isoform of the GH receptor (GHR). Horm Res 2005;64(Suppl 1):1–397 [Abstracts of the ESPE/LWPES 7th Joint Meeting Paediatric Endocrinology. Lyon, France, September 21–24, 2005. p. 3–71].
[71] Thomas-Teinturier C, Dos Santos C, Bougneres P. The growth hormone receptor polymorphism influences the response of growth hormone-deficient children to GH treatment. Horm Res 2005;64(Suppl 1):1–397 [Abstracts of the ESPE/LWPES 7th Joint Meeting Paediatric Endocrinology. Lyon, France, September 21–24, 2005. p. 1–152].
[72] Hirschhorn JN, Altshuler D. Once and again-issues surrounding replication in genetic association studies. J Clin Endocrinol Metab 2002;87:4438–41.
[73] Freedman ML, Reich D, Penney KL, et al. Assessing the impact of population stratification on genetic association studies. Nat Genet 2004;36:388–93.
[74] Farrall M. Quantitative genetic variation: a post-modern view. Hum Mol Genet 2004;13(1): R1–7.

Genetics of Growth Hormone Deficiency

Primus E. Mullis, MD

Paediatric Endocrinology and Diabetology and Metabolism, University Children's Hospital, University of Berne, Inselspital, CH-3010 Bern, Switzerland

The most fundamental characteristic of infancy and childhood is that it is a time of growth. Although the process of growth is quite multifactorial and complex, the growth pattern of children is rather predictable. Any deviation from a normal pattern of growth can be the first manifestation of a wide variety of disease processes, including endocrine and nonendocrine disorders and, importantly, may involve any organ system of the human body.

Over the last decades, growth disorders were managed on the basis of a growth hormone (GH)–oriented classification system, from historical development. Nowadays, however, clinicians are well aware that (1) GH is not the major mediator of skeletal growth, (2) skepticism and criticism are adequate and accepted when analyzing the variable results of the various GH stimulation tests, and (3) many genetic defects have been described and have presented important insights into the molecular basis of non–GH-deficient growth failure.

When a child is not following the normal, predicted growth curve, however, an evaluation for underlying illness and central nervous system abnormalities is required. Where appropriate, genetic defects causing GH deficiency (GHD) should be considered. Because insulin-like growth factor-I (IGF-I) plays a pivotal role in growth, where it mediates most, if not all, of the effects of GH, GHD could also be considered somehow as IGF-I deficiency (Box 1). Although IGF-I deficiency can develop at any level of the growth hormone–releasing hormone (GH-RH)–GH–IGF axis, one should differentiate, however, between GHD (absent to low GH in circulation) and IGF-I deficiency (normal to high GH in circulation). Without ignoring the fact that classifications and taxonomies are decisions imposed on nature providing a chronicle of historical changes and not accepting

---

The study was supported by a grant of the Swiss National Science Foundation (SNF 3200-064623.01).

E-mail address: primus.mullis@insel.ch

> **Box 1. Alteration of the growth hormone-releasing hormone—growth hormone axis affecting growth in humans (differential diagnosis of insulin-like growth factor-I deficiency)**
>
> *Hypothalamus*
> Transcription factors
> GHRH-gene
>
> *Pituitary gland*
> Transcription factors
>   TPIT
>   SOX2
>   SOX3
>   HESX1
>   LHX3
>   LHX4
>   PROP1
>   POU1F1
> GH-RH–receptor
> GH-gene cluster
>   GH-deficiency/bioinactivity
>
> *GH-target organs*
> GH-receptor (primary: extracellular, transmembrane, intracellular)
>   GH-insensitivity
>   Signaling (JAK2/STAT5b/ERK)
> GH-insensitivity (secondary)
>   Malnutrition (eg, anorexia)
>   Liver disease (eg, Byler's disease)
>   Chronic illness
>   Anti-GH antibodies
>
> *IGF-I defects*
> *IGF-I transport/metabolism/clearance*
> *IGF-I resistance*
> IGF-I receptor defect (type I)
> IGF-I signaling (postreceptor defect)

the medieval rule of parsimony, well known as Ockham's razor (Pluritas non est ponenda sine necissitate—plurality should not be assumed without necessity; in modern English, keep it simple) the main focus of this article is on the *GH* gene, the various gene alterations, and their possible impact on the pituitary gland.

The application of the powerful tool of molecular biology has made it possible to ask questions not only about hormone production and action but also to characterize many of the receptor molecules that initiate responses to the hormones. Clinicians are beginning to understand how cells regulate the expression of genes and how hormones intervene in regulatory processes to modulate the expression of individual genes. In addition, great strides have been made in understanding how individual cells talk to each other through locally released factors to coordinate growth, differentiation, secretion, and other responses within a tissue. When focusing on regulatory systems playing a most important role defining a physiologic or pathophysiologic mechanism one has to be aware of having entered already the postgenomic area, in which not only a defect at the molecular level but also its functional impact at the cellular level becomes important.

Transcription factors of clinical importance for pituitary gland development

In Table 1 the various phenotypes caused by a defect of various transcription factors of the pituitary gland are summarized. All these disorders caused by any altered transcription factor do end up in various forms of combined pituitary hormone deficiency. At the very beginning, however, GHD might be the only hormonal deficiency, and these factors should be taken into account when examining and following-up these patients [1]. The two most important transcription factors, namely POU1F1 (PIT 1) and PROP 1, are discussed in detail.

POU1F1 (PIT1)

The pituitary transcription factor PIT-1 is a member of the POU-family of homeoproteins, which regulates important steps during embryologic development of the pituitary gland and regulates target gene function during postnatal life. It is a 291-amino acid peptide, containing a transactivation domain and two conserved DNA-binding domains: the POU-homeo domain and the POU-specific domain. Because PIT1 is confined to the nuclei of somatotropes, lactotrops, and thyrotropes in the anterior pituitary gland, the target genes of PIT1 include the *GH*, *prolactin* (PRL), and the *thyroid-stimulating hormone* (TSH) subunit, and the *POU1F1* gene itself. The defects in the human *POU1F1* gene result in a total deficiency of GH and PRL, whereas a variable hypothyroidism caused by insufficient TSH secretion, at least during childhood, has been described (see Table 1). Although it is important to stress that the clinical variability is caused by other factors than the exact location of the mutation reported, the type of inheritance, however, seems to correlate well with the genotype. The first mutation within the *POU1F1* gene was identified by Tatsumi [2]. Most of the mutations reported so far are recessive; however, a number of heterozygous point mutations have been reported [3]. Of those the amino acid substitution

Table 1
Transcription factors of clinical importance

Gene	Phenotype	Inheritance

Pit1/POU1F1	*Hormonal deficiencies:* GH, PRL, TSH
	Imaging
	  anterior pituitary gland: normal to hypo
	  posterior pituitary gland: normal
	Other manifestation: none
		R/D

PROP 1	*Hormonal deficiencies:* GH, PRL, TSH, LH, FSH, (ACTH)
	Imaging
	  anterior pituitary gland: hypo to hyper
	  posterior pituitary gland: normal
	Other manifestation: none
		R

HESX1	*Hormonal deficiencies:* GH, PRL, TSH, LH, FSH, ACTH, IGHD, CPHD
	Imaging
	  anterior pituitary gland: hypo
	  posterior pituitary gland: ectopic
	Other manifestation: eyes, brain, septo-optic dysplasia
		R/D

LHX3	*Hormonal deficiencies:* GH, PRL, TSH, LH, FSH (ACTH)
	Imaging
	  anterior pituitary gland: hypo
	  posterior pituitary gland: normal
	Other manifestation: neck rotation 75°–85° (?) (no:160°–180°)
		R

LHX4	*Hormonal deficiencies:* GH, TSH, ACTH
	Imaging
	  anterior pituitary gland: hypo
	  posterior pituitary gland: normal
	Other manifestation: sella turcica/skull defects, cerebellar defects
		D

SOX2	*Hormonal deficiencies:* GH, LH, FSH
	Imaging
	  anterior pituitary gland: hypo
	  posterior pituitary gland: normal/hypo
	Other manifestation: bilateral anophthalmia, spastic, altered brain development esophage atresia
		?/D

SOX3	*Hormonal deficiencies:* GH
	Imaging
	  anterior pituitary gland: normal to hypo
	  posterior pituitary gland: normal, ectopic
	Other manifestation: mental retardation, abnormality of corpus callosum, absent infundibulum
		XL

Abbreviations: ACTH, adrenocorticotropic hormone; CPHD, combined pituitary hormone deficiency; D, autosomal dominantly inherited; FSH, follicle-stimulating hormone; GH, growth hormone; IGHD, isolated growth hormone deficiency; LH, luteinising hormone; PRL, prolactin; R, autosomal recessively inherited; TSH, thyroid-stimulating hormone; XL, x-linked.

R271W (Arg271Trp) seems to be a hot spot. Further, the dominant-negative effect of the R271W POU1F1 form has been recently challenged by Kishimoto and coworkers [4]. Although most cases with R271W are sporadic and present with an autosomal-dominant mode of inheritance, Okamoto and coworkers [5] reported a family with normal family members who were clearly heterozygous for that mutation. Further in vitro expression studies were performed, which could not confirm its dominant negative effect that is well in contrast with the original report using identical experimental conditions [1,4].

PROP1

Wu and coworkers [6] described four families in which combined pituitary hormone deficiency was associated with homozygosity or compound heterozygosity for inactivating mutations of the *PROP1* gene. PROP1 (prophet of Pit1) is a paired-like homeodomain transcription factor and, originally, a mutation in this gene (Ser83Pro) was found causing the Ames dwarf (*df*) mouse phenotype [7]. In mice, *Prop1* gene mutation primarily causes GH, PRL, and TSH deficiency, and in humans, *PROP1* gene defects also seem to be a major cause of combined pituitary hormone deficiency. In agreement with the model of Prop1 playing a role in commitment of dorsal lineages (GH, PRL, and TSH), Prop1 mutant mice exhibit a dorsal expansion of gonadotrophs that normally arise on the ventral side.

To date, many different missense, frameshift, and splice site mutations, deletions, and insertions have been reported. The clinical phenotype varies not only among the different gene mutations but also among the affected siblings with the same mutation [8,9]. Although the occurrence of the hormonal deficiency varies from patient to patient [8], the affected patients as adults are not only GH, PRL, and TSH deficient, but also gonadotropin deficient (see Table 1). The three tandem repeats of the dinucleotides GA at location 296-302 in the *PROP1* gene represent a hot spot for combined pituitary hormone deficiency [8–10]. Low levels of cortisol have also been described in some patients with *PROP1* gene mutations [11]. In addition, pituitary enlargement with subsequent involution has been reported in patients with PROP1 mutations [11]. The mechanism underlying this phenomenon remains unknown.

Classification of isolated growth hormone deficiency

Structure and function of GH and CS genes

The *GH* gene cluster consists of five structurally similar genes in the order 5′ (GH-1, CSHP [chorionic somatomammotropin pseudogene], CSH-1 [chorionic somatomammotropin gene], GH-2, and CSH-2) 3′ encompassing a distance of about 65,000 bp (65 kb) on the long arm of chromosome 17 at bands q22-24 [12]. The *GH-1* gene encodes the mature human GH, a 191-amino acid peptide, and consists of five exons and four introns [12–14]. Approximately 75% of circulating GH is expressed in the anterior pituitary gland as a major 22-kd product, whereas alternative splicing can give rise to minor forms [13–15]. The most prominent minor form (5%–10%) is a bioactive 20-kd GH peptide that results from the use of a cryptic 3' splice site in exon 3, deleting amino acid 32-46 [15–18]. The *GH-2* gene encodes a protein (GH-V) that is expressed in the placenta rather than in the pituitary gland and differs from the primary sequence of GH-N (product of *GH-1* gene) by 13 amino acids. This hormone replaces pituitary GH in the maternal circulation during the second half of pregnancy. The *CSH-1*, *CSH-2* genes encode proteins of identical sequences, whereas the CSHP encodes a protein that differs by 13 amino acids and contains a mutation (donor splice site of its second intron) that should alter its pattern of mRNA splicing and the primary sequence of the resulting protein. The extensive homology (92%–98%) between the immediate flanking, intervening, and coding sequences of these five genes suggests that this multigene family arose through a series of duplicational events. With the exception of *CSHP*, each gene encodes a 217-amino acid prehormone that is cleaved to yield a mature hormone with 191-amino acids and a molecular weight of 22 kd. The expression of *GH-1* gene is further controlled by *cis*- and *trans*-acting elements and -factors, respectively [1,13].

Familial isolated growth hormone deficiency

Short stature associated with GH deficiency has been estimated to occur in about 1 in 4000 to 1 in 10,000 in various studies [19–21]. Although most cases are sporadic and are believed to result from environmental cerebral insults or developmental anomalies, 3% to 30% of cases have an affected first-degree relative suggesting a genetic etiology. Because MRI examinations detect only about 12% to 20% anomalies of either hypothalamus or pituitary gland in isolated growth hormone deficiency (IGHD), it can be assumed that many genetic defects may not be diagnosed and a significantly higher proportion of sporadic cases may have indeed a genetic cause [22]. Familial IGHD, however, is associated with at least four mendelian disorders [1,13]. These include two forms that have autosomal-recessive inheritance (IGHD type IA, IB) and autosomal-dominant (IGHD type II) and X-linked (IGHD III) forms. Table 2 depicts the mutational spectrum of GHD, which is discussed in greater detail later.

Isolated growth hormone deficiency type IA

In 1970, IGHD type IA was first described by Illig [23] in three Swiss children with unusually severe growth impairment and apparent deficiency of

Table 2
Mutational spectrum of growth hormone-deficiency

Microdeletions

Deficiency type	Deletion	Codon	GH-antibodies on treatment	References
IA	TGcCTG	-10	Yes	74
IA	GGCcTGC	-12	Yes	Mullis unpublished
II	CGGggatggggagacctgtaGT	5'IVS-3 del +28 to +45	No	42
IA	GagTCTAT	55	No	75

Single base-pair substitutions in the *GH-1* gene coding region

Deficiency type	Mutation	Codon nucleotide	AB on treatment	References
IA	TGG -> TAG Trp -> stop	-7	Yes	76
IA	GAG -> TAG Glu -> stop	-4	No	77
II	R183H	G6664A	No	52
II	V110F	G6191T	No	53
II	P89L	C6129T	No	54
II/bioinactivity	CGC-TGC Arg -> Cys	77	No	78,79

Single base-pair substitutions affecting mRNA splicing

Deficiency type	5'IVS-3	Δ exon 3	Origin	References
II	GTGAGT -> GTGAAT	Yes	Chile	80
II	GTGAGT -> GTGACT	Yes	Turkey	Mullis unpublished
II	GTGAGT -> GTGAGC	Yes	Turkey, Asia	81
II	GT -> AT	Yes	Europe, America, Africa	43
II	GT -> CT	Yes	Turkey	39
II	GT -> TT	Yes	India	Mullis unpublished
II	GT -> GC	Yes	Germany, Holland	53
	Exon splice enhancer	Yes		
II	ESE1m1: +1G -> T	Yes	Japan	44
II	ESE1m2: +2A -> C	Yes	Switzerland	Mullis unpublished
II	ESE1m3: +5A -> G	Yes		45
	Intron splice enhancer	Yes		
II	ISEm1:IVS-3 +28 G -> A	Yes		42
II	ISEm3:IVS-3 del28-45	Yes		42
	Length of the intron	Yes		
II	IVS3 del56-77	Yes	Italy	51
	5'IVS-4			
IB	GT -> CT	No	Saudi Arabia	76
IB	GT -> TT	No	Saudi Arabia	82

GH. Affected individuals occasionally have short length at birth and hypoglycemia in infancy but uniformly develop severe growth retardation by the age of 6 months. Their initial good response to exogenous GH is hampered by the development of anti-GH antibodies leading to dramatic slowing of growth [13,24].

GH-1 gene deletions

In 1981, Phillips and coworkers [24], examining the genomic DNA from these Swiss children, reported and discovered using Southern blotting technique that the *GH-1* gene was missing. Subsequently, additional cases of *GH-1* gene deletions have been described responding well to the GH treatment. The development of anti-GH antibodies is an inconsistent finding in IGHD IA patients despite having identical molecular defects (homozygosity for *GH-1* gene deletions) [25]. The frequency of *GH-1* gene deletions as a cause of GH deficiency varies among different populations and the criteria and definition of short stature chosen [1]. The sizes of the deletions are heterogeneous with the most frequent (70%–80%) being 6.7 kb [1,13]. The remaining deletions described include 7.6, 7, and 45 kb, and double deletions within the *GH* gene cluster [1,13]. At the molecular level, these deletions involve unequal recombination and crossing over within the *GH*-gene cluster at meiosis [13].

GH-1 gene frameshift- and nonsense mutations

Single base pair deletions and non-sense mutations of the signal peptide may result in an absent production of mature GH and in the production of anti-GH antibodies on exogenous replacement therapy [1].

Isolated growth hormone deficiency type IB

Patients with IGHD type IB are characterized by low but detectable levels of GH (<7 mU/L; <2.5 ng/mL); short stature (<-2 standard deviation score [SDS] for age and sex); growth deceleration and height velocity less than 25th percentile for age and sex; significantly delayed bone age; an autosomal-recessive inheritance (two parents of normal height, two siblings affected); no demonstrable direct or endocrine cause for IGHD; and a positive response and immunologic tolerance to treatment with exogenous GH. This subgroup of IGHD has been broadened and reclassified on the basis of the nature of their *GH* gene defects and includes splicing sites mutations of the *GH* gene; even an apparent lack of GH has been found by radioimmunoassay. The phenotype of IGHD type IB is more variable than IA. In one family, the children may resemble IGHD type IA, whereas in other families, growth during infancy is relatively normal and growth failure is not noted until mid-childhood. Similarly, GH may be nearly lacking or simply low following stimulation test. This heterogeneous phenotype suggests that there is more than one candidate gene causing the disorder as summarized recently [1].

Candidates genes in isolated growth hormone deficiency type IB

Some of the components of the GH pathway are unique to GH, whereas many others are shared. In patients with IGHD, mutational changes in genes specific to the GH-RH–GH axis are of importance and there is a need to focus on them.

GHRH-gene

Many laboratories put a lot of energy to define any *GHRH* gene alterations. To date, GHRH gene mutations or deletions causing IGHD have been reported [26,27].

GHRH-receptor gene

In 1992, Mayo [28] cloned and sequenced the rat and human *GHRH-receptor* gene, which provided the opportunity to examine the role of GH-RH–receptor in growth abnormalities that involve the GH-axis. Sequencing of the *GHRH-receptor* gene in the *little*-mouse (*lit/lit*) showed a single nucleotide substitution in codon 60 that changed aspartic acid to glycine (D60G) eliminating the binding of GH-RH to its own receptor [29]. Because the phenotype of IGHD type IB in humans has much in common with the phenotype of homozygous *lit/lit* mice including autosomal-recessive inheritance, time of onset of growth retardation, diminished secretion of GH and IGF-I, proportional reduction in weight and skeletal size, and delay in sexual maturation, the *GHRH-receptor* gene was searched for alteration in these patients suffering from IGHD type IB [30,31]. Wajnrajch and coworkers [31] reported a non-sense mutation similar to the *little* mouse in an Indian Moslem kindred. Furthermore, in two villages in the Sindh area of Pakistan, Baumann and Maheshwari [32] reported another form of severe short stature caused by a point mutation in the *GHRH-receptor* gene resulting in a truncation of the extracellular domain of this receptor. Individuals who are homozygous for this mutation are very short (-7.4 SDS) but normally proportioned. They seem of normal intelligence, and at least some are fertile. Biochemical testing revealed that they have normal levels of GH-RH and GHBP, but undetectable levels of GH and extremely low levels of IGF-I. Later, families from Sri Lanka, Brazil, United States, Spain, and Pakistan were reported [33–36]. Mutations in the *GHRH-receptor* gene have been described as the basis for a syndrome characterized by autosomal-recessive IGHD and anterior pituitary hypoplasia, defined as pituitary height more than 2 SD below age-adjusted normal, which is likely caused by depletion of the somatotroph cells (OMIM: 139,190). In a most recent report, however, certain variability in anterior pituitary size even in siblings with the same mutation was described [37].

Specific trans-acting factor to GH-gene

Any alteration to the specific transcriptional regulation of the *GH-1* gene may produce IGHD type IB. Mullis and coworkers [38] have reported a heterozygous 211-bp deletion within the *retinoic acid receptor-α* gene causing the phenotype of IGHD type IB.

Isolated growth hormone deficiency type II

Focusing on the autosomal-dominant form of IGHD, type II (IGHD II) is mainly caused by mutations within the first six base pairs of intervening sequences 3 (5′IVS-3) [1], which result in a missplicing at the mRNA level and the subsequent loss of exon 3, producing a 17.5-kd human GH (hGH) isoform [1,39]. This GH product lacks amino acid 32-71 (del32-71 GH), which is the entire loop that connects helix 1 and helix 2 in the tertiary structure of hGH [40,41]. Skipping of exon 3 caused by *GH-1* gene alterations other than those at the donor splice site in 5′IVS-3 has also been reported in other patients with IGHD II. These include mutations in exon 3 (E3) splice enhancer ESE1 (E3 + 1G-> T:ESE1m1; E3 + 2A-> C:ESE1m2, E3 + 5A-> G:ESE1m3) and ESE2 (downstream of the cryptic splice site in E3; ESE2: Δ721-735) and within suggested intron splice enhancers (ISE) (IVS-3 + 28 G-> A: ISEm1; IVS-3del + 28-45: ISEm2) sequences [1,42–49]. Such mutations lie within purine-rich sequences and cause increased levels of exon 3 skipped transcripts [42–45,47–49], suggesting that the usage of the normal splicing elements (ESE1 at the 5′ end of exon 3 and ISE in intron 3) may be disrupted [47–49]. Importantly, the first seven nucleotides in exon 3 (ESE1) are crucial for the splicing of GH mRNA [49] such that some nonsense mutations might cause skipping of one or more exons during mRNA splicing in the nucleus. This phenomenon is called "nonsense-mediated altered splicing"; its underlying mechanisms are still unknown [50]. Furthermore, there is a recent report of Vivenza and coworkers [51] presenting a patient with a specific deletion within intron 3 leading to exon 3 skipping, which underlines the importance of intron length on the splicing machinery, as it was previously suggested by the elegant work by Ryther and coworkers [49]. In addition to the previously described splice site mutations that result in the production of del32-71 GH, three other mutations within the *GH-1* gene (missense mutations) are reported to be responsible for IGHD II: the substitution of leucine for proline, histidine for arginine, and phenylalanine for valine at amino acid positions 89 (P89V), 183 (R183H), and 110 (V110F), respectively [52–54].

At the functional level, the 17.5-kd isoform exhibits a dominant-negative effect on the secretion of the 22-kd isoforms in both tissue cultures and in transgenic animals [55–57]. The 17.5-kd isoform is initially retained in the endoplasmic reticulum, disrupts the Golgi apparatus, impairs both GH and other hormonal trafficking [58], and partially reduces the stability of the 22-kd isoform [55]. Furthermore, transgenic mice overexpressing the 17.5-kd isoform exhibit a defect in the maturation of GH secretory vesicles and the anterior pituitary gland is hypoplastic because of a loss of most somatotropes [47,55,56]. Trace amounts of the 17.5-kd isoforms, however, are normally present in children and adults of normal growth and stature [59], and heterozygosity for A731G mutation (K41R) within the newly defined ESE2 (which is important for exon 3 inclusion) led to approximately 20% exon 3 skipping resulting in both normal and short stature [47,49,60]. From the clinical point of view, severe short stature (<-4.5 SDS) is not present in all affected individuals, indicating that in some forms growth of IGHD II, growth failure is less severe than one might expect [53]. It has been hypothesized that children with splice site mutations may be younger and shorter at diagnosis than their counterparts with missense mutations [53]. Furthermore, more recent in vitro and animal data suggest that both a quantitative and qualitative difference in phenotype may result from variable splice site mutations causing differing degrees of exon 3 skipping [1,60–64]. These data suggest that the variable phenotype of autosomal-dominant GHD may reflect a threshold and a dose dependency effect of the amount of 17.5-kd relative to 22-kd hGH [56,57,60]. Specifically, this has a variable impact on pituitary size, and on onset and severity of GHD and, unexpectedly, the most severe, rapid onset forms of GHD might be subsequently associated with the evolution of other pituitary hormone deficiencies [65,66].

Isolated growth hormone deficiency type III

This reported type is X-linked recessively inherited. In these families, the affected males were immunoglobulin and GH deficient [67,68]. Recent studies have shown that the long arm of chromosome X may be involved and that the disorder may be caused by mutations or deletions of a portion of the X chromosome containing two loci, one necessary for normal immunoglobulin production, and the other for GH expression [69]. In addition, Duriez and coworkers [70] reported an exon-skipping mutation in the btk-gene of a patient with X-linked agammaglobulinemia and IGHD.

Isolated growth hormone deficiency II disrupts secretory vesicles in vitro and in vivo in transgenic mice

To determine the mechanisms of the dominant negative effects McGuinness and coworkers [56] used a combination of transgenic and morphologic approaches in both in vitro and in vivo models. In the transgenic mouse model, McGuinness and coworkers [56] drove exon-3-skipped hGH expression specifically in somatotrophs, using an 5′IVS-3 + 1G-> A mutant construct that could only generate an hGH product lacking exon 3, this in a normal mouse background (ie, with two copies of the mouse GH gene). The results showed that the most severely affected lines rapidly developed severe GHD and dwarfism, with profound pituitary hypoplasia and almost total loss of somatotrophs. Importantly, however, the onset and severity of IGHD II was transgene copy number dependent; high-copy lines rapidly developed profound GHD by weaning, whereas low-copy mice showed milder adult-onset GHD. These data were supported by an in vitro splicing study showing that other human IGHD II mutations within splice-enhancer or repressor sequences generated a variable ratio of mutant–wild-type GH (*wt*-hGH) transcripts, which might help to explain some variability seen in IGHD II patients with a similar range of mutations [47,49]. In detail, an unexpected observation, however, was that in the most severely affected high-copy number IGHD II transgenic mice, the marked anterior pituitary gland hypoplasia was associated with deficits in other pituitary derived hormones [47,56]. Prolactin, TSH, and luteinizing hormone in males only were all significantly reduced in adult high-copy transgenic mice, whereas these deficits were absent or mild, with a much later onset in low-copy lines. This could simply reflect an artificially high ratio of exon-3 skipped hGH to mouse GH, but raised the question of whether other hormone deficits might evolve spontaneously in the more severe forms of human IGHD II.

Furthermore, analysis of the different lines of transgenic mice was consistent with the notion that the onset, severity, and rate of progression are proportional to the relative amounts of del32-71-GH versus *wt*-GH and somehow predictable. The most severely affected lines were already GH deficient at weaning and developed proportionate reductions in weight and length and bone growth, presented with other pituitary gland–derived hormonal deficiencies over time and with a concomitant effect on pituitary size.

Isolated growth hormone deficiency II: an evolving pituitary deficit in humans?

General aspects of the dominant-negative nature of isolated growth hormone deficiency type II

IGHD II is a rare autosomal-dominant form of GHD, usually caused by hGH splicing mutations that generate internally truncated GH forms that block the secretion of *wt*-GH produced from the normal allele [1,49,71]. From the clinical point of view there is some evidence that there is great variability of the IGHD II phenotype in terms of onset, severity, and rate of progression according to the mutation (splice site versus missense) within the *GH-1* gene [52–54,60,61]. Severe short stature was only present in one third of the affected individuals at diagnosis in the study by Binder and coworkers [53]; in addition, children with splice site mutations were younger and shorter at diagnosis than those with missense mutations [53].

Furthermore, it has to be stressed that in a recent study the authors described five families, clinically clearly autosomal dominantly GH deficient, where they could not find any *GH-1* gene alterations [72]. In these subjects not only the whole *GH-1* gene but also *HESX-1* was sequenced; as in mice and humans monoallelic *hesx-1/HESX-1* alterations may cause mild forms of GHD.

Variable clinical course depending on the GH-1 *gene alteration*

The authors studied a total 89 subjects on a clinical basis, and sequenced the whole *GH-1*, introns and exons included, in the index cases of 32 families with IGHD II [65,66]. Sixty-nine subjects belonging to 27 families suffered from different splice site and missense mutations within the *GH-1* gene. The subjects presenting with a 5'IVS-3 + 1/+ 2-bp splice site mutation causing a skipping of exon 3 were more likely to harbor other pituitary hormone deficiencies [65]. In addition, although the patients with missense mutations have previously been reported to be less affected, a number of patients presenting with the P89L missense GH form showed some pituitary hormone impairment [65,66]. The development of multiple hormonal deficiencies is not age dependent and there is a clear variability in onset, severity, and progression, even within the same families presenting with identical *GH-1* gene alterations [65,66].

Diagnosis

Focusing on the time of diagnosis during infancy and childhood, there is clearly a tendency to diagnose the IGHD II children presenting a 5'IVS-3 + 1/+ 2-bp mutations earlier than those with a splice site mutation within 5'IVS-3 + 5/+ 6-bp and with a missense mutation. Further, there was no difference in age at diagnosis among the various forms of missense mutated GH. One should interpret this with caution, however, because awareness of a familial GHD increases the likelihood of an earlier diagnosis in a second affected child, and it is interesting that on GH stimulation testing, peak GH values did not vary among the different groups.

Compared with patients suffering from IGHD II caused by splice site mutations leading to a skipping of exon 3 (median age at diagnosis, 3 years) [65], patients with P89L GH were rather older (median age, 7.35 years) when GHD was diagnosed by their physician (mean height velocity, 2.7 cm/y; -5.8 SDS). Although this time of primary diagnosis was comparable with that in subjects presenting with R183H GH missense mutations (median, 7.2 years) and much later than in subjects with splice site mutations, examination of their growth charts revealed that subjects with P89L GH were falling off the centiles much earlier. Their response to GH replacement therapy (15-20 $IU/m^2/wk$) seems to be judged as good because all subjects surpassed their estimated parental target heights. The authors interpret this result with caution, however, because in an autosomal-dominant growth disorder one parental height may be compromised.

Follow-up of endocrine parameters

Hormonal data splice-site mutation

There is evidence that other pituitary hormone deficiencies may develop in patients with GH-1 gene alterations. In two families with (5'IVS-3 + 1;

GT-> CT) splice mutation, the authors identified either adrenocorticotropic hormone (ACTH) or combined ACTH-TSH deficiency, requiring replacement therapy. In this study the authors also took the opportunity to analyze the impact of age or GH treatment on the disorder by studying the affected family members that did not receive any rhGH-treatment. Of these 24 older subjects (mean age, 44 years; range, 21–73 years), four (two males, two females) with 5'IVS-3 + 5/+ 6-bp splice site mutations could be retested and other than GHD, no further pituitary hormonal deficiencies were present. In contrast, the 5'IVS-3 + 1/+ 2-bp subgroup (N = 7) contained two women (5'IVS-3 + 1G- > A; 5'IVS-3 + 1G-> T) who exhibited partial TSH and ACTH deficiency, which needed replacement therapy.

Hormonal data missense mutations

The most striking difference noted in patients with P89L GH is the longer-term impact on other pituitary hormonal axes. Out of 12 subjects, eight were partially ACTH or TSH deficient either at the median age of 16.5 years following GH replacement therapy (N = 3) or in all but one untreated affected adult family members at the median adult age of 47 years (N = 5). This contrasts markedly with data from 17 patients with IGHD II caused by the R183H missense mutated GH form.

Size of pituitary gland

Patients who develop any additional pituitary hormone deficiency show a more hypoplastic pituitary gland with time. The authors believe it is noteworthy that whereas no difference in pituitary size was evident at diagnosis in the two groups of 5'IVS-3 splice site mutated subjects, a significant difference clearly emerged at final height, with the pituitaries significantly smaller in the 5'IVS-3 + 1/+ 2-bp than in the 5'IVS-3 + 5/+ 6-bp patients. Furthermore, in the affected family members that did not receive rhGH-treatment the size of the pituitary gland is in line with the hormonal deficiencies, with the 5'IVS-3 + 1/+ 2-bp group having smaller pituitary size than the 5'IVS-3 + 5/+ 6-bp group, although the numbers are small. The mechanism affecting other pituitary cell types is not obvious, but it might reflect bystander damage from activated macrophages clearing dying somatotroph debris, as observed in the transgene mouse study [56].

Pituitary hypoplasia probably evolves slowly. Several studies have reported no differences in pituitary sizes assessed by MRI scans between the various forms of IGHD II. At re-evaluation, however, the pituitary gland sizes of P89L GH affected subjects were significantly smaller when compared with the normative data reported by Tsunoda and coworkers [73]. The combination of severity of phenotype, pituitary hypoplasia, and incidence of pituitary endocrine deficits of the most severe splice site mutations is shared by this particular P89L missense mutation.

Possible mechanism causing variable phenotypes depending on the splice site mutation

It is of interest that different splice site mutations generating the same exon 3–deleted product may end up with a variable phenotype. The authors' clinic-derived data, however, are broadly in line with the results reported in the transgenic mice by McGuinness and coworkers [56]. The high-copy lines showed compromised numbers of corticotrophs, gonadotrophs, and lactotrophs by electron microscopy; thyrotroph deficits were not evaluated directly, but were probably reduced because the TSH contents were markedly lower [56]. In the most severely affected mouse lines, luteinizing hormone was reduced; however, in the milder lines with later-onset GHD, the mice were fully fertile. In the authors' subjects, clinically the gonadotroph axis seems to be normal, whereas the corticotroph and thyrotroph axes are compromised in some patients mainly suffering from the 5′IVS-3 + 1-bp caused exon-3 skipped IGHD II variant.

Based on the fact that the splice sites in *GH-1*, particularly those flanking exon 3, are supposed to be rather weak, Ryther and coworkers [47,49] developed a model of IGHD II pathogenesis in which mutations that weaken exon 3 definition, either at the splice sites or within defined splice enhancers (ESE, ISE), lead to an enhanced exon 3 skipping. In that sense, it seems that a splice site mutation at position 5′IVS-3 + 1/+ 2-bp has a greater impact on spliceosome assembly and splicing efficiency than a splice site mutation at position 5′IVS-3 + 5/+ 6-bp. This may possibly increase the exon 3 skipping and, consequently, may be responsible for the higher quantity of 17.5-kd GH variant, which apparently exerts a dominant-negative effect on the packaging and secretion of 22-kd GH, and leads eventually to a more severe phenotype.

In addition, McGuinness and coworkers [56] raised the concern that untreated GH deficiency could exacerbate pituitary damage because the lack of GH feedback clearly increased GH-RH expression in the mice, which might drive the exhaustion of pituitary somatotrophs, directly or indirectly damaging other pituitary axes. Data comparing treated with untreated groups suggest this is not of clinical concern.

Analysis at the cellular level

Because all of these variable *GH-1* gene alterations and defects, which either lead to exon 3 skipping and or an altered GH peptide, have a major impact on GH secretion and seem dissimilarly to affect the clinical phenotype, both in vitro and in animal models of IGHD II, a more detailed analysis of the secretory pathway is crucial. The subcellular distribution and colocalization of *wt*-GH with the mutant GH peptide can be studied by applying quantitative confocal analysis. Further, using the immunofluorescent technique, cells can be double stained for GH plus one of the following organelles: endoplasmic reticulum (ER; anti-Grp94); Golgi (anti-βCOP); or secretory granules (anti-Rab3a). This is a novel approach, which allows quantitative assessment of the intracellular localization of a particular peptide and, specifically, the analysis of the impact of the del32-71 GH/mutant GH peptide form in comparison with the *wt-* GH at the cellular and subcellular level in terms of GH synthesis, transport, storage, and secretion. Furthermore, while using AtT-20 cells, which endogenously secrete ACTH, the significance of the del32-71 GH on ACTH secretion as an internal control for the controlled secretory pathway could be investigated. Moreover, GH secretion and cell viability can be studied and compared in detail [65,66].

Clinical impact of GH gene mutations leading to isolated growth hormone deficiency II

These findings clearly support the notion that depending on the *GH-1* gene alteration there is a clinical variability in the severity of the IGHD II phenotype. Furthermore, subjects suffering from IGHD II caused by a 5′IVS-3 + 1/+ 2-bp splice site mutations leading to a skipping of exon 3 may also present with other pituitary hormone deficiencies, mainly in the ACTH and TSH, but not in the gonadotroph axis. Although missense mutations were reported to generate a less severe phenotype, this generalization may not hold true for patients presenting the P89L GH form. Because the occurrence of multiple hormonal deficiencies does not obviously increase with age or prior GH treatment, the concern that enhanced GH-RH drive in the absence of GH aggravates and supports the development of other hormonal abnormalities at the pituitary level does not seem to be a problem in human IGHD II. The imaging data strongly suggest, however, that pituitary growth deficits continue to develop with time on GH treatment, so this should be borne in mind when considering such patients for GH replacement in adulthood. The analysis also suggests that variability in onset, severity, and progression, evident between mutation genotypes, may also occur within families with the same mutation. Perhaps the most important message is that other hormone deficits can develop in IGHD II patients, underscoring the clinical importance of maintaining vigilance for the development of other hormonal deficiencies over the years.

Acknowledgment

These studies are only possible because of the most successful collaboration with many friends: Paul Czernichow, Dominique Simon, and Serge Amselem from Paris, France; Gerhard Binder, Tübingen Germany; Iain CAF Robinson, Mill Hill, London, England.

References

[1] Mullis PE. Genetic control of growth. Eur J Endocrinol 2005;152:11–31.
[2] Tatsumi K, Miyai K, Notomi T, et al. Cretinism with combined hormone deficiency caused by a mutation in the PIT-1 gene. Nat Genet 1992;1:56–8.
[3] Cohen LE, Wondisford FE, Radovick S. Role of Pit-1 in the gene expression of growth hormone, prolactin, and thyrotropin. Endocrinol Metab Clin North Am 1996;25:523–40.
[4] Kishimoto M, Okimura Y, Fumoto M, et al. The R271W mutant form of Pit-1 does not act as a dominant inhibitor of Pit-1 action to activate the promoters of GH and prolactin genes. Eur J Endocrinol 2003;148:619–25.
[5] Okamoto N, Wada Y, Ida S, et al. Monoallelic expression of normal mRNA in the PIT1 mutation heterozygotes with normal phenotype and biallelic expression in the abnormal phenotype. Hum Mol Genet 1994;3:1565–8.
[6] Wu W, Cogan JD, Pfaffle RW, et al. Mutations in PROP1 cause familial combined pituitary hormone deficiency. Nat Genet 1998;18:147–9.
[7] Sornson MW, Wu W, Dasen JS, et al. Pituitary lineage determination by the *Prophet of Pit-1* homeodomain factor defective in Ames dwarfism. Nature 1996;384:327–33.
[8] Fluck C, Deladoey J, Rutishauser K, et al. Phenotypic variability in familial combined pituitary hormone deficiency caused by a PROP1 gene mutation resulting in the substitution of Arg→Cys at codon 120 (R120C). J Clin Endocrinol Metab 1998;83:3727–34.
[9] Duquesnoy P, Roy A, Dastot F, et al. Human Prop-1: cloning, mapping, genomic structure. Mutations in familial combined pituitary hormone deficiency. FEBS Lett 1998;437:216–20.
[10] Deladoey J, Fluck C, Buyukgebiz A, et al. Hot spot in the PROP1 gene responsible for combined pituitary hormone deficiency. J Clin Endocrinol Metab 1999;84:1645–50.
[11] Mendonca BB, Osorio MG, Latronico AC, et al. Longitudinal hormonal and pituitary imaging changes in two females with combined pituitary hormone deficiency due to deletion of A301,G302 in the PROP1 gene. J Clin Endocrinol Metab 1999;84:942–5.
[12] Chen EY, Liao YC, Smith DH, et al. The human growth hormone locus: nucleotide sequence, biology, and evolution. Genomics 1989;4:479–87.
[13] Phillips JA III. Inherited defects in growth hormone synthesis and action. In: Scriver CR, Beaudet AL, Sly WS, et al, editors. The metabolic and molecular bases of inherited disease, vol. 2. 7th edition. New York: McGraw-Hill; 1995. p. 3023–44.
[14] Rosenfeld RG, Cohen P. Disorders of growth hormone / insulin-like growth factor secretion and action. In: Sperling MA, editor. Pediatric endocrinology. 2nd edition. Philadelphia: WB Saunders; 2002. p. 211–88.
[15] Baumann G. Growth hormone heterogeneity: genes, isohormones, variants, and binding proteins. Endocr Rev 1991;12:424–49.
[16] Lewis UJ, Bonewald LF, Lewis LJ. The 20,000-dalton variant of human growth hormone: location of the amino acid deletions. Biochem Biophys Res Commun 1980;92:511–6.
[17] DeNoto FM, Moore DD, Goodman HM. Human growth hormone DNA sequence and mRNA structure: possible alternative splicing. Nucleic Acids Res 1981;9:3719–30.
[18] Nuoffer JM, Flück C, Deladoëy J, et al. Regulation of human GH receptor gene transcription by 20- and 22-kDa GH in a human hepatoma cell line. J Endocrinol 2000;165:313–20.
[19] Lacey KA, Parkin JM. Causes of short stature: a community study of children in Newcastle upon Tyne. Lancet 1974;1:42–5.
[20] Rona RJ, Tanner JM. Aetiology of idiopathic growth hormone deficiency in England and Wales. Arch Dis Child 1977;52:197–208.
[21] Vimpani GV, Vimpani AF, Lidgard GP, et al. Prevalence of severe growth hormone deficiency. BMJ 1977;2:427–30.
[22] Cacciari E, Zucchini S, Carla G, et al. Endocrine function and morphological findings in patients with disorders of the hypothalamo-pituitary area: a study with magnetic resonance. Arch Dis Child 1990;65:1199–202.

[23] Illig R. Growth hormone antibodies in patients treated with different preparations of human growth hormone (hGH). J Clin Endocrinol Metab 1970;31:679–88.
[24] Phillips JA III, Hjelle B, Seeburg PH, et al. Molecular basis for familial isolated growth hormone deficiency. Proc Natl Acad Sci U S A 1981;78:6372–5.
[25] Laron Z, Kelijman M, Pertzelan A, et al. Human growth hormone deletion without antibody formation or growth arrest during treatment: a new disease entity? J Med Sci 1985;21: 999–1006.
[26] Mullis PE, Patel M, Brickell PM, et al. Isolated growth hormone deficiency: analysis of the growth hormone (GH) releasing hormone gene and the GH gene cluster. J Clin Endocrinol Metab 1990;70:187–91.
[27] Perez-Jurado LA, Phillips JA III, Francke U. Exclusion of growth hormone (GH)-releasing hormone gene mutations in family isolated GH deficiency by linkage and single strand conformation analysis. J Clin Endocrinol Metab 1994;78:622–8.
[28] Mayo K. Molecular cloning and expression of a pituitary-specific receptor for growth hormone-releasing hormone. Mol Endocrinol 1992;6:1734–44.
[29] Lin SC, Lin CR, Gukovsky I. Molecular basis of the *little* mouse phenotype and implications for cell type-specific growth. Nature 1993;364:208–13.
[30] Cao Y, Wagner JK, Hindmarsh PC, et al. Isolated growth hormone deficiency: testing the *little* mouse hypothesis in man and exclusion of mutations within the extracellular domain of the growth hormone-releasing hormone receptor. Pediatr Res 1995;38:962–6.
[31] Wajnrajch MP, Gertner JM, Harbison MD, et al. Nonsense mutation in the human growth hormone receptor causes growth failure analogous to the little (*lit*) mouse. Nat Genet 1996; 12:88–90.
[32] Baumann G, Maheshwari H. The dwarfs of Sindh: severe growth hormone (GH) deficiency caused by a mutation in the GH-releasing hormone receptor gene. Acta Paediatr Suppl 1997; 423:33–8.
[33] Maheshwari HG, Silverman BL, Dupuis J, et al. Phenotype and genetic analysis of a syndrome caused by an inactivating mutation in the growth hormone-releasing hormone receptor: dwarfism of Sindh. J Clin Endocrinol Metab 1998;83:4065–74.
[34] Netchine I, Talon P, Dastot F, et al. Extensive phenotypic analysis of a family with growth hormone (GH) deficiency caused by a mutation in the GH-releasing hormone receptor gene. J Clin Endocrinol Metab 1998;83:432–6.
[35] Salvatori R, Hayashida CY, Aguiar-Oliveira MH, et al. Familial dwarfism due to a novel mutation of the growth hormone-releasing hormone receptor gene. J Clin Endocrinol Metab 1999;84:917–23.
[36] Salvatori R, Fan X, Phillips JA III, et al. Three new mutations in the gene for the growth hormone (GH)-releasing hormone receptor in familial isolated GH deficiency type Ib. J Clin Endocrinol Metab 2001;86:273–9.
[37] Alba M, Hall CM, Whatmore AJ, et al. Variability in anterior pituitary size within members of a family with GH deficiency due to a new splice mutation in the GHRH receptor gene. Clin Endocrinol (Oxf) 2004;60:470–5.
[38] Mullis PE, Eblé A, Wagner JK. Isolated growth hormone deficiency is associated with a 211bp deletion within RAR α gene [abstract]. Horm Res 1994;41:61.
[39] Binder G, Ranke MB. Screening for growth hormone (GH) gene splice-site mutations in sporadic cases with severe isolated GH deficiency using ectopic transcript analysis. J Clin Endocrinol Metab 1995;80:1247–52.
[40] de Vos AM, Ultsch M, Kossiakoff AA. Human growth hormone and extracellular domain of its receptor: crystal structure of the complex. Science 1992;255:306–12.
[41] Cunningham BC, Ultsch M, De Vos AM, et al. Dimerization of the extracellular domain of the human growth hormone receptor by a single hormone molecule. Science 1991;254: 821–5.
[42] Cogan JD, Prince MA, Lekhakula S, et al. A novel mechanism of aberrant pre-mRNA splicing in humans. Hum Mol Genet 1997;6:909–12.

[43] Cogan JD, Ramel B, Lehto M, et al. A recurring dominant negative mutation causes autosomal dominant growth hormone deficiency: a clinical research center study. J Clin Endocrinol Metab 1995;80:3591–5.
[44] Takahashi I, Takahashi T, Komatsu M, et al. An exonic mutation of the GH-1 gene causing familial isolated growth hormone deficiency type II. Clin Genet 2002;61:222–5.
[45] Moseley CT, Mullis PE, Prince MA, et al. An exon splice enhancer mutation causes autosomal dominant GH deficiency. J Clin Endocrinol Metab 2002;87:847–52.
[46] Mullis PE, Deladoey J, Dannies PS. Molecular and cellular basis of isolated dominant-negative growth hormone deficiency, IGHD type II: Insights on the secretory pathway of peptide hormones. Horm Res 2002;58:53–66.
[47] Ryther RC, McGuinness LM, Phillips JA III, et al. Disruption of exon definition produces a dominant-negative growth hormone isoform hat causes somatotroph death and IGHD II. Hum Genet 2003;113:140–8.
[48] McCarthy EMS, Phillips JA III. Characterization of an intron splice enhancer that regulates alternative splicing of human GH pre-mRNA. Hum Mol Genet 1998;7:1491–6.
[49] Ryther RC, Flynt AS, Harris BD, et al. GH1 splicing is regulated by multiple enhancers whose mutation produces a dominant-negative GH isoform that can be degraded by allele-specific siRNA. Endocrinology 2004;145:2988–96.
[50] Dietz HC. Nonsense mutations and altered splice-site selections. Am J Hum Genet 1997;60: 729–30.
[51] Vivenza D, Guazzarotti L, Godi M, et al. A novel deletion in the GH1 gene including the IVS3 branch site responsible for autosomal dominant isolated growth hormone deficiency. J Clin Endocrinol Metab 2006;91:980–6.
[52] Deladoey J, Stocker P, Mullis PE. Autosomal dominant GH deficiency due to an Arg183His GH-1 gene mutation: clinical and molecular evidence of impaired regulated GH secretion. J Clin Endocrinol Metab 2001;86:3941–7.
[53] Binder G, Keller E, Mix M, et al. Isolated GH deficiency with dominant inheritance: new mutations, new insights. J Clin Endocrinol Metab 2001;86:3877–81.
[54] Duquesnoy P, Simon D, Netchine I, et al. Familial isolated growth hormone deficiency with slight height reduction due to a heterozygote mutation in GH gene. In: Program of the 80th Annual Meeting of the Endocrine Society. New Orleans: Endocrine Society; 1998. p. 2–202.
[55] Lee MS, Wajnrajch MP, Kim SS, et al. Autosomal dominant growth hormone (GH) deficiency type II: the Del32-71-GH deletion mutant suppresses secretion of wild-type GH. Endocrinology 2000;141:883–90.
[56] McGuinness L, Magoulas C, Sesay AK, et al. Autosomal dominant growth hormone deficiency disrupts secretory vesicles in vitro and in vivo n transgenic mice. Endocrinology 2003;144:720–31.
[57] Hayashi Y, Yamamoto M, Ohmori S, et al. Inhibition of growth hormone (GH) secretion by a mutant GH-I gene product in neuroendocrine cells containing secretory granules: an implication for isolated GH deficiency inherited in an autosomal dominant manner. J Clin Endocrinol Metab 1999;84:2134–9.
[58] Graves TK, Patel S, Dannies PS, et al. Misfolded growth hormone causes fragmentation of the Golgi apparatus and disrupts endoplasmic reticulum-to-Golgi traffic. J Cell Sci 2001; 114(Pt 20):3685–94.
[59] Lewis UJ, Sinha YN, Haro LS. Variant forms and fragments of human growth hormone in serum. Acta Paediatr Suppl 1994;399:29–31.
[60] Millar DS, Lewis MD, Horan M, et al. Novel mutations of the growth hormone 1 (GH1) gene disclosed by modulation of the clinical selection criteria for individuals with short stature. Hum Mutat 2003;21:424–40.
[61] Fofanova OV, Evgrafov OV, Polyakov AV, et al. A novel IVS2 -2A > T splicing mutation in the GH-1 gene in familial isolated growth hormone deficiency type II in the spectrum of other splicing mutations in the Russian population. J Clin Endocrinol Metab 2003;88: 820–6.

[62] Katsumata N, Matsuo S, Sato N, et al. A novel and de novo splice-donor site mutation in intron 3 of the GH-1 gene in a patient with isolated growth hormone deficiency. Growth Horm IGF Res 2001;11:378–83.
[63] Kamijo T, Hayashi Y, Seo H, et al. Hereditary isolated growth hormone deficiency caused by GH1 gene mutations in Japanese patients. Growth Horm IGF Res 1999;9(Suppl B):31–4.
[64] Kamijo T, Hayashi Y, Shimatsu A, et al. Mutations in intron 3 of GH-1 gene associated with isolated GH deficiency type II in three Japanese families. Clin Endocrinol (Oxf) 1999;51: 355–60.
[65] Mullis PE, Robinson IC, Salemi S, et al. Isolated autosomal dominant growth hormone deficiency: an evolving pituitary deficit? A multicenter follow-up study. J Clin Endocrinol Metab 2005;90:2089–96.
[66] Salemi S, Yousefi S, Baltensperger K, et al. Variability of isolated autosomal dominant GH deficiency (IGHD II): impact of the P89L GH mutation on clinical follow-up and GH secretion. Eur J Endocrinol 2005;153:791–802.
[67] Fleisher TA, White RM, Broder S, et al. X-linked hypogamma-globulinemia and isolated growth hormone deficiency. N Engl J Med 1980;302:1429–34.
[68] Sitz KV, Burks AW, Williams LW, et al. Confirmation of X-linked hyogammaglobuliemia with isolated growth hormone deficiency as a disease entity. J Pediatr 1990;116:292–4.
[69] Conley ME, Burks AW, Herrod HG, et al. Molecular analysis of X-linked agammaglobulinemia and isolated growth hormone deficiency. J Pediatr 1991;119:392–7.
[70] Duriez B, Duquesnoy P, Dastot F, et al. An exon-skipping mutation in the btk gene of a patient with x-linked agammaglobulinemia and isolated growth hormone deficiency. FEBS Lett 1994;346(2–3):165–70.
[71] Binder G, Brown M, Parks J. Mechanisms responsible for dominant expression of human growth hormone gene mutations. J Clin Endocrinol Metab 1996;81:4047–50.
[72] Fintini D, Salvatori R, Salemi S, et al. Autosomal dominant growth hormone deficiency (IGHD II) with normal GH-1 gene. Horm Res 2006;65:76–82.
[73] Tsunoda A, Okuda O, Sato K. MR height of the pituitary gland as a function of age and sex: especially physiological hypertrophy in adolescence and in climacterium. AJNR Am J Neuroradiol 1997;18:551–4.
[74] Duquesnoy P, Amselem S, Gourmelen M, et al. A frameshift mutation causing isolated growth hormone deficiency type IA. Am J Med Genet 1990;47:A110.
[75] Igarashi Y, Ogawa M, Kamijo T, et al. A new mutation causing inherited growth hormone deficiency: a compound heterozygote of a 6.7 kb deletion and a two base deletion in the third exon of the GH-1 gene. Hum Mol Genet 1993;2:1073–4.
[76] Cogan JD, Phillips JA III, Sakati N, et al. Heterogeneous growth hormone (GH) gene mutations in familial GH deficiency. J Clin Endocrinol Metab 1993;76:1224–8.
[77] Wagner JK, Eble A, Cogan JD, et al. Allelic variations in the human growth hormone-1 gene promoter of growth hormone-deficient patients and normal controls. Eur J Endocrinol 1997; 137:474–81.
[78] Takahashi Y, Kaji H, Okimura Y, et al. Brief report: short stature caused by a mutant growth hormone. N Engl J Med 1996;334:432–6.
[79] Chihara K. Identification of a growth hormone mutation responsible for short stature. Acta Paediatr Suppl 1996;417:49–50.
[80] Missarelli C, Herrera L, Mericq V, et al. Two different 5′ splice site mutations in the growth hormone gene causing autosomal dominant growth hormone deficiency. Hum Genet 1997; 101:113–7.
[81] Cogan JD, Phillips JA III, Schenkman SS, et al. Familial growth hormone deficiency: a model of dominant and recessive mutations affecting a monomeric protein. J Clin Endocrinol Metab 1994;79:1261–5.
[82] Phillips JA III, Cogan JD. Genetic basis of endocrine disease. 6. Molecular basis of familial human growth hormone deficiency. J Clin Endocrinol Metab 1994;78:11–6.

Factors Regulating Growth Hormone Secretion in Humans

Naila Goldenberg, MD, Ariel Barkan, MD*

Division of Metabolism, Endocrinology and Diabetes, 3920 Taubman, Box 0354, University of Michigan Medical Center, Ann Arbor, MI 48109, USA

Growth hormone (GH) secretion is pulsatile in nature in all species. The periodic pattern of GH release plays an important role in transmitting the GH message in a tissue-specific manner. For example, only pulsatile GH can normalize muscle and cartilage insulin-like growth factor (IGF)-1 mRNA levels [1] and only the continuous component of GH's secretory profile induces hepatic mRNAs for certain cytochrome P-450 enzymes [2]. The question of what regulates the pulsatile GH secretion pattern is an issue of not only theoretical interest but of considerable practical importance for designing different GH therapies for a variety of human diseases.

GH synthesis and secretion are regulated primarily by the hypothalamic neuropeptides growth hormone–releasing hormone (GH-RH) and somatotropin release–inhibiting factor (somatostatin [SRIF]). Similar to other endocrine systems, the end product of GH's action, IGF-1, exerts a negative feedback effect on GH secretion. The amount of GH secreted and the pattern of its release is also subject to the nutritional state and to nutrients themselves and to the prevailing gonadal steroid milieu. Additionally, the recently discovered gastric hormone, ghrelin, may play a role. All these factors interact with each other in a precise and coordinated manner and the interplay between them is necessarily complex.

This article provides a brief introductory overview of the different regulators of GH secretion and concentrates primarily on human studies.

Growth hormone–releasing hormone

The stimulatory hypothalamic influence on GH secretion was first suggested by Reichlin [3], who demonstrated that hypothalamic destruction

---

* Corresponding author.

E-mail address: abarkan@umich.edu (A. Barkan).

abolished somatic growth in young rats. Subsequent studies by Frohman and coworkers [4,5] extended this observation by showing that destruction of the ventromedial hypothalamus reduced pituitary and plasma GH, whereas electrical stimulation of the ventromedial and arcuate nuclei acutely increased GH release. The long search for the putative GH-RH culminated in the isolation of this neurohormone from two malignant extrapituitary tumors producing acromegaly in humans [6,7] and the subsequent identification of the same peptide in human hypothalamus [8]. Administration of synthetic GH-RH reliably increases GH release in humans.

Studies by Plotsky and Vale [9] in male rats showed that pituitary-portal GH-RH is grossly augmented at the time of GH pulse. Moreover, immunoneutralization of GH-RH in these animals abolished the occurrence of GH pulses without altering interpulse GH levels. Interestingly, the same methodologic approach in female rats not only abolished the pulses but also decreased tonic, interpulse GH levels [10], and suggested sexual dimorphism of GH regulation. Finally, Nakamura and coworkers [11] demonstrated pulsatile GH-RH release from the stalk-median eminence in female rhesus monkeys and studies by Cataldi and coworkers [12] found correlation between GH pulses and GH-RH pulses in the pituitary-portal blood in conscious sheep. These studies have firmly established the causative role of GH-RH as a GH pulse generator. Indeed, only a pulsatile mode of GH-RH administration increased growth in normal and GH-RH–deficient rats [13] and upregulated GH mRNA and protein content in the pituitaries of young female rats [14]. Interestingly, however, continuous GH-RH infusion augmented pulsatile GH release in normal and GH-RH–deficient humans [15]. The mechanisms observed in animals may differ from the human model.

To elucidate the role of endogenous GH-RH in the regulation of GH secretion in humans, the authors administered a competitive GH-RH antagonist, (N-Ac1, D-Tyr2) GH-RH 1-29 [16]. Administration of this compound, either as a bolus dose at night [17] or as a continuous 24-hour intravenous infusion [18], severely impaired pulsatile GH release and proportionately suppressed acute GH responses to bolus doses of exogenous GH-RH. This is in complete accord with studies performed in human dwarfs with inactivating mutations of GH-RH receptor [19]. Additionally, the authors studied the role of endogenous GH-RH in the genesis of acute GH responses to pharmacologic stimuli. Previous studies, using indirect approaches, attributed GH responses to clonidine and levodopa to GH-RH release. GH responses to pyridostigmine were thought to result, however, from acute suppression of hypothalamic SRIF secretion. Responses after insulin hypoglycemia or arginine infusion are attributed to both SRIF and GH-RH. The difficulty with interpretation of such studies was best demonstrated by the studies of Magnan and coworkers [20] in sheep. A combined injection of GH-RH and neostigmine in sheep elicited much more powerful GH release than either of these compounds given separately. This suggested that neostigmine, a parenteral analogue of pyridostigmine, suppressed hypothalamic SRIF secretion. Direct measurements of GH-RH and SRIF in the hypothalamic-pituitary portal blood after a neostigmine challenge revealed powerful release of GH-RH, however, but no decline in SRIF secretion. Similarly, passive immunoneutralization of GH-RH in sheep abolished endogenous GH pulsatility and responses to neostigmine, whereas anti-SRIF serum was ineffective [20]. These data indicated that a reliance on a pharmacologic approach of synergy and potentiation may be misleading in attributing the effect of a particular GH-releaser to SRIF inhibition.

Pretreatment of humans with GH-RH antagonist [21] severely impaired acute GH responses to exogenous GH-RH but did not attenuate the response to acute termination of SRIF infusion (an analogue of acute inhibition of endogenous SRIF release). This suggests that if the GH response to a particular secretogogue was not inhibited by GH-RH receptor antagonist, the mechanism of that secretogogue's action did not include GH-RH release and was likely to be caused by inhibition of SRIF release. If, on the contrary, the response was blocked by GH-RH receptor antagonist, the involvement of endogenous GH-RH was evident.

In subsequent studies, GH-RH receptor antagonist severely inhibited acute GH responses to all stimuli tested: clonidine; levodopa; arginine; insulin hypoglycemia; pyridostigmine; and to a synthetic ghrelin agonist, GHRP-6 [21,22]. GH-RH is crucial for the genesis of acute GH responses to all of these stimuli.

In another series of experiments, the authors have developed an approach for a semiquantification of hypothalamic GH-RH output in humans. Direct sampling of hypothalamic-pituitary portal blood in humans is not feasible and the peripheral GH-RH concentrations do not reflect portal concentrations because of an approximately 5000-fold dilution of the hypothalamic effluent within the peripheral circulation. Only indirect approaches are possible. The authors used a pharmacologic approach of graded GH-RH infusions over a 1000-fold range (0.033–33 mcg/kg/h) and tested the ability of graded GH-RH boluses to affect an acute GH rise under those circumstances. In young men and women, there was a dose-dependent correlation between the dose of GH-RH antagonist and the suppression of acute GH responses to GH-RH, so that the dose-inhibition curve to the lowest dose of GH-RH (0.1 mcg/kg/h) was shifted to the left. This indicated that the model allowed differentiation between different degrees of GH-RH levels over a 3- to 10-fold range [16].

Aging-related attenuation of GH secretion (somatopause) is thought to represent a model of diminished GH-RH secretion. This is suggested indirectly, because attenuated GH pulse amplitude in aging is accompanied by intact pituitary responsiveness to exogenous GH-RH [16], and directly, because GH-RH output is decreased in the hypothalamus of elderly monkeys [11]. In the authors' studies, the dose-inhibition curve of GH output with graded doses of GH-RH antagonist was markedly shifted to the left in elderly men [16], showing that the somatopause is a GH-RH–deficient state.

Sexual dimorphism of GH secretion is seen in another interesting model. In men, most GH is secreted at the early hours of night, whereas daytime GH secretion is relatively hypopulsatile and has low baseline component [23]. In contrast, young women have similar total 24-hour GH output, but the structure of their GH secretion pattern is different. They have multiple GH secretory episodes and high basal GH output during the day and relatively attenuated nocturnal GH release [23]. The authors have shown that, unlike males, women have higher sensitivity to GH-RH antagonist during the night but also suppress daytime basal GH levels [18]. This is in complete agreement with the data in male and female rats [9,10] and indicates higher GH-RH output at night in men and the involvement of endogenous GH-RH in the maintenance of basal GH secretion in women.

Endogenous GH-RH is the principal regulator of pulsatile GH secretion in humans and is indispensable in the generation of GH responses to a variety of pharmacologic stimuli. The age-related GH decline in humans is a GH-RH–mediated phenomenon, as is the sexual dimorphism of the structure of pulsatile GH secretion during both daytime and nighttime (Fig. 1).

Somatostatin

SRIF, a 14–amino acid peptide, is the main negative regulator of GH secretion. It also suppresses thyroid-stimulating hormone and, to a lesser degree, adrenocorticotropic hormone and prolactin. Five subtypes of SRIF receptors have been identified, of which types 2 and 5 are the main mediators of the suppressive GH effect. SRIF powerfully antagonizes the mitogenic effect of GH-RH on somatotrophs, but does not inhibit GH synthesis [24]. It suppresses spontaneous GH release, however, and GH responses to all stimuli tested: GH-RH, hypoglycemia, arginine, exercise, and so forth. SRIF has a very short circulating half-life (approximately 3 minutes) and termination of its infusion elicits modest rebound GH rise and significantly augments GH response to GH-RH. This is likely caused by the accumulation of the readily releasable GH in the somatotrophs during SRIF exposure. A continuous infusion of GH-RH almost completely desensitizes pituitary responses to GH-RH in humans, but actually augments pulsatile GH release [15], and the rebound GH release post-SRIF withdrawal does not require GH-RH [21].

This led to a hypothesis that the periodic, pulsatile GH release is the result of a coordinated, but 180-degree out-of-phase, SRIF and GH-RH secretion. Direct measurements of SRIF in the pituitary-portal circulation in conscious sheep, however, failed to disclose any relations between SRIF and GH pulses [12].

Administration of SRIF antibody to rats does not alter pulsatile GH release but elevates the interpulse GH levels [9]. In humans, continuous infusion of a selective type 2 SRIF receptor agonist, octreotide, created circulating octreotide levels in excess of 1 to 2 ng/mL [25,26], at least 50 times

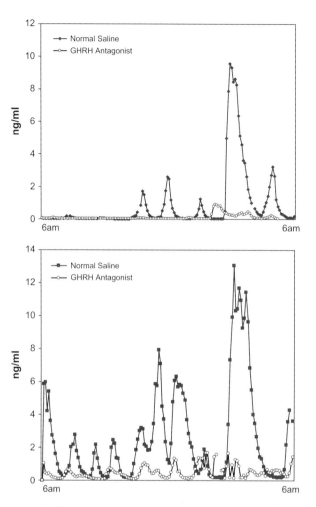

Fig. 1. Suppression of GH pulsatility in a young man (*upper panel*) and in a young woman (*lower panel*) by growth hormone–releasing hormone (GH-RH) receptor antagonist. Also note the easily discernible sexually dimorphic pattern of GH secretion at baseline.

higher than the pituitary-portal SRIF concentrations in rats [9] and sheep [12]. Even complete cessation of the endogenous hypothalamic SRIF release in these subjects decreases the pituitary exposure to this hormone by only 1% to 2%. This paradigm effectively created a model of an unvarying pituitary exposure to SRIF. Both in men and in women, GH pulse occurrence remained unmodified, but both the interpulse GH levels and the amplitude of GH pulses were powerfully suppressed. Based on this information, it seems that the role of SRIF in humans may be limited to the adjustment of the magnitude of basal and pulsatile GH release but not to affect generation of GH pulsatility (Fig. 2).

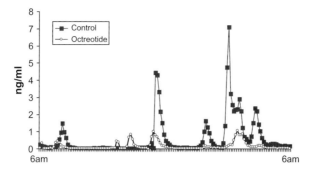

Fig. 2. Suppression of GH secretion by continuous octreotide infusion in a young man.

SRIF receptor type-2 knock-out mice [27] or animals with constitutive hypersecretion of SRIF [28] grew at a normal rate and passive immunization to SRIF in domestic livestock did not increase somatic growth [29].

The potential role of SRIF hypersecretion had been postulated to explain the diminished GH secretion in obesity and aging [30]. At least in the latter model, the decline in GH secretion is more likely to be caused by loss of GH-RH rather than gain of SRIF [11,16]. The physiologic role of SRIF in human physiology and pathology remains enigmatic. The resolution of this problem must await development of SRIF receptor antagonists.

Ghrelin

Ghrelin is a peptide of a primarily gastric origin, although ghrelin mRNA had been found in the hypothalamus [31]. Both circulating and hypothalamic ghrelin need to be considered when discussing the regulation of GH secretion. Ghrelin consists of 28 amino acids and circulates in two forms: the biologically active octanoylated one and the inactive deoctanoylated one. Synthetic analogues of ghrelin (GHS) had been produced as early as 25 years ago and were instrumental in discovering ghrelin receptor and, subsequently, ghrelin itself [32]. Most of the information regarding the GH-promoting effects of ghrelin had been obtained with these synthetic peptides.

Acute administration of GHS produces an immediate and massive release of GH [33]. Co-administration of GHS and GH-RH results in powerful GH rise that is greater than the effect of either peptide administered alone [33]. GHS were shown to act in a manner compatible with a functional SRIF antagonist [34], which may explain its synergism with GH-RH. GHS act directly at the pituitary level, but are much more powerful when applied to combined hypothalamic-pituitary segments in vitro [33] or in intact animals [33], suggesting that the presence of hypothalamic GH-RH is essential for their full action. Indeed, pretreatment of humans with GH-RH receptor antagonist attenuates their responsiveness to GHS [22]. Interestingly, however,

GHS retains full activity in a human model of homologous desensitization to GH-RH [35]. Functional occupancy of GH-RH receptors is indispensable for the full GHS action.

Continuous infusion of GHS induces homologous desensitization, but amplifies the magnitude of GH pulses and GH responsiveness to GH-RH [36], suggesting a cross-talk between GH-RH and ghrelin receptors. Studies aimed at the elucidation of the potential role of endogenous ghrelin failed, however, to find a significant role for this peptide as a regulator of spontaneous GH secretion. Neither total nor octanoylated ghrelin increased during fasting in parallel to the massive increase in GH secretion, and there was no concordance between the level of either of these peptides and GH pulses [37]. Similarly, GH secretion was not augmented in patients with ghrelin-producing tumors [38,39]. Even more importantly, in several animal models, knock-outs of either ghrelin or ghrelin receptor genes were not associated with any attenuation of growth [40–43]. Neither circulating nor hypothalamic ghrelin seems to be involved in meaningful physiologic regulation of GH secretion.

Negative feedback regulation of growth hormone secretion

As in all other endocrine systems, GH secretion is a target of multiple negative feedback loops at multiple levels. First, there are ultrashort feedback loops. GH-RH acutely inhibits its own secretion. This was shown in elegant studies by Lumpkin and coworkers [44,45] in rats demonstrating that intracerebroventricular injections of low doses of GH-RH inhibited, whereas those of GH-RH receptors antagonist stimulated, GH secretion. Similarly, hypothalamic SRIF also suppresses its own neuronal release both in vitro and in vivo [46].

The so-called "short-feedback" loops are also operative. Hypothalamic SRIF secretion is directly stimulated by GH-RH [47] and GH-RH secretion is inhibited by SRIF [48]. Pituitary GH inhibits hypothalamic GH-RH secretion [49,50] and stimulates SRIF release from the hypothalamic neurons [51,52]. Any stimulus altering GH secretion by acting at the hypothalamic levels automatically extinguishes its own effect. This likely participates in the creation of pulsatile GH-RH and SRIF secretion patterns and, ultimately, in the maintenance of pulsatile GH release from the pituitary.

These mechanisms are likely to be operative on a short-term basis. The long-term negative feedback of GH secretion is accomplished mainly through a classical "trophic-target" loop involving the ultimate product of GH action (ie, IGF-1).

The earliest and the most obvious evidence of this effect was obtained by Laron and coworkers [53] who identified a group of dwarfs with low-to-undetectable plasma IGF-1 levels and grossly augmented GH secretion. These patients were subsequently shown to have mutant GH receptors [54,55] and were unable to generate IGF-1.

The site of IGF-1 negative feedback on GH secretory mechanisms is still uncertain. Continuous infusion of IGF-1 in young men and women increased plasma IGF-1 concentrations three to four times above the upper limit of the normal range [23,56]. This reliably suppressed plasma GH concentrations by approximately 50% to 80% in both sexes, at the expense of grossly diminished GH pulse amplitude (Fig. 3). Administration of exogenous GH-RH to the same individuals, however, had a distinct sexually dimorphic effect. In men, IGF-1 infusion grossly suppressed plasma GH response to GH-RH suggesting that the negative feedback of IGF-1 was expressed either directly at the pituitary level or at the hypothalamic level by stimulating SRIF secretion. In contrast, administration of exogenous IGF-1 to women was completely ineffective in suppressing their GH response to GH-RH. This suggests that, in women, elevated IGF-1 suppresses GH secretion by selective suppression of the hypothalamic GH-RH output.

The reverse model (ie, assessment of GH secretion during an artificially induced decline in serum IGF-1) was also used. A study by Chapman and coworkers [57] has shown that the increase in GH secretion after termination of IGF-1 infusion is very rapid, occurring almost simultaneously with the first evidence of the decline of free IGF-1 in the peripheral circulation. Fasting has been shown to lower free and total [57,58] IGF-1 concentrations in the peripheral circulation. Whereas free IGF-1 declines rapidly, a statistically measurable decline in total IGF-1 may take 3 to 5 days because of a prolonged half-life of the bound hormone. Within a single day of fasting, plasma GH increases twofold to threefold and reaches a plateau after approximately 3 days of fasting [58]. The increase in GH secretion is accomplished almost exclusively at the expense of augmented GH pulse amplitude,

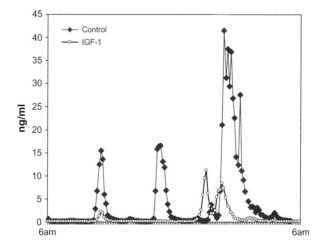

Fig. 3. Suppression of GH secretion by continuous insulin-like growth factor (IGF)-1 infusion in a young man.

whereas GH pulse frequency and the interpulse GH concentrations contribute only minimally to the total effect. Interestingly, plasma GH responsiveness to exogenous GH-RH does not change appreciably [58] suggesting the increase in GH-RH pulse amplitude is the underlying mechanism of fast-induced GH increase. Fasting also augments GH secretion in patients with hypothalamic hypopituitarism secondary to radiation therapy [59], but the degree of that effect is much lower than in healthy subjects suggesting that the negative IGF-1 feedback requires an intact hypothalamus.

It has been suggested that GH exhibits negative feedback on its own secretion. This was suggested primarily by the demonstration of decreased levels of the 22k GH isoform in subjects infused with a biologically active, but not cross-reacting 20k GH [60] and in pregnant women with high levels of endogenous placental GH-variant [61]. Because in both situations plasma IGF-1 also rises, it is difficult to attribute the decline in GH secretion to GH itself rather than to its end product, circulating IGF-1.

Sleep

One of the hallmarks of pulsatile GH architecture in humans is the sleep-related augmentation of GH secretion. It usually occurs around midnight and the GH levels at that time are, as a rule, at their highest during the 24-hour period. Partially, this phenomenon is time-entrained and partially related to sleep itself [62,63]. It is associated with a slow wave sleep, and the maximal GH levels occur within minutes of the onset of slow wave sleep [64]. γ-Hydroxybutyrate enhances both slow wave sleep and GH release [65]. There is marked sexual dimorphism of the nocturnal GH augmentation in humans, constituting only a fraction of the total daily GH release in women, but the bulk of GH output in men [23]. The neuroendocrine mechanisms of nocturnal GH augmentation are not known, and several potential mechanisms have been advanced.

The simplest explanation was a cortisol-mediated mechanism. Cortisol is secreted mostly in the morning and its levels reach the nadir around midnight. Because glucocorticoids had been shown to suppress GH secretion [66] one could hypothesize that low cortisol levels in the early hours of the night remove this inhibition and allow GH to be secreted in an unimpeded fashion. Indeed, this had been shown to be the case in patients with Addison's disease who lose their nocturnal thyroid-stimulating hormone surge [67]. Administration of cortisol in a physiologic manner (high pulses in the morning and no cortisol in the late-evening or early morning hours) fully restored the physiologic profile of thyroid-stimulating hormone secretion in these patients [67]. Continuous infusion of cortisol or a "reverse" pattern (low morning–high nighttime), however, was completely ineffective in this regard. When GH profiles were examined in the same group of patients, the results were negative. GH exhibited normal nocturnal augmentation during placebo administration and during all three cortisol regimens: (1) continuous, (2) physiologic, and (3) reverse [68]. The nocturnal GH rhythmicity does not depend on cortisol secretory patterns. The possibility that the nocturnal GH rise is caused by nutritional stimuli had been excluded by the demonstration of its persistence in healthy humans after a week-long fasting [58]. Another possibility is mediation by hypothalamic SRIF. Jaffe and coworkers [69] have shown that bolus doses of exogenous GH-RH, 0.33 mcg/kg/h every 2 hours, maintain pulsatile, one-on-one, GH responses in healthy men. The magnitude of these responses varied, however, throughout the day. They were at their lowest between 0800 and 1400 hours, after which they started to increase gradually. The magnitude of GH responses reached its peak between 2200 and 0200 hours, and this augmentation was of the same timing, magnitude, and duration as the naturally occurring nocturnal GH release. After that, GH responses returned to the relatively low magnitude toward the morning hours. This suggested that the nocturnal GH augmentation may be a manifestation of increased pituitary sensitivity to GH-RH and implied a wave-form diurnal SRIF pattern: the highest in the morning and the lowest around midnight. If this was the case, the abolition of the apparent nocturnal SRIF decrease would be expected to prevent the nocturnal GH rise. To this end, Dimaraki and coworkers [25,26] infused the synthetic SRIF receptor agonist, octreotide, at high doses to young men and postmenopausal women. In both groups, octreotide infusions significantly suppressed overall GH output accompanied by gross diminution of GH pulse amplitude. In both groups, however, the nocturnal GH rise clearly persisted, suggesting that SRIF decline may not be the main mechanism of the nocturnal GH rise.

Continuous infusion of GHRP-6, an analogue of endogenous ghrelin, dramatically augmented GH pulse amplitude and the nocturnal GH rise increased proportionately [36]. Ghrelin measurements of the peripheral blood failed to detect any diurnal changes, however, including the time of the nocturnal GH rise [37].

Whether endogenous GH-RH mediates the nocturnal GH surge is equally uncertain. Maheshwari and coworkers [19] detected minutely increased GH "pulses" around midnight in a small number of dwarfs with inactivating mutations of the GH-RH receptor, suggesting that the nocturnal pulses were not caused by GH-RH. In a larger group of similar patients, however, they also found small GH responses to exogenous GH-RH [70]. The functional ability of mutant GH-RH receptors in these subjects was not completely absent and the nocturnal GH pulses might have been caused by the exposure to presumably elevated GH-RH levels in the pituitary-portal blood. The difficulty in interpreting these studies is caused by the role of GH-RH as a major inducer of GH synthesis. Chronically impaired pituitary exposure to GH-RH likely resulted in grossly diminished intrapituitary stores of readily releasable GH. Ideally, such studies would be better performed in healthy subjects with acute GH-RH deficiency or insensitivity to GH-RH. Indeed, continuous infusion of GH-RH antagonist reliably suppressed nighttime GH secretion by 77% ± 4% and acute GH response to GH-RH by 93% ± 2% [71]. This indicated that GH-RH played a major role in the genesis of the nocturnal GH surge in humans; however, small nocturnal GH pulses were still present. Whether they occurred because of only partial GH-RH receptor blockade or because of a non–GH-RH related mechanism is still uncertain.

The genesis of the nocturnal GH augmentation in humans is still unknown. Most likely hypothalamic GH-RH release is a major contributing component, but an additional role of another factor, presumably augmenting GH-RH responsiveness of the somatotrophs, is likely.

Gonadal steroids

Activation of the gonadal system during puberty is accompanied by increased GH and IGF-1 concentrations [72,73]. Conversely, attenuation of gonadal activity with aging is accompanied by declines in both GH and IGF-1 concentrations [16]. Whether gonadal steroids are positive regulators of the somatotropic axis, or the relationship between the two systems are temporarily linked but causally unrelated, had been a topic of much interest.

The sexual dimorphism of GH secretion suggests that androgens and estrogens may play different roles in the regulation of somatotrope axis [74]. There are conflicting data on the content and release of hypothalamic GH-RH between male and female rats and overall it does not seem that these parameters are regulated by either androgens or estrogens. The same is true of hypothalamic SRIF synthesis and secretion.

In contrast, there is major sexual dimorphism in pituitary GH synthesis. Male rats have two to four higher levels of GH mRNA and protein. Also, male rats exhibit higher sensitivity to exogenous GH-RH. Administration of androgens, but not estrogens, increased GH responses to GH-RH in castrated animals. Coupled with the previously mentioned data on the stability of both GH-RH and SRIF hypothalamic secretions, it seems that the high GH pulses in male rats may be caused entirely by the sexually dimorphic effects of gonadal steroids at the pituitary level.

Gonadal-somatotropic relations in humans are often diametrically opposed to those found in animal models. Women secrete more GH in response to GH-RH than men. Studies with androgen or estrogen receptor blockers revealed that puberty-related augmentation of GH secretion is an estrogen-related, but not an androgen-related, phenomenon [75,76]. Similarly, only aromatizable (testosterone) but not nonaromatizable (dihydrotestosterone or oxandrolone) androgens increase GH pulse amplitude in boys [77].

Administration of exogenous transdermal testosterone at physiologic concentrations to boys with constitutional delay in growth or in adolescence powerfully increased their GH pulse amplitude and plasma IGF-1 concentrations, but did not alter their set point for the inhibition of GH secretion by GH-RH receptor antagonist [78]. This suggested that androgens did not exert their effect through augmentation of GH-RH secretion. In contrast, when the same levels of testosterone were achieved in elderly men, their plasma GH actually fell twofold, whereas their plasma IGF-1 minimally increased [79]. Again, no difference between the effects of GH-RH receptor antagonist was noted both before and after testosterone administration. The parallelism between gonadal and somatotropic parameters during puberty and senescence does not imply that similar mechanisms are involved.

The gonadal-somatotropic interrelations are further complicated by the direct effects of gonadal steroids at the level of hepatic IGF-1 synthesis and the resultant alterations in the negative feedback effects of IGF-1 on GH. Castrated female rats given physiologic male concentrations of dihydrotestosterone exhibited increased hepatic IGF-1 mRNA concentrations, plasma IGF-1 levels, and somatic growth [80]. Pituitary GH content in these animals was grossly increased and plasma GH levels were low. In contrast, estradiol-treated animals exhibited virtually abolished somatic growth, high plasma GH concentrations, low pituitary GH content, low hepatic IGF-1 mRNA, and plasma IGF-1 levels [80]. Taken together, these data imply that, in rats, androgens increase hepatic IGF-1 synthesis, which subsequently suppresses pituitary GH release. Estrogen, in contrast, suppresses hepatic IGF-1 synthesis and the lower circulating IGF-1 level removes the negative feedback at the pituitary level.

The same mechanism may be operative in humans, as evidenced by the inhibitory action of estrogen on IGF-1 synthesis [81–83], probably as a result of interference with intracellular GH signaling mechanisms [84].

Nutritional influences

In man, hyperglycemia causes transient GH suppression for 1 to 3 hours, followed by GH rise 3 to 5 hours after oral glucose administration [85,86]. The GH response to GH-RH and GHS infusion after glucose load is attenuated in healthy volunteers [87,88], suggesting the rapid inhibitory effect of glucose on GH release may be caused by a discharge of SRIF from the hypothalamus. When SRIF release declines, endogenous GH-RH secretion is activated reciprocally, and available pituitary stores of GH are released, leading to the rebound increase in serum GH levels [89]. Hypoglycemia, however, leads to acute GH secretion, which is the basis for the insulin-induced hypoglycemia test, a gold standard evaluation of pituitary function [90]. This response requires intact GH-RH signaling [21].

Elevation of free fatty acid levels is a strong inhibitor of GH release in normal humans [91]. The response of GH to GH-RH is markedly decreased when lipid infusion is maintained, suggesting significant elevation of SRIF tone [91]. Acipimox, which suppresses plasma concentrations of free fatty acids by blocking their release from adipose tissue [92], concomitantly increases GH secretion [91,93] and the GH-secretory response to GH-RH in normal subjects [94].

Conditions associated with chronic elevation of free fatty acid, such as obesity, present with suppressed GH level [95]. The degree of GH attenuation correlates with amount of total and visceral fat [96]. Acipimox treatment or weight loss leads to improvement of free fatty acid levels in obese people and to normalization of GH secretion, pointing out an important role of increased free fatty acid flux in this relatively GH-deficient state [93,97,98]. In obesity, however, free IGF-I levels are elevated because of insulin-dependent IGFBP-1 reduction, and this may suppress pituitary GH secretion directly [99], as suggested in humans and sheep [100,101]. The response of somatotrophs to GH-RH and GHS stimulations is reduced [102,103] in obese people, suggesting increased SRIF tone associated with this condition or a direct pituitary effect of free fatty acids. Contrary to the state of energy excess, fasting powerfully increases GH secretion, despite high free fatty acid levels, most likely as a result of direct stimulation by decreasing IGF-1 levels [104,105].

Summary

Of all the pituitary hormones, GH is regulated in the most complex and multifactorial fashion (Fig. 4). Whereas the reason for the augmentation of

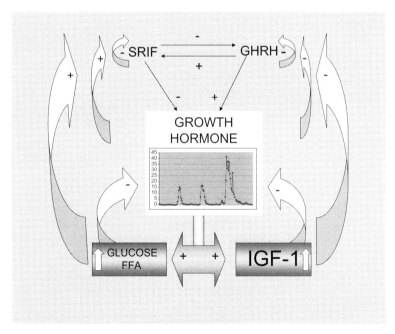

Fig. 4. Schematic diagram of GH regulation in humans. FFA, free fatty acid; GNRH, growth hormone–releasing hormone; IGF, insulin-like growth factor; SRIF, somatostatin.

GH secretion during puberty is obvious, there are no readily apparent teleologic explanations for the phenomenon of nocturnal GH augmentation, sexual dimorphism of GH secretion, or the attenuation of GH secretion in obesity or during aging. Similarly, the physiologic purpose of GH pulsatility has begun to be understood only recently. GH-RH seems to be the main regulator of the GH pulsatile pattern and of the total GH. The roles of SRIF and, especially, ghrelin are still poorly understood. Elucidation of their roles requires the availability of specific receptor antagonists. Indirect studies assessing GH responsiveness to GH-RH cannot be easily interpreted because of the potential involvement of these peptides in modifying or determining GH release. Many novel investigative tools and many additional studies are still required to understand better the mechanisms and the physiologic purpose of pulsatile GH secretion in health and disease.

References

[1] Isgaard J, Carlsson L, Isaksson OG, et al. Pulsatile intravenous growth hormone (GH) infusion to hypophysectomized rats increases insulin-like growth factor 1 messenger ribonucleic acid in skeletal tissues more effectively than continuous GH infusion. Endocrinology 1988;123:2605–10.
[2] Cheung C, Yu AM, Chen CS, et al. Growth hormone determines sexual dimorphism of hepatic cytochrome P450 3A4 expression in transgenic mice. J Pharmacol Exp Ther 2006;316: 1328–34.
[3] Reichlin S. Growth and the hypothalamus. Endocrinology 1960;67:760–3.
[4] Frohman LA, Bernardis LL, Kant KJ. Hypothalamic stimulation of growth hormone secretion. Science 1968;162:580–2.
[5] Frohman LA, Bernardis LL. Growth hormone and insulin levels in weanling rats with ventromedial hypothalamic lesions. Endocrinology 1968;82:1125–32.
[6] Thorner MO, Perryman RL, Cronin MJ, et al. Somatotroph hyperplasia: successful treatment of acromegaly by removal of a pancreatic islet tumor secreting growth hormone-releasing factor. J Clin Invest 1982;70:967–77.
[7] Sassolas G, Chayvialle JA, Partensky C, et al. Acromegaly, clinical expression of the production of growth hormone releasing factor in pancreatic tumors. Ann Endocrinol (Paris) 1983;44:347–54.
[8] Ling N, Esch F, Bohlen P, et al. Isolation, primary structure, and synthesis of human hypothalamic somatocrine: growth hormone-releasing factor. Proc Natl Acad Sci U S A 1984;81:4302–6.
[9] Plotsky PM, Vale W. Patterns of growth hormone-releasing factor and somatostatin secretion into the hypophysial-portal circulation of the rat. Science 1985;230:461–3.
[10] Ono M, Miki N, Demura H. Effect of antiserum to rat growth hormone (GH)-releasing factor on physiological GH secretion in the female rat. Endocrinology 1991;129:1791–6.
[11] Nakamura S, Mizuno M, Katakami H, et al. Aging-related changes in in vivo release of growth hormone-releasing hormone and somatostatin from the stalk-median eminence in female rhesus monkeys (Macaca mulatta). J Clin Endocrinol Metab 2003;88:827–33.
[12] Cataldi M, Magnan E, Guillaume V, et al. Relationship between hypophyseal portal GH-RH and somatostatin and peripheral GH levels in the conscious sheep. J Endocrinol Invest 1994;17:717–22.
[13] Clark RG, Robinson IC. Growth induced by pulsatile infusion of an amidated fragment of human growth hormone releasing factor in normal and GHRF-deficient rats. Nature 1985; 314:281–3.

[14] Borski RJ, Tsai W, Demott-Friberg R, et al. Induction of growth hormone (GH) mRNA by pulsatile GH-releasing hormone in rats is pattern specific. Am J Physiol Endocrinol Metab 2000;278:885–91.
[15] Vance ML, Kaiser DL, Martha PM Jr, et al. Lack of in vivo somatotroph desensitization or depletion after 14 days of continuous growth hormone (GH)-releasing hormone administration in normal men and a GH-deficient boy. J Clin Endocrinol Metab 1989; 68:22–8.
[16] Russell-Aulet M, Dimaraki EV, Jaffe CA, et al. Aging-related growth hormone (GH) decrease is a selective hypothalamic GH-releasing hormone pulse amplitude mediated phenomenon. J Gerontol A Biol Sci Med Sci 2001;56:124–9.
[17] Jaffe CA, Friberg RD, Barkan AL. Suppression of growth hormone (GH) secretion by a selective GH-releasing hormone (GH-RH) antagonist: direct evidence for involvement of endogenous GH-RH in the generation of GH pulses. J Clin Invest 1993;92:695–701.
[18] Jessup SK, Dimaraki EV, Symons KV, et al. Sexual dimorphism of growth hormone (GH) regulation in humans: endogenous GH-releasing hormone maintains basal GH in women but not in men. J Clin Endocrinol Metab 2003;88:4776–80.
[19] Maheshwari HG, Pezzoli SS, Rahim A, et al. Pulsatile growth hormone secretion persists in genetic growth hormone-releasing hormone resistance. Am J Physiol Endocrinol Metab 2002;282:943–51.
[20] Magnan E, Cataldi M, Guillaume V, et al. Neostigmine stimulates growth hormone-releasing hormone release into hypophysial portal blood of conscious sheep. Endocrinology 1993;132:1247–51.
[21] Jaffe CA, DeMott-Friberg R, Barkan AL. Endogenous growth hormone (GH)-releasing hormone is required for GH responses to pharmacological stimuli. J Clin Invest 1996;97: 934–40.
[22] Pandya N, DeMott-Friberg R, Bowers CY, et al. Growth hormone (GH)-releasing peptide-6 requires endogenous hypothalamic GH-releasing hormone for maximal GH stimulation. J Clin Endocrinol Metab 1998;83:1186–9.
[23] Jaffe CA, Ocampo-Lin B, Guo W, et al. Regulatory mechanisms of growth hormone secretion are sexually dimorphic. J Clin Invest 1998;102:153–64.
[24] Billestrup N, Swanson LW, Vale W. Growth hormone-releasing factor stimulates proliferation of somatotrophs in vitro. Proc Natl Acad Sci U S A 1986;83:6854–7.
[25] Dimaraki EV, Jaffe CA, DeMott-Friberg R, et al. Generation of growth hormone pulsatility in women: evidence against somatostatin withdrawal as pulse initiator. Am J Physiol Endocrinol Metab 2001;280:489–95.
[26] Dimaraki EV, Jaffe CA, Bowers CY, et al. Pulsatile and nocturnal growth hormone secretions in men do not require periodic declines of somatostatin. Am J Physiol Endocrinol Metab 2003;285:163–70.
[27] Zheng H, Bailey A, Jiang MH, et al. Somatostatin receptor subtype 2 knockout mice are refractory to growth hormone-negative feedback on arcuate neurons. Mol Endocrinol 1997;11:1709–17.
[28] Low MJ, Hammer RE, Goodman RH, et al. Tissue-specific posttranslational processing of pre-prosomatostatin encoded by a metallothionein-somatostatin fusion gene in transgenic mice. Cell 1985;41:211–9.
[29] Bass JJ, Gluckman PD, Fairclough RJ, et al. Effect of nutrition and immunization against somatostatin on growth and insulin-like growth factors in sheep. J Endocrinol 1987;112:27–31.
[30] Maccario M, Grottoli S, Procopio M, et al. The GH/IGF-1 axis in obesity: influence of neuroendocrine and metabolic factors. Int J Obes Relat Metab Disord 2000;24(Suppl 2):S96–9.
[31] Mozid AM, Tringali G, Forsling ML, et al. Ghrelin is released from rat hypothalamic explants and stimulates corticotrophin-releasing hormone and arginine-vasopressin. Horm Metab Res 2003;35:455–9.
[32] Smith RG, Jiang H, Sun Y. Developments in ghrelin biology and potential clinical relevance. Trends Endocrinol Metab 2005;16:436–42.

[33] Bowers CY, Sartor AO, Reynolds GA, et al. On the actions of the growth hormone-releasing hexapeptide, GHRP. Endocrinology 1991;128:2027–35.
[34] Herrington J, Hille B. Growth hormone-releasing hexapeptide elevates intracellular calcium in rat somatotropes by two mechanisms. Endocrinology 1994;135:1100–8.
[35] Robinson BM, Friberg RD, Bowers CY, et al. Acute growth hormone (GH) response to GH-releasing hexapeptide in humans is independent of endogenous GH-releasing hormone. J Clin Endocrinol Metab 1992;75:1121–4.
[36] Jaffe CA, Ho PJ, DeMott-Friberg R, et al. Effects of a prolonged growth hormone (GH)-releasing peptide infusion on pulsatile GH secretion in normal men. J Clin Endocrinol Metab 1993;77:1631–7.
[37] Avram AM, Jaffe C, Symons KV, et al. Endogenous circulating ghrelin does not mediate growth hormone rhythmicity or response to fasting. J Clin Endocrinol Metab 2005;90:2982–7.
[38] Corbetta S, Peracchi M, Cappiello V, et al. Circulating ghrelin levels in patients with pancreatic and gastrointestinal neuroendocrine tumors: identification of one pancreatic ghrelinoma. J Clin Endocrinol Metab 2003;88:3117–20.
[39] Tsolakis AV, Portela-Gomes GM, Stridsberg M, et al. Malignant gastric ghrelinoma with hyperghrelinemia. J Clin Endocrinol Metab 2004;89:3739–44.
[40] Sun Y, Ahmed S, Smith RG. Deletion of ghrelin impairs neither growth nor appetite. Mol Cell Biol 2003;23:7973–81.
[41] Wortley KE, Anderson KD, Garcia K, et al. Genetic deletion of ghrelin does not decrease food intake but influences metabolic fuel preference. Proc Natl Acad Sci U S A 2004;101: 8227–32.
[42] Wortley KE, del Rincon JP, Murray JD, et al. Absence of ghrelin protects against early-onset obesity. J Clin Invest 2005;115:3393–7.
[43] Zigman JM, Nakano Y, Coppari R, et al. Mice lacking ghrelin receptors resist the development of diet-induced obesity. J Clin Invest 2005;115:3564–72.
[44] Lumpkin MD, McDonald JK. Blockade of growth hormone-releasing factor (GRF) activity in the pituitary and hypothalamus of the conscious rat with a peptidic GRF antagonist. Endocrinology 1989;124:1522–31.
[45] Lumpkin MD, Mulroney SE, Haramati A. Inhibition of pulsatile growth hormone (GH) secretion and somatic growth in immature rats with a synthetic GH-releasing factor antagonist. Endocrinology 1989;124:1154–9.
[46] Peterfreund RA, Vale WW. Somatostatin analogs inhibit somatostatin secretion from cultured hypothalamus cells. Neuroendocrinology 1984;39:397–402.
[47] Mitsugi N, Arita J, Kimura F. Effects of intracerebroventricular administration of growth hormone-releasing factor and corticotrophin-releasing factor on somatostatin secretion into rat hypophysial portal blood. Neuroendocrinology 1990;51:9306.
[48] Yamauchi N, Shibasaki T, Ling N, et al. In vitro release of growth hormone-releasing factor (GRF) from the hypothalamus: somatostatin inhibits GRF release. Regul Pept 1991;33: 71–8.
[49] Frohman MA, Downs TR, Chomczynski P, et al. Cloning and characterization of mouse growth hormone-releasing hormone (GRH) complementary DNA: increased GRH messenger RNA levels in the growth hormone-deficient lit/lit mouse. Mol Endocrinol 1989;3: 1529–36.
[50] Chomczynski P, Downs TR, Frohman LA. Feedback regulation of growth hormone (GH)-releasing hormone gene expression by GH in rat hypothalamus. Mol Endocrinol 1988;2: 236–41.
[51] Chihara K, Minamitani N, Kaji H, et al. Intraventricularly injected growth hormone stimulates somatostatin release into rat hypophysial portal blood. Endocrinology 1981;109: 2279–81.
[52] Berelowitz M, Firestone SL, Frohman LA. Effects of growth hormone excess and deficiency on hypothalamic somatostatin content and release and on tissue somatostatin distribution. Endocrinology 1981;109:714–9.

[53] Laron Z, Pertzelan A, Karp M. Pituitary dwarfism with high serum levels of growth hormone. Isr J Med Sci 1968;4:883–94.
[54] Rosenfeld RG, Buckway CK. Growth hormone insensitivity syndromes: lessons learned and opportunities missed. Horm Res 2001;55:36–9.
[55] Rosenfeld RG, Hwa V. New molecular mechanisms of GH resistance. Eur J Endocrinol 2004;151:11–5.
[56] Bermann M, Jaffe CA, Tsai W, et al. Negative feedback regulation of pulsatile growth hormone secretion by insulin-like growth factor I. Involvement of hypothalamic somatostatin. J Clin Invest 1994;94:138–45.
[57] Chapman IM, Hartman ML, Pieper KS, et al. Recovery of growth hormone release from suppression by exogenous insulin-like growth factor I (IGF-I): evidence for a suppressive action of free rather than bound IGF-I. J Clin Endocrinol Metab 1998;83: 2836–42.
[58] Ho PJ, Friberg RD, Barkan AL. Regulation of pulsatile growth hormone secretion by fasting in normal subjects and patients with acromegaly. J Clin Endocrinol Metab 1992;75: 812–9.
[59] Darzy KH, Murray RD, Gleeson HK, et al. The impact of short-term fasting on the dynamics of 24-hour growth hormone (GH) secretion in patients with severe radiation-induced GH deficiency. J Clin Endocrinol Metab 2005;91:987–94.
[60] Hashimoto Y, Kamioka T, Hosaka M, et al. Exogenous 20K growth hormone (GH) suppresses endogenous 22K GH secretion in normal men. J Clin Endocrinol Metab 2000;85: 601–6.
[61] Beckers A, Stevenaert A, Foidart JM, et al. Placental and pituitary growth hormone secretion during pregnancy in acromegalic women. J Clin Endocrinol Metab 1990;71: 725–31.
[62] Obal F Jr, Krueger JM. GH-RH and sleep. Sleep Med Rev 2004;8:367–77.
[63] Van Cauter E, Plat L, Copinschi G. Interrelations between sleep and the somatotropic axis. Sleep 1998;21:553–66.
[64] Holl RW, Hartman ML, Veldhuis JD, et al. Thirty-second sampling of plasma growth hormone in man: correlation with sleep stages. J Clin Endocrinol Metab 1991;72:854–61.
[65] Van Cauter E, Plat L, Scharf MB, et al. Simultaneous stimulation of slow-wave sleep and growth hormone secretion by gamma-hydroxybutyrate in normal young men. J Clin Invest 1997;100:745–53.
[66] Hochberg Z. Mechanisms of steroid impairment of growth. Horm Res 2002;58:33–8.
[67] Samuels MH. Effects of variations in physiological cortisol levels on thyrotropin secretion in subjects with adrenal insufficiency: a clinical research center study. J Clin Endocrinol Metab 2000;85:1388–93.
[68] Barkan AL, DeMott-Friberg R, Samuels MH. Growth hormone (GH) secretion in primary adrenal insufficiency: effects of cortisol withdrawal and patterned replacement of GH pulsatility and circadian rhythmicity. Pituitary 2000;3:175–9.
[69] Jaffe CA, Turgeon DK, Friberg RD, et al. Nocturnal augmentation of growth hormone (GH) secretion is preserved during repetitive bolus administration of GH-releasing hormone: potential involvement of endogenous somatostatin—a clinical research center study. J Clin Endocrinol Metab 1995;80:3321–6.
[70] Maheshwari HG, Silverman BL, Dupuis J, et al. Phenotype and genetic analysis of a syndrome caused by an inactivating mutation in the growth hormone-releasing hormone receptor: dwarfism of Sindh. J Clin Endocrinol Metab 1998;83:4065–74.
[71] Jessup SK, Malow BA, Symons KV, et al. Blockade of endogenous growth hormone-releasing hormone receptors dissociates nocturnal growth hormone secretion and slow-wave sleep. Eur J Endocrinol 2004;151:561–6.
[72] Veldhuis JD, Roemmich JN, Richmond EJ, et al. Somatotropic and gonadotropic axes linkages in infancy, childhood, and the puberty-adult transition. Endocr Rev 2006;27: 101–40.

[73] Giordano R, Lanfranco F, Bo M, et al. Somatopause reflects age-related changes in the neural control of GH/IGF-I axis. J Endocrinol Invest 2005;28:94–8.
[74] Wehrenberg WB, Giustina A. Feedback regulation of growth hormone secretion. In: Kostyo JL, Goodman HM, editors. Handbook of physiology, vol. 5. Hormonal control of growth. Oxford: Oxford University Press; 1999. p. 299–327.
[75] Metzger DL, Kerrigan JR. Estrogen receptor blockade with tamoxifen diminishes growth hormone secretion in boys: evidence for stimulatory role of endogenous estrogens during male adolescence. J Clin Endocrinol Metab 1994;79:513–8.
[76] Metzger DL, Kerrigan JR. Androgen receptor blockade with flutamide enhances growth hormone secretion in late pubertal males: evidence for independent actions of estrogen and androgen. J Clin Endocrinol Metab 1993;76:1147–52.
[77] Veldhuis JD, Metzger DL, Martha PM Jr, et al. Estrogen and testosterone, but not a non-aromatizable androgen, direct network integration of the hypothalamosomatotrope (growth hormone)-insulin-like growth factor I axis in the human: evidence from pubertal pathophysiology and sex-steroid hormone replacement. J Clin Endocrinol Metab 1997; 82:3414–20.
[78] Racine MS, Symons KV, Foster CM, et al. Augmentation of growth hormone secretion after testosterone treatment in boys with constitutional delay of growth and adolescence: evidence against an increase in hypothalamic secretion of growth hormone-releasing hormone. J Clin Endocrinol Metab 2004;89:3326–31.
[79] Orrego JJ, Dimaraki E, Symons K, et al. Physiological testosterone replenishment in healthy elderly men does not normalize pituitary growth hormone output: evidence against the connection between senile hypogonadism and somatopause. J Clin Endocrinol Metab 2004;89:3255–60.
[80] Borski RJ, Tsai W, DeMott-Friberg R, et al. Regulation of somatic growth and the somatotropic axis by gonadal steroids: primary effect of insulin-like growth factor I gene expression and secretion. Endocrinology 1996;137:3253–9.
[81] Dimaraki EV, Symons KV, Barkan AL. Raloxifene decreases serum IGF-1 in male patients with active acromegaly. Eur J Endocrinol 2004;150:481–7.
[82] Cozzi R, Barausse M, Lodrini S, et al. Estroprogestinic pill normalizes IGF-1 levels in acromegalic women. J Endocrinol Invest 2003;26:347–52.
[83] Leung KC, Johannsson G, Leong GM, et al. Estrogen regulation of growth hormone action. Endocr Rev 2004;25:693–721.
[84] Leung KC, Doyle N, Ballesteros M, et al. Estrogen inhibits GH signaling by suppressing GH-induced JAK2 phosphorylation, an effect mediated by SOCS-2. Proc Natl Acad Sci U S A 2003;100:1016–21.
[85] Roth A, Cligg SM, Yallow CP, et al. Secretion of human growth hormone: physiological and experimental modification. Metabolism 1963;12:557–9.
[86] Yalow RS, Goldsmith SJ, Berson SA. Influence of physiologic fluctuations in plasma growth hormone on glucose tolerance. Diabetes 1969;18:402–8.
[87] Masuda A, Shibasaki T, Nakahara M, et al. The effect of glucose on growth hormone (GH)-releasing hormone-mediated GH secretion in man. J Clin Endocrinol Metab 1985; 60:523–6.
[88] Broglio F, Benso A, Gottero C, et al. Effects of glucose, free fatty acids or arginine load on the GH-releasing activity of ghrelin in humans. Clin Endocrinol (Oxf) 2002;57: 265–71.
[89] Valcavi R, Zini M, Davoli S, et al. The late growth hormone rise induced by oral glucose is enhanced by cholinergic stimulation with pyridostigmine in normal subjects. Clin Endocrinol (Oxf) 1992;37:360–4.
[90] Gharib H, Cook DM, Saenger PH, et al. American Association of Clinical Endocrinologists medical guidelines for clinical practice for growth hormone use in adults and children: 2003 update. Endocr Pract 2003;9:64–76.

[91] Imaki T, Shibasaki T, Shizume K, et al. The effect of free fatty acids on growth hormone (GH)-releasing hormone-mediated GH secretion in man. J Clin Endocrinol Metab 1985; 60:290–3.
[92] Fuccella L, Golaniga G, Lovisolo P. Inhibition of lipolysis by nicotinic acid and acipimox. Clin Pharmacol Ther 1980;28:790–5.
[93] Cordido F, Peino R, Penalva A, et al. Impaired growth hormone secretion in obese subjects is partially reversed by acipimox-mediated plasma free fatty acid depression. J Clin Endocrinol Metab 1996;81:914–8.
[94] Pontiroli AE, Lanzi R, Monti LD, et al. Effect of acipimox, a lipid lowering drug, on growth hormone (GH) response to GH-releasing hormone in normal subjects. J Endocrinol Invest 1990;13:539–42.
[95] Maccario M, Grottoli S, Procopio M, et al. The GH/IGF-I axis in obesity: influence of neuro-endocrine and metabolic factors. Int J Obes Relat Metab Disord 2000;24(Suppl 2):S96–9.
[96] Clasey JL, Weltman A, Patrie J, et al. Abdominal visceral fat and fasting insulin are important predictors of 24-hour GH release independent of age, gender, and other physiological factors. J Clin Endocrinol Metab 2001;86:3845–52.
[97] Kok P, Buijs MM, Kok SW, et al. Acipimox enhances spontaneous growth hormone secretion in obese women. Am J Physiol Regul Integr Comp Physiol 2004;286:R693–8.
[98] Douyon L, Schteingart DE. Effect of obesity and starvation on thyroid hormone, growth hormone, and cortisol secretion. Endocrinol Metab Clin North Am 2002;31:173–89.
[99] Sandhu MS, Gibson JM, Heald AH, et al. Association between insulin-like growth factor-I: insulin-like growth factor-binding protein-1 ratio and metabolic and anthropometric factors in men and women. Cancer Epidemiol Biomarkers Prev 2004;13:166–70.
[100] Fletcher TP, Thomas GB, Dunshea FR, et al. IGF feedback effects on growth hormone secretion in ewes: evidence for action at the pituitary but not the hypothalamic level. J Endocrinol 1995;144:323–31.
[101] Hartman ML, Clayton PE, Johnson ML, et al. A low dose euglycemic infusion of recombinant human insulin-like growth factor I rapidly suppresses fasting-enhanced pulsatile growth hormone secretion in humans. J Clin Invest 1993;91:2453–62.
[102] Ghigo E, Procopio M, Boffano GM, et al. Arginine potentiates but does not restore the blunted growth hormone response to growth hormone-releasing hormone in obesity. Metabolism 1992;41:560–3.
[103] Alvarez-Castro P, Isidro ML, Garcia-Buela J, et al. Marked GH secretion after ghrelin alone or combined with GH-.releasing hormone (GH-RH) in obese patients. Clin Endocrinol (Oxf) 2004;61:250–5.
[104] Koutkia P, Schurgin S, Berry J, et al. Reciprocal changes in endogenous ghrelin and growth hormone during fasting in healthy women. Am J Physiol Endocrinol Metab 2005;289: E814–22.
[105] Katz LE, DeLeon DD, Zhao H, et al. Free and total insulin-like growth factor (IGF)-I levels decline during fasting: relationships with insulin and IGF-binding protein-1. J Clin Endocrinol Metab 2002;87:2978–83.

Regulation of Growth Hormone Action by Gonadal Steroids

Udo J. Meinhardt, MD[a],
Ken K.Y. Ho, FRACP, MD[a,b,c,*]

[a]*Pituitary Research Unit, Garvan Institute of Medical Research, 384 Victoria Street, Darlinghurst, NSW 2010, Sydney, Australia*
[b]*University of New South Wales, Kensington, NSW 2010, Australia*
[c]*Department of Endocrinology, St. Vincent's Hospital, Darlinghurst, NSW 2010, Sydney, Australia*

Sex steroids modulate growth hormone (GH) action at a number of levels, centrally at the hypothalamus and the pituitary by regulating secretion, and peripherally by modifying GH responsiveness. Regulation of GH action by gonadal steroids has been an active subject of investigation in the basic and the clinical arena over the last several decades. This article addresses data of potential clinical relevance from studies in human models on interactions between sex steroids at the level of neurosecretion and action, encompassing effects on the insulin-like growth factor (IGF)-I system, substrate metabolism, and body composition. For clinicians, a better understanding of sex steroid–GH interactions leads to improved management of children with growth and developmental disorders and of adults with hypopituitarism.

Neurosecretory modulation

At the central level GH secretion is controlled by three peptide hormones: (1) growth hormone–releasing hormone (GH-RH), (2) somatostatin, and (3) ghrelin. IGF-I mediates a negative feedback control of GH secretion by acting directly on the somatotroph and on hypothalamic GH-RH and somatostatin neurons. GH secretion is also regulated by metabolic cues: insulin, glucose, and nonesterified fatty acids inhibit secretion, whereas certain

---

Dr. Udo Meinhardt is supported by grants from the Swiss National Foundation and Novo Nordisk. Professor Ken Ho is supported in part by the National Health and Medical Council of Australia.

* Corresponding author. Pituitary Research Unit, Garvan Institute of Medical Research, 384 Victoria Street, Darlinghurst, NSW 2010, Sydney, Australia.
E-mail address: k.ho@garvan.org.au (K.K.Y. Ho).

amino acids, such as arginine, stimulate secretion [1–4]. GH secretion is affected by adiposity; obesity and in particular visceral obesity is a powerful negative regulator of GH output and probably accounts for a substantial component of the age-related decline in GH status [5]. GH secretion can be increased in two ways. Secretion can be pushed by factors that activate central drive enhancing the GH–IGF-I system. GH secretion can be stimulated or pulled indirectly by reduction of feedback inhibitory signals. The change in GH secretion induced by fasting is an example of stimulation through a pulling effect [6].

Sex steroids modulate secretion within this framework. Evidence of a regulatory role of sex steroids on GH comes from association studies in children and adults. In children there is a positive correlation between sex steroid and GH status during puberty. These studies report a twofold to threefold increase in GH secretion along with an increase in gonadal steroid concentrations during puberty [7]. GH secretion is reduced in hypogonadal children [8,9] and increased in precocious puberty, successful therapy of which normalizes GH secretion [10,11]. In pubertal children, there is a significant relationship between IGF-I and GH secretion [12]. These associations provide strong evidence that endogenous sex steroids drive GH secretion during puberty in boys and girls.

In adults, the relationships between gonadal and GH status are strong for men but seem less so for women. GH secretion is reduced in men with hypogonadism regardless of cause and is normalized by androgen replacement or by pulsatile gonadotropin-releasing hormone treatment in those with hypogonadotrophic hypogonadism [13–16]. In women the relationship between estrogen and GH levels is inconsistent. During the menstrual cycle, estradiol levels during the periovulatory phase are positively related with GH concentration [17]. Mean 24-hour GH levels are higher in premenopausal than in postmenopausal women [18]. The suggestion, however, of a positive influence of estrogen is not supported by the observation that GH-RH–stimulated GH secretion is similar between normal women and women with premature ovarian failure either replaced or unreplaced with estrogen [19]. Studies in adult women reveal no clear positive correlation between estrogen status and GH secretion in contrast to the situation with androgen status in men.

Androgens

Insights into the regulation of GH secretion have come from intervention studies manipulating sex steroid status in males and females. In prepubertal and peripubertal boys, testosterone in physiologic and pharmacologic doses increases spontaneous and stimulated GH secretion [8,20,21]. In hypogonadal and elderly men, testosterone and gonadotropin-releasing hormone treatment, respectively, increase spontaneous and stimulated GH secretion [13–16,22]. The stimulatory action of testosterone on the GH axis seems to be mediated at the hypothalamic level by GH-RH [15]. In all of these situations increased GH secretion is accompanied by increased IGF-I levels, indicating a "push" effect of testosterone on the GH-IGF axis.

The stimulation of GH secretion by testosterone requires prior aromatization to estrogen. Evidence comes from studies reporting that administration of an estrogen receptor antagonist abrogates the GH-stimulatory effect of testosterone in hypogonadal men and reduces GH secretion in normal healthy men [14]. In boys with delayed puberty, dihydrotestosterone, an androgen that cannot be aromatized to an estrogen, does not stimulate GH secretion [21]. The central push mechanism of testosterone involves estrogen receptor mediation.

Estrogens

Early studies convincingly demonstrated that GH secretion is stimulated by estrogen administration. Most of these studies did not report IGF-I data, however, because many were performed before the availability of IGF-I assays [16,23,24] and before the realization that the route of administration is a major determinant of the effect of estrogen on the GH–IGF-I axis [25]. When administered orally, the liver is exposed to pharmacologic estrogen concentrations inhibiting IGF-I production, an effect that is avoided by the parenteral route [18]. In postmenopausal women, estrogen replacement by the oral route increases 24-hour serum GH but significantly reduces circulating IGF-I [18,26,27]. If estrogen activates GH–IGF-I axis in a similar manner to testosterone, then estrogen treatment should increase both GH and IGF-I levels.

To the authors' knowledge only five recent human studies have addressed whether estrogen exerts a central effect by administration of this sex steroid transdermally to hypogonadal women. The study in Turner's girls reported increased urinary GH excretion but no change in mean blood IGF-I concentration [9]. Three studies in menopausal women used physiologic replacement doses of 100 μg daily and observed no significant change in GH secretion. Of these three studies, the IGF-I concentration was unchanged in two [26,28] but increased in one [18]. One of these studies also included women with premature ovarian failure, who also failed to demonstrate a change in mean GH or IGF-I after transdermal estrogen administration [28]. A fourth study in postmenopausal women administered suprarphysiologic doses transdermally, resulting in systemic estradiol levels threefold to fourfold higher than observed from conventional dosages. This high-dose transdermal regimen increased mean 24-hour GH but reduced IGF-I levels [27]. The data in menopausal women provide little support for a push effect on GH secretion but strong evidence for a pull effect by reducing IGF-I feedback inhibition. This seems intriguing, in view of the apparent estrogen receptor mediated push effect of testosterone on GH secretion in men.

Stimulated GH secretion may differ between women and men, a phenomenon probably reflecting the modulatory central effects of sex steroids. In response to GH-RH, most recent studies report a greater GH release in women than men [29]. In women, the GH-RH–stimulated GH release is positively related to estrogen levels during the menstrual cycle [30]. Similarly, L-arginine elicits higher GH release in women than in men [3]. Conversely, GH release is greater in men than women in response to GHRP-2, somatostatin-induced rebound [31], and clonidine [32]. Interestingly, the GH response to glucagon [33], insulin-induced hypoglycemia [34], or galanin [35] does not seem to be different between men and women. The influence of gender on stimulated GH secretion is dependent on the stimulus type and likely reflects the complex interactions between sex steroids and the stimulus-generated neuroendocrine pathways regulating GH secretion. Regardless of the mechanisms, the collective data indicate that clinicians should be mindful of the influence of sex steroids in evaluating GH status in children and adults.

Modulation of the peripheral biologic effects of growth hormone

Insulin-like growth factor-I

IGF-I is the mediator of the anabolic action of GH. The liver is a major target tissue of GH action and the principal source of circulating IGF-I. Because the liver is also a sex steroid responsive organ it represents a potential site of regulatory interaction between gonadal steroids and GH.

Observational studies in children have reported a correlation between IGF-I and sex steroid levels in both sexes during puberty [36–38] related closely to a concomitant increase in GH levels. Coutant and coworkers [39] showed that progression through puberty was significantly associated with increases in mean sex steroid, GH, and IGF-I concentrations and with increased IGF-I response to an injection of GH. These observations suggest that the endogenous gonadal steroid milieu increases GH sensitivity in girls and boys during puberty. Few studies have explored whether androgens or estrogens alter GH sensitivity.

Androgens

Testosterone and IGF-I levels are positively correlated in young, middle aged [40], and elderly men [41] and androgens positively modulate GH-induced stimulation of IGF-I levels as indicated by studies administering testosterone to normal men [42], hypogonadal men [16], or female-to-male transsexuals [43]. Because androgens stimulate GH secretion, the peripheral effect of testosterone on GH responsiveness must be distinguished from its central stimulatory effect. To address whether testosterone enhances the IGF-I response to GH, it is important first to determine whether androgens increase serum IGF-I in the absence of GH.

A study in 10 hypopituitary boys reported that testosterone treatment minimally elevated IGF-I levels but that the addition of GH resulted in a greater than threefold increase [44]. Another study compared the effects of a human chorionic gonadotropin-β injection in GH-deficient children with and without GH replacement. Although testosterone increased significantly following human chorionic gonadotropin-β, IGF-I concentrations increased only in boys on GH replacement therapy [45]. The findings indicate that testosterone does not regulate circulating IGF-I but requires GH to exert a stimulatory effect. The question as to whether androgens enhance the IGF-I responsiveness to GH has been addressed in a recent study of hypopituitary men. Gibney and coworkers [46] compared the effects of GH alone with combined GH and testosterone treatment. They observed that although GH increased IGF-I levels, co-administration of testosterone to GH induced a further rise in circulating IGF-I. In the same study, treatment with testosterone alone failed to elevate IGF-I levels, confirming the observations in children. Androgens alone exert no effect but enhance the IGF-I response to GH.

Estrogens

The question as to whether estrogens influence GH action had until recently been controversial. The realization that oral estrogens impair the metabolic and endocrine functions of the liver, however, has clarified some of the previous confusion in the literature.

Studies in girls have been conducted almost without exception with oral administration. A large French study [39] found that a 2-mg oral dose of 17β-estradiol reduced IGF-I response to GH in girls with delayed puberty. In contrast, testosterone priming increased the IGF-1 response. In girls with constitutional tall stature, high doses of oral estrogen (ethinyl-estradiol 100 µg/day) reduced IGF-I levels [47]. In Turner's girls, oral or transdermal estrogen replacement therapy increased GH but did not change IGF-I levels [9]. Present data in girls indicate that oral estrogen especially at higher doses impairs the IGF-I response to GH. Effects of endogenous estrogen or of a nonparenteral route in childhood are less clear, although evidence that estrogen pushes the GH–IGF-I axis is weak.

In adulthood, observational studies suggest that the presence of estrogen levels is associated with GH resistance. GH but not IGF-I levels are higher in young women than in age-matched men [48]. IGF-I levels are lower in GH-deficient women and the IGF-I increase in response to GH treatment is about half that of their male counterparts resulting in women requiring a higher replacement dose of GH than men [49]. In acromegaly, there is a gender difference in the relationship between GH output and IGF-I with IGF-I being lower in women for a given GH concentration [50]. Indirect evidence supporting a role of estrogens comes from studies exploring changes in the GH–IGF-I axis during the female menstrual cycle. In women, some [17] studies report a periovulatory rise in GH and IGF-I levels. The

IGF-I response to GH administration is reduced, however, suggesting that the rises are caused by a temporal association rather than a causal effect [51]. These observations suggest some modulatory effect of estrogens on the GH–IGF-I axis, the manner and mechanisms of which have been clarified from intervention studies, which must control for indirect (IGF-I feedback) effects of estrogen on GH secretion.

Estrogen reduces IGF-I when used to prime GH stimulation tests in prepubertal children [39], taken as oral contraceptive [52], or as hormone-replacement therapy in menopausal [18] or hypopituitary women [53]. The estrogen effect is independent of formulation [54]. The dose and route dependency has been explored further in a number of situations. In hypopituitary and postmenopausal women oral administration of 17β-estradiol reduced IGF-I levels in a dose-dependent manner [55]. GH elevates IGF-I to a higher level with transdermal than oral estrogen treatment in hypopituitary women [53]. A number of studies have demonstrated that estrogens administered by the parenteral route can reduce circulating IGF-I if sufficiently high concentrations are achieved in the systemic circulation. In one study involving an IGF-I generation test, the response to GH with high-dose (200 μg/day) was significantly lower than that after low-dose (50 μg/day) transdermal treatment [56]. In a study of postmenopausal women, transdermal estrogen administered at seven times the therapeutic dose increased GH but reduced IGF-I levels, a response indistinguishable from that observed after oral administration [27]. In a third study, incremental doses of transdermal estradiol administration (up to 200 μg/day) reduced IGF-I significantly at the highest dose [57]. These data establish unequivocally that estrogen inhibits hepatic IGF-I production dose dependently regardless of whether this is achieved by the portal or systemic circulation. At the cellular level, estrogen inhibits GH signaling in a concentration-dependent manner [25].

Progestogens

The question as to whether progestogens influence GH action has been studied indirectly by investigating how they modulate the hepatic effects of estrogens. Progestogens attenuate the effects of estrogen on circulating lipid levels according to their androgenic properties, decreasing high-density lipoprotein cholesterol levels [58] and increasing low-density lipoprotein cholesterol levels [59]. Progestogens oppose the estrogen antagonistic effect on IGF-I in relation to their androgenic properties [60]. In postmenopausal women receiving estrogen, co-administration of a neutral progestogen (eg, dihydrogesterone) exerted no effect on IGF-I, whereas the more androgenic progestogens (eg, medroxyprogesterone acetate and norethisterone) opposed the fall in IGF-I induced by oral estrogen administration [60]. The relationship between the progestogen androgenicity and IGF-I may explain why increases in IGF-I during progestogen and transdermal estrogen administration are not reported by all studies [18].

The failure of progestogens in general to oppose estrogen-induced changes in IGF-I indicates that this class of sex steroid does not play a significant role in modulating GH action compared with estrogen. Nevertheless, the androgenic progestogens co-administered with estrogen on the GH-IGF system may impart potentially significant effects on substrate metabolism and body composition.

The mechanism by which progestogens exert these modulatory effects on the liver is not clear. Human liver expresses estrogen and androgen receptors [61,62]. Whether the liver expresses any progesterone receptors, however, is uncertain [61,63]. The limited information points to a paucity of progesterone receptors in human hepatic tissue. It is likely that the effects of androgenic progestogens are exerted through hepatic androgen receptors. The negative effects of androgenic progestogens on lipids do not occur when administered by a nonoral route [64], indicating that their actions arise from a first pass hepatic effect, a phenomenon similar to that of estrogens administered by the oral route.

Selective estrogen receptor modulators

Selective estrogen receptor modulators are synthetic estrogen compounds with tissue-specific estrogen agonistic and antagonistic properties. They are used increasingly in the therapy of breast cancer and of osteoporosis. Classical selective estrogen receptor modulators, such as tamoxifen and raloxifene, exert estrogen-like effects on the liver reducing circulating IGF-I levels in women [65,66]. Therapeutic doses of raloxifene seem to be less potent than estrogens in reducing IGF-I levels. This was shown in a study comparing effects of raloxifene (60 and 120 mg daily) and 17β-estradiol (2 and 4 mg daily) in GH-deficient and normal postmenopausal women (Fig. 1). In both groups, estradiol reduced IGF-I levels in a dose-dependent manner to a greater extent than raloxifene. Raloxifene reduced IGF-I levels in the GH-deficient group, but not in the postmenopausal group [55].

Estrogen and estrogen-like compounds attenuate GH action by reducing IGF-I. This is a concentration-dependent phenomenon, which arises invariably from oral administration. From the mechanistic perspective the divergent effects of estrogen and testosterone on the modulation of IGF-I suggest that even if the neurosecretory action of testosterone might be at least partly mediated through the estrogen receptor, its seems likely that the peripheral effect on GH responsiveness is mediated through the androgen receptor.

Growth

The onset of puberty and its associated growth spurt are triggered by interplay between the somatotrophic and the gonadotrophic hormone axes [67]. Sex steroids and GH interact closely to regulate pubertal growth [68]. In girls, height velocity is positively correlated with levels of GH,

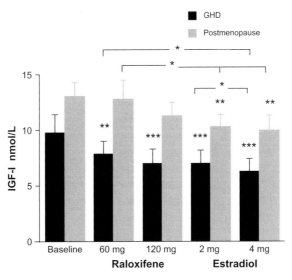

Fig. 1. GH and IGF-I action and the effect of oral estrogen. Mean serum IGF-1 concentrations in GH-deficient and postmenopausal women treated with 17β estradiol (2 mg and 4 mg/day) and raloxifene (60 mg and 120 mg). $*P < .05$, $**P < .01$, and $***P < .001$ compared with baseline. (*From* Gibney J, Johannsson G, Leung KC, et al. Comparison of the metabolic effects of raloxifene and oral estrogen in postmenopausal and growth hormone-deficient women. J Clin Endocrinol Metab 2005;90:3899; with permission.)

estradiol, and androstenedione, whereas in boys GH, testosterone, and estradiol are correlated with peak height velocity [69].

Studying how testosterone and estrogen modulate GH effects on linear growth is complicated by the fact that sex steroids act directly on the growth plate independently of GH. The effect of estrogen is particularly relevant because it is responsible for the fusion of the growth plate [70]. Given its clear GH antagonistic effects, it may be predicted that estrogen impairs (GH dependent) growth. The extent to which estrogen reduces growth independent of its direct effect on the growth plate, however, has been difficult to quantify.

Studies in children suggest that GH and testosterone interact positively to enhance growth. Testosterone enhances the growth of boys with hypogonadism, and with hypopituitarism during GH treatment. The effect of testosterone on somatic growth is poor in boys with hypopituitarism without concomitant GH treatment [71], and androgens alone exert no effect on IGF-I but enhance its response to GH [46]. These collective observations suggest that the growth-promoting and anabolic effects of testosterone may be dependent on GH and possibly mediated in part by IGF-I.

Substrate metabolism and body composition

Gender-related differences in body composition are likely to be mediated by sex steroids modulating the GH–IGF-I axis. This is supported by the observation that body composition differences between the sexes emerge at the time of pubertal growth. GH treatment induces a greater increase in lean mass and decrease in fat mass in GH-deficient male compared with female patients [72]. GH replacement of GH-deficient adults also induces a greater increase in indices of bone turnover and in bone mass in men than in women [73]. GH supplementation in the elderly results in greater body compositional change in men than women [74]. This section discusses how sex steroids modulate GH-induced changes in substrate metabolism and body composition.

Estrogens

The dissociation induced by the oral route on the GH–IGF-I axis suggests that estrogens inhibit GH-regulated endocrine and metabolic function of the liver. Studies undertaken in hypopituitary and postmenopausal women have provided strong evidence that estrogens impair GH action when administered by the oral route.

In GH-deficient women, the effects of graded doses of GH on IGF-I, fat oxidation, and whole-body protein metabolism were compared in a group of hypopituitary women replaced with therapeutic doses of estrogen administered by the oral and transdermal routes [53]. Mean IGF-I concentrations and lipid oxidation were significantly lower during oral estrogen treatment, both before and during GH administration, than during transdermal treatment (Fig. 2). Whole-body protein synthesis was also lower during oral estrogen both before and during GH administration. Estrogen at a therapeutic dose exerted significant route-dependent effects on GH action in women with organic GH deficiency. In comparison with the transdermal route, oral estrogen markedly attenuated the metabolic effects of GH.

O'Sullivan and coworkers [75] compared the metabolic and body compositional effects of oral and transdermal estrogen treatment in postmenopausal women. When compared with transdermal estrogen, oral estrogen therapy reduced IGF-I and induced a transient suppression of lipid oxidation and a reciprocal elevation of carbohydrate oxidation. Significantly different effects on fat mass and lean body mass were observed between the two routes of estrogen therapy. Oral therapy led to a significant increase in fat mass equivalent to a 5% change in body fat and a loss of lean body mass equivalent to a 3% change compared with that observed during transdermal estrogen therapy. The effects on fat oxidation and IGF-I induced by the oral route of administration are opposite to the effects of GH and consistent with an antagonistic effect on GH action.

These studies in GH-deficient and postmenopausal women provide compelling evidence that estrogen levels achieved in the portal circulation after ingestion of therapeutic doses of oral estrogen impair the GH-regulated function of the liver. The molecular mechanisms by which estrogen inhibits GH action have been elucidated by Leung and coworkers [76] who reported

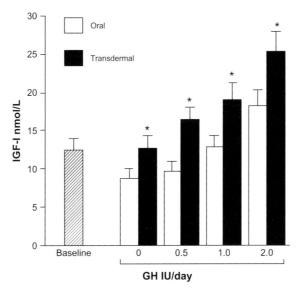

Fig. 2. Mean serum IGF-1 concentration before and during incremental dosages of GH (0.5, 1, and 2 IU/d) during oral and transdermal estrogen therapy. *$P < .05$ by analysis of covariance, oral versus transdermal. (*From* Wolthers T, Hoffman DM, Nugent AG, et al. Oral estrogen antagonizes the metabolic actions of growth hormone in growth hormone-deficient women. Am J Physiol Endocrinol Metab 2001;281:E193; with permission.)

that estrogen directly inhibits the function of the GH receptor by inducing the expression of proteins called SOCS (suppressors of cytokine signaling), which are negative regulators of GH signaling.

The influence of selective estrogen receptor modulators on the metabolic process has not been studied in detail. In the previously mentioned study comparing effects of therapeutic doses of raloxifene and estrogen in GH-deficient and postmenopausal women, fat oxidation was reduced by both treatments by a similar extent [55]. To date, there are no published data comparing the effects of selective estrogen receptor modulators and estrogen on the metabolic actions GH.

Androgens

There is evidence that both GH and testosterone are necessary to exert an optimal anabolic effect. Even following adequate androgen replacement, lean body mass is reduced in GH-deficient men and increases by about 4 kg on GH replacement therapy [72]. The observation that the effects of GH replacement are more marked in men compared with women [77] provides support that testosterone might enhance the anabolic effects of GH.

Gibney and coworkers [46] studied the independent and combined effects of testosterone and GH on IGF-I, energy, and protein metabolism in a group of hypopituitary men. Although testosterone enhanced the IGF-I response to GH, testosterone alone did not change IGF-I concentrations. GH alone stimulated resting energy expenditure and fat oxidation, whereas co-administration of testosterone amplified these metabolic changes (Fig. 3A, B). GH treatment significantly reduced protein oxidation and increased synthesis, and these changes were enhanced by combined treatment with testosterone (Fig. 3C). The results provide strong evidence that testosterone enhances the metabolic actions of GH.

Of particular interest is whether testosterone alone is anabolic, because it failed to elevate IGF-I levels in hypopituitary men. Gibney and coworkers [46] observed that testosterone alone reduced whole-body protein oxidation and stimulated synthesis, and these effects were enhanced by cotreatment with GH. Testosterone also stimulated resting energy expenditure and lipid oxidation, and these effects were increased by combined GH treatment. Testosterone exerts anabolic effects independent of IGF-I and GH while interacting positively with GH to enhance energy metabolism and protein anabolism. Mauras and coworkers [44] have also reported that testosterone alone reduces whole-body protein oxidation in prepubertal boys with GH deficiency and that GH cotreatment enhances this effect. Protein synthesis was significantly stimulated only after combined treatment. The change in fat-free mass was greater after combined treatment than after testosterone alone.

In addition to effects on the metabolic process, GH regulates sodium and fluid retention [78]. This property of GH is frequently observed as soft tissue swelling and edema during initiation of GH treatment. In studies of hypopituitary men, Johannsson and coworkers [79] have recently reported that the antinatriuretic action of GH is enhanced by testosterone. Testosterone itself also causes sodium retention and GH amplifies this effect. These changes occurred without a significant effect on components of the renin-angiotensin system or on atrial-natriuretic peptide concentration. These observations suggest that the antinatriuretic effects of GH and its influence by androgens may be exerted at the renal tubular level. Testosterone and GH exert independent and additive effects on energy metabolism, protein anabolism, and extracellular fluid volume.

Summary

Estrogens and androgens modulate the secretion and the action of GH. These interactions occur centrally and peripherally in the liver. Sex steroids regulate GH secretion directly and indirectly through IGF-I modulation. Testosterone stimulates GH secretion centrally, pushing the GH–IGF-I axis, an effect dependent on prior aromatization to estrogen. Estrogen stimulates GH secretion indirectly by reducing IGF-I feedback inhibition. This pulling effect is a consequence of a first pass hepatic effect from oral administration. Whether estrogen stimulates GH secretion centrally in females is unresolved; however, evidence that it pushes secretion is weak. Testosterone enhances GH action. It amplifies GH stimulation of IGF-I, sodium

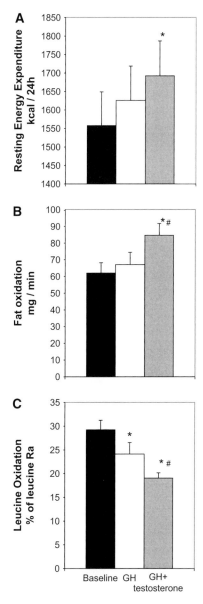

Fig. 3. Study in hypopituitary men investigating the impact of GH alone and combined GH and testosterone replacement on (*A*) resting energy expenditure, (*B*) fat oxidation, and (*C*) leucine oxidation, an index of whole-body protein oxidation expressed as percent leucine rate of appearance. Black boxes for baseline, white boxes for GH, and gray boxes for GH plus testosterone; all panels showing mean (+ standard error of the mean). *$P < .05$ versus baseline. #$P < .05$ versus GH. (*Data from* Gibney J, Wolthers T, Johannsson G, et al. Growth hormone and testosterone interact positively to enhance protein and energy metabolism in hypopituitary men. Am J Physiol Endocrinol Metab 2005;289:E266–71.)

retention, substrate metabolism, and protein anabolism while possessing similar but independent actions of its own. Estrogen attenuates GH action by inhibiting GH-regulated endocrine function of the liver. This is a concentration-dependent phenomenon, which arises invariably from oral administration of therapeutic doses of estrogen, an effect that can be avoided by using a parenteral route. Estrogen levels achieved in the female menstrual cycle and during pregnancy, however, affect hepatic metabolic function.

The strong modulatory action of gonadal steroids on GH responsiveness provides insights into the biologic basis of sexual dimorphism in growth, development, and body composition. GH is a major regulator of growth, somatic development, and body composition in later adult life. As GH secretion increases equally in boys and girls during puberty with arguably gender difference in adult years, it is likely that sex steroids, interacting with the GH–IGF-1 axis play a major part in determining growth and body composition differences between the sexes.

The physiologic regulation by sex steroids of GH secretion and action also has practical implications for the clinical endocrinologist. In pediatric endocrinology, it calls for an appraisal of the diagnostic criteria for GH deficiency of GH stimulation tests, which currently are based on arbitrary cut-offs that do not take into account the shifting baseline from the changing gonadal steroid milieu of puberty. The rationale and interpretation of sex steroid priming for GH stimulation tests requires a similar re-evaluation. In adult endocrinology, diagnostic tests for GH deficiency need to take into account the effects of gonadal steroids, particularly in the female with pituitary disease already replaced with oral estrogen. In the management of GH deficiency in the hypopituitary female, estrogen should be administered by a nonoral route. In hypopituitary men, androgens should be replaced concurrently to maximize the benefits of GH. In the general population, the metabolic consequences of long-term treatment of women with oral estrogens (ie, oral contraceptive steroids, hormone-replacement therapy) and with estrogen compounds (ie, selective estrogen receptor modulators) are largely unknown and deserve study.

References

[1] Roth J, Glick SM, Yalow RS, et al. Secretion of human growth hormone: physiologic and experimental modification. Metabolism 1963;12:577–9.
[2] Imaki T, Shibasaki T, Shizume K, et al. The effect of free fatty acids on growth hormone (GH)-releasing hormone-mediated GH secretion in man. J Clin Endocrinol Metab 1985;60:290–3.
[3] Merimee TJ, Rabinowtitz D, Fineberg SE. Arginine-initiated release of human growth hormone; factors modifying the response in normal man. N Engl J Med 1969;280:1434–8.
[4] Melmed S. Insulin suppresses growth hormone secretion by rat pituitary cells. J Clin Invest 1984;73:1425–33.
[5] Vahl N, Jorgensen JO, Skjaerbaek C, et al. Abdominal adiposity rather than age and sex predicts mass and regularity of GH secretion in healthy adults. Am J Physiol 1997;272(6 Pt 1): E1108–16.

[6] Ho KY, Veldhuis JD, Johnson ML, et al. Fasting enhances growth hormone secretion and amplifies the complex rhythms of growth hormone secretion in man. J Clin Invest 1988;81: 968–75.
[7] Kerrigan JR, Rogol AD. The impact of gonadal steroid hormone action on growth hormone secretion during childhood and adolescence. Endocr Rev 1992;13:281–98.
[8] Giustina A, Scalvini T, Tassi C, et al. Maturation of the regulation of growth hormone secretion in young males with hypogonadotropic hypogonadism pharmacologically exposed to progressive increments in serum testosterone. J Clin Endocrinol Metab 1997; 82:1210–9.
[9] Jospe N, Orlowski CC, Furlanetto RW. Comparison of transdermal and oral estrogen therapy in girls with Turner's syndrome. J Pediatr Endocrinol Metab 1995;8:111–6.
[10] Harris DA, Van Vliet G, Egli CA, et al. Somatomedin-C in normal puberty and in true precocious puberty before and after treatment with a potent luteinizing hormone-releasing hormone agonist. J Clin Endocrinol Metab 1985;61:152–9.
[11] Mansfield MJ, Rudlin CR, Crigler JF Jr, et al. Changes in growth and serum growth hormone and plasma somatomedin-C levels during suppression of gonadal sex steroid secretion in girls with central precocious puberty. J Clin Endocrinol Metab 1988;66:3–9.
[12] Blum WF, Albertsson-Wikland K, Rosberg S, et al. Serum levels of insulin-like growth factor I (IGF-I) and IGF binding protein 3 reflect spontaneous growth hormone secretion. J Clin Endocrinol Metab 1993;76:1610–6.
[13] Giusti M, Torre R, Cavagnaro P, et al. The effect of long-term pulsatile GnRH administration on the 24-hour integrated concentration of GH in hypogonadotropic hypogonadic patients. Acta Endocrinol (Copenh) 1989;120:724–8.
[14] Weissberger AJ, Ho KK. Activation of the somatotropic axis by testosterone in adult males: evidence for the role of aromatization. J Clin Endocrinol Metab 1993;76:1407–12.
[15] Bondanelli M, Ambrosio MR, Margutti A, et al. Activation of the somatotropic axis by testosterone in adult men: evidence for a role of hypothalamic growth hormone-releasing hormone. Neuroendocrinology 2003;77:380–7.
[16] Liu L, Merriam GR, Sherins RJ. Chronic sex steroid exposure increases mean plasma growth hormone concentration and pulse amplitude in men with isolated hypogonadotropic hypogonadism. J Clin Endocrinol Metab 1987;64:651–6.
[17] Ovesen P, Vahl N, Fisker S, et al. Increased pulsatile, but not basal, growth hormone secretion rates and plasma insulin-like growth factor I levels during the periovulatory interval in normal women. J Clin Endocrinol Metab 1998;83:1662–7.
[18] Weissberger AJ, Ho KK, Lazarus L. Contrasting effects of oral and transdermal routes of estrogen replacement therapy on 24-hour growth hormone (GH) secretion, insulin-like growth factor I, and GH-binding protein in postmenopausal women. J Clin Endocrinol Metab 1991;72:374–81.
[19] Hartmann BW, Kirchengast S, Albrecht A, et al. Effect of hormone replacement therapy on growth hormone stimulation in women with premature ovarian failure. Fertil Steril 1997;68: 103–7.
[20] Loche S, Colao A, Cappa M, et al. The growth hormone response to hexarelin in children: reproducibility and effect of sex steroids. J Clin Endocrinol Metab 1997;82:861–4.
[21] Keenan BS, Richards GE, Ponder SW, et al. Androgen-stimulated pubertal growth: the effects of testosterone and dihydrotestosterone on growth hormone and insulin-like growth factor-I in the treatment of short stature and delayed puberty. J Clin Endocrinol Metab 1993;76:996–1001.
[22] Veldhuis JD, Keenan DM, Mielke K, et al. Testosterone supplementation in healthy older men drives GH and IGF-I secretion without potentiating peptidyl secretagogue efficacy. Eur J Endocrinol 2005;153:577–86.
[23] Moll GW Jr, Rosenfield RL, Fang VS. Administration of low-dose estrogen rapidly and directly stimulates growth hormone production. Am J Dis Child 1986;140:124–7.

[24] Schober E, Frisch H, Waldhauser F, et al. Influence of estrogen administration on growth hormone response to GHRH and L-Dopa in patients with Turner's syndrome. Acta Endocrinol (Copenh) 1989;120:442–6.
[25] Leung KC, Johannsson G, Leong GM, et al. Estrogen regulation of growth hormone action. Endocr Rev 2004;25:693–721.
[26] Bellantoni MF, Vittone J, Campfield AT, et al. Effects of oral versus transdermal estrogen on the growth hormone/insulin-like growth factor I axis in younger and older postmenopausal women: a clinical research center study. J Clin Endocrinol Metab 1996;81:2848–53.
[27] Friend KE, Hartman ML, Pezzoli SS, et al. Both oral and transdermal estrogen increase growth hormone release in postmenopausal women–a clinical research center study. J Clin Endocrinol Metab 1996;81:2250–6.
[28] Lieman HJ, Adel TE, Forst C, et al. Effects of aging and estradiol supplementation on GH axis dynamics in women. J Clin Endocrinol Metab 2001;86:3918–23.
[29] Soares-Welch C, Farhy L, Mielke KL, et al. Complementary secretagogue pairs unmask prominent gender-related contrasts in mechanisms of growth hormone pulse renewal in young adults. J Clin Endocrinol Metab 2005;90:2225–32.
[30] Benito P, Avila L, Corpas MS, et al. Sex differences in growth hormone response to growth hormone-releasing hormone. J Endocrinol Invest 1991;14:265–8.
[31] Veldhuis JD, Patrie JT, Brill KT, et al. Contributions of gender and systemic estradiol and testosterone concentrations to maximal secretagogue drive of burst-like growth hormone secretion in healthy middle-aged and older adults. J Clin Endocrinol Metab 2004;89:6291–6.
[32] Kimber J, Sivenandan M, Watson L, et al. Age- and gender-related growth hormone responses to intravenous clonidine in healthy adults. Growth Horm IGF Res 2001;11:128–35.
[33] Rao RH, Spathis GS. Intramuscular glucagon as a provocative stimulus for the assessment of pituitary function: growth hormone and cortisol responses. Metabolism 1987;36:658–63.
[34] Qu XD, Gaw Gonzalo IT, Al Sayed MY, et al. Influence of body mass index and gender on growth hormone (GH) responses to GH-releasing hormone plus arginine and insulin tolerance tests. J Clin Endocrinol Metab 2005;90:1563–9.
[35] Giustina A, Licini M, Bussi AR, et al. Effects of sex and age on the growth hormone response to galanin in healthy human subjects. J Clin Endocrinol Metab 1993;76:1369–72.
[36] Juul A, Dalgaard P, Blum WF, et al. Serum levels of insulin-like growth factor (IGF)-binding protein-3 (IGFBP-3) in healthy infants, children, and adolescents: the relation to IGF-I, IGF-II, IGFBP-1, IGFBP-2, age, sex, body mass index, and pubertal maturation. J Clin Endocrinol Metab 1995;80:2534–42.
[37] Clayton PE, Hall CM. Insulin-like growth factor I levels in healthy children. Horm Res 2004;62(Suppl 1):2–7.
[38] Lofqvist C, Andersson E, Gelander L, et al. Reference values for insulin-like growth factor-binding protein-3 (IGFBP-3) and the ratio of insulin-like growth factor-I to IGFBP-3 throughout childhood and adolescence. J Clin Endocrinol Metab 2005;90:1420–7.
[39] Coutant R, de Casson FB, Rouleau S, et al. Divergent effect of endogenous and exogenous sex steroids on the insulin-like growth factor I response to growth hormone in short normal adolescents. J Clin Endocrinol Metab 2004;89:6185–92.
[40] Erfurth EM, Hagmar LE, Saaf M, et al. Serum levels of insulin-like growth factor I and insulin-like growth factor-binding protein 1 correlate with serum free testosterone and sex hormone binding globulin levels in healthy young and middle-aged men. Clin Endocrinol (Oxf) 1996;44:659–64.
[41] Pfeilschifter J, Scheidt-Nave C, Leidig-Bruckner G, et al. Relationship between circulating insulin-like growth factor components and sex hormones in a population-based sample of 50- to 80-year-old men and women. J Clin Endocrinol Metab 1996;81:2534–40.
[42] Hobbs CJ, Plymate SR, Rosen CJ, et al. Testosterone administration increases insulin-like growth factor-I levels in normal men. J Clin Endocrinol Metab 1993;77:776–9.

[43] van Kesteren P, Lips P, Deville W, et al. The effect of one-year cross-sex hormonal treatment on bone metabolism and serum insulin-like growth factor-1 in transsexuals. J Clin Endocrinol Metab 1996;81:2227–32.
[44] Mauras N, Rini A, Welch S, et al. Synergistic effects of testosterone and growth hormone on protein metabolism and body composition in prepubertal boys. Metabolism 2003;52:964–9.
[45] Saggese G, Cesaretti G, Franchi G, et al. Testosterone-induced increase of insulin-like growth factor I levels depends upon normal levels of growth hormone. Eur J Endocrinol 1996;135:211–5.
[46] Gibney J, Wolthers T, Johannsson G, et al. Growth hormone and testosterone interact positively to enhance protein and energy metabolism in hypopituitary men. Am J Physiol Endocrinol Metab 2005;289:E266–71.
[47] Rooman RP, De Beeck LO, Martin M, et al. Ethinylestradiol and testosterone have divergent effects on circulating IGF system components in adolescents with constitutional tall stature. Eur J Endocrinol 2005;152:597–604.
[48] Landin-Wilhelmsen K, Lundberg PA, Lappas G, et al. Insulin-like growth factor I levels in healthy adults. Horm Res 2004;62(Suppl 1):8–16.
[49] Span JP, Pieters GF, Sweep CG, et al. Gender difference in insulin-like growth factor I response to growth hormone (GH) treatment in GH-deficient adults: role of sex hormone replacement. J Clin Endocrinol Metab 2000;85:1121–5.
[50] Parkinson C, Ryder WDJ, Trainer PJ. The relationship between serum GH and serum IGF-I in acromegaly is gender-specific. J Clin Endocrinol Metab 2001;86:5240–4.
[51] Gleeson HK, Shalet SM. GH responsiveness varies during the menstrual cycle. Eur J Endocrinol 2005;153:775–9.
[52] Eden Engstrom B, Burman P, Johansson AG, et al. Effects of short-term administration of growth hormone in healthy young men, women, and women taking oral contraceptives. J Intern Med 2000;247:570–8.
[53] Wolthers T, Hoffman DM, Nugent AG, et al. Oral estrogen antagonizes the metabolic actions of growth hormone in growth hormone-deficient women. Am J Physiol Endocrinol Metab 2001;281:E1191–6.
[54] Kelly JJ, Rajkovic IA, O'Sullivan AJ, et al. Effects of different oral oestrogen formulations on insulin-like growth factor-I, growth hormone and growth hormone binding protein in post-menopausal women. Clin Endocrinol (Oxf) 1993;39:561–7.
[55] Gibney J, Johannsson G, Leung KC, et al. Comparison of the metabolic effects of raloxifene and oral estrogen in postmenopausal and growth hormone-deficient women. J Clin Endocrinol Metab 2005;90:3897–903.
[56] Lissett CA, Shalet SM. The impact of dose and route of estrogen administration on the somatotropic axis in normal women. J Clin Endocrinol Metab 2003;88:4668–72.
[57] Liu PY, Hoey KA, Mielke KL, et al. A randomized placebo-controlled trial of short-term graded transdermal estradiol in healthy gonadotropin-releasing hormone agonist-suppressed pre- and postmenopausal women: effects on serum markers of bone turnover, insulin-like growth factor-I, and osteoclastogenic mediators. J Clin Endocrinol Metab 2005;90:1953–60.
[58] Campagnoli C, Colombo P, De Aloysio D, et al. Positive effects on cardiovascular and breast metabolic markers of oral estradiol and dydrogesterone in comparison with transdermal estradiol and norethisterone acetate. Maturitas 2002;41:299–311.
[59] Lobo RA. Clinical review 27: effects of hormonal replacement on lipids and lipoproteins in postmenopausal women. J Clin Endocrinol Metab 1991;73:925–30.
[60] Nugent AG, Leung KC, Sullivan D, et al. Modulation by progestogens of the effects of oestrogen on hepatic endocrine function in postmenopausal women. Clin Endocrinol (Oxf) 2003;59:690–8.
[61] Duffy MJ, Duffy GJ. Estradiol receptors in human liver. J Steroid Biochem 1978;9:233–5.
[62] Nagasue N, Ito A, Yukaya H, et al. Androgen receptors in hepatocellular carcinoma and surrounding parenchyma. Gastroenterology 1985;89:643–7.

[63] Nagasue N, Kohno H, Yamanoi A, et al. Progesterone receptor in hepatocellular carcinoma: correlation with androgen and estrogen receptors. Cancer 1991;67:2501–5.
[64] Lindgren R, Berg G, Hammar M, et al. Plasma lipid and lipoprotein effects of transdermal administration of estradiol and estradiol/norethisterone acetate. Eur J Obstet Gynecol Reprod Biol 1992;47:213–21.
[65] Duschek EJ, de Valk-de Roo GW, Gooren LJ, et al. Effects of conjugated equine estrogen vs. raloxifene on serum insulin-like growth factor-I and insulin-like growth factor binding protein-3: a 2-year, double-blind, placebo-controlled study. Fertil Steril 2004;82:384–90.
[66] Helle SI, Anker GB, Tally M, et al. Influence of droloxifene on plasma levels of insulin-like growth factor (IGF)-I, Pro-IGF-IIE, insulin-like growth factor binding protein (IGFBP)-1 and IGFBP-3 in breast cancer patients. J Steroid Biochem Mol Biol 1996;57:167–71.
[67] Rogol AD. Gender and hormonal regulation of growth. J Pediatr Endocrinol Metab 2004;17(Suppl 4):1259–65.
[68] Bourguignon JP. Linear growth as a function of age at onset of puberty and sex steroid dosage: therapeutic implications. Endocr Rev 1988;9:467–88.
[69] Delemarre-van de Waal HA, van Coeverden SC, Rotteveel J. Hormonal determinants of pubertal growth. J Pediatr Endocrinol Metab 2001;14(Suppl 6):1521–6.
[70] Weise M, De-Levi S, Barnes KM, et al. Effects of estrogen on growth plate senescence and epiphyseal fusion. Proc Natl Acad Sci U S A 2001;98:6871–6.
[71] Zachmann M, Prader A. Anabolic and androgenic affect of testosterone in sexually immature boys and its dependency on growth hormone. J Clin Endocrinol Metab 1970;30:85–95.
[72] Span JP, Pieters GF, Sweep FG, et al. Gender differences in rhGH-induced changes in body composition in GH-deficient adults. J Clin Endocrinol Metab 2001;86:4161–5.
[73] Valimaki MJ, Salmela PI, Salmi J, et al. Effects of 42 months of GH treatment on bone mineral density and bone turnover in GH-deficient adults. Eur J Endocrinol 1999;140:545–54.
[74] Blackman MR, Sorkin JD, Munzer T, et al. Growth hormone and sex steroid administration in healthy aged women and men: a randomized controlled trial. JAMA 2002;288:2282–92.
[75] O'Sullivan AJ, Crampton LJ, Freund J, et al. The route of estrogen replacement therapy confers divergent effects on substrate oxidation and body composition in postmenopausal women. J Clin Invest 1998;102:1035–40.
[76] Leung KC, Doyle N, Ballesteros M, et al. Estrogen inhibits GH signaling by suppressing GH-induced JAK2 phosphorylation, an effect mediated by SOCS-2. Proc Natl Acad Sci U S A 2003;100:1016–21.
[77] Burman P, Johansson AG, Siegbahn A, et al. Growth hormone (GH)-deficient men are more responsive to GH replacement therapy than women. J Clin Endocrinol Metab 1997;82:550–5.
[78] Ho KY, Weissberger AJ. The antinatriuretic action of biosynthetic human growth hormone in man involves activation of the renin-angiotensin system. Metabolism 1990;39:133–7.
[79] Johannsson G, Gibney J, Wolthers T, et al. Independent and combined effects of testosterone and growth hormone on extracellular water in hypopituitary men. J Clin Endocrinol Metab 2005;90:3989–94.

Effects of Growth Hormone on Glucose and Fat Metabolism in Human Subjects

Jens O.L. Jørgensen, MD, DMSc[a],*,
Louise Møller, MD[a], Morten Krag, MD[a],
Nils Billestrup, PhD[b], Jens S. Christiansen, MD, DMSc[a]

[a]*Medical Department M (Endocrinology and Diabetes) and Institute of Experimental Clinical Research, Aarhus University Hospital, Norrebrogade 44, DK-800C, Aarhus, Denmark*
[b]*Steno Diabetes Center, Niels Steensens Vej 2, 2820 Gentofte, Denmark*

A link between the pituitary gland and glucose metabolism was originally observed by Houssay [1], who recorded increased sensitivity to insulin in hypophysectomized animals that was reversed by administration of anterior pituitary extracts. Forty years ago Rabinowitz and colleagues [2] showed that infusion of high-dose growth hormone (GH) into the brachial artery of healthy adults reduced forearm glucose uptake in both muscle and adipose tissue. This reduction is paralleled by increased muscular uptake and oxidation of free fatty acids (FFA). Moreover, GH completely blocked the effects of insulin when the two hormones were administered together. Despite years of subsequent research, the mechanisms whereby GH causes insulin resistance and stimulates lipolysis remain unclear. Several lines of evidence, however, suggest a close link between these two important effects of GH. This article focuses on in vivo data from tests performed in normal subjects and in patients who had abnormal GH status. The effect of GH on protein metabolism is discussed elsewhere in this issue.

Effects of growth hormone on glucose and fat metabolism

Healthy subjects

Using the human forearm as an experimental model, Rabinowitz and colleagues [2] performed several classic studies on the effects of insulin and GH,

---

* Corresponding author.
  E-mail address: jolj@dadlnet.dk (J.O.L. Jørgensen).

singly and together, on muscle and adipose tissue metabolism in situ. Intra-arterial administration of insulin alone caused a 10-fold increase in skeletal muscle glucose uptake, whereas the regional release of FFA from adipose tissue was obliterated. By contrast, GH alone stimulated the release of FFA from adipose tissue and its uptake into muscle, whereas glucose uptake was decreased. Administration of GH together with insulin virtually blocked forearm uptake of glucose as well as FFA release from fat. The observation that the antilipolytic effects of insulin prevail during concomitant GH exposure has been challenged subsequently, but the authors were the first to provide a comprehensive theory to explain the physiologic and pathophysiologic interaction between insulin and GH.

By combining the infusion of radiolabeled glucose with the euglycemic, hyperinsulinemic clamp, it later was demonstrated that GH infusion impairs hepatic as well as peripheral insulin sensitivity [3,4]. Møller and colleagues [5] studied the time course of these events using similar methods in addition to the forearm model; insulin resistance in muscle, as measured by the glucose infusion rate, and forearm uptake of glucose were detectable after 2 hours of GH infusion and were paralleled by a reduced insulin-induced suppression of hepatic glucose output. In the same study GH infusion also increased the circulating levels of lipid intermediates as well as the forearm uptake of FFA, but these lipolytic effects were delayed compared with the effects of GH on glucose fluxes. It subsequently was demonstrated that administration of a physiologic GH pulse was associated with distinct time-course effects on substrate metabolism with a rapid decrease in skeletal muscle glucose uptake followed by stimulation of lipolysis and a net muscle uptake of lipid intermediates [6]. At first glance these observations may seem to challenge the theory that FFA is the cause of insulin resistance by means of substrate competition with glucose in the glycolytic pathways (Fig. 1) [7].

Acromegaly before and after treatment

Active acromegaly is a hypermetabolic condition characterized by elevated resting energy expenditure, increased lipolysis, and hepatic as well as peripheral insulin resistance [8]. An increased prevalence of impaired glucose tolerance and overt diabetes mellitus also is well documented [9]. It is assumed, albeit not formally demonstrated, that these abnormalities contribute to the observed increase in cardiovascular morbidity and mortality [9]. In indirect support of this assumption, successful surgery, which has been shown to normalize the mortality rate, also is associated with a correction of the metabolic aberrations. Møller and colleagues [8] studied substrate metabolism and insulin sensitivity in six newly diagnosed acromegalic patients before and several months after successful transsphenoidal adenomectomy. The methods included indirect calorimetry, use of a glucose tracer, measurement of substrate exchanges across the forearm, and a hyperinsulinemic, euglycemic glucose clamp. In the basal state plasma

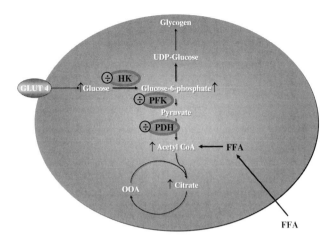

Fig. 1. The glucose–fatty acid (Randle) cycle in muscle. Oxidation of fatty acids (FFA) inhibits pyruvate dehydrogenase (PDH). Citrate inhibits phosphofructokinase (PFK). The rise in glucose-6-phosphate inhibits hexokinase. GLUT 4, glucose transporter 4; UDP, uridine diphosphate.

levels of insulin and glucose were significantly elevated before surgery and normalized afterwards. This normalization was associated with reduced forearm uptake of glucose and increased hepatic glucose output. Moreover, the rate of lipid oxidation also was increased in active acromegaly and normalized after surgery. The glucose infusion rate during the clamp was abnormally low in active acromegaly and was normalized with surgery. The latter finding suggests GH-induced resistance to insulin-stimulated glucose uptake in skeletal muscle compatible with the observations made in the basal state. In addition, failure of insulin to suppress lipid oxidation was observed in the untreated state. More recently, pegvisomant, which is a specific GH antagonist, has been approved for treating acromegaly [10]. Drake and colleagues [11] and the present authors [12] have reported that administration of pegvisomant improves glucose tolerance in acromegalic patients and also have observed, by means of the glucose clamp technique, that pegvisomant improves insulin sensitivity in acromegaly (unpublished data).

Growth hormone–deficient patients

Fasting hypoglycemia is a frequent occurrence in naive GH-deficient children, and, in keeping with the known effects of GH, β-cell function and insulin sensitivity in GH-deficient adults were assumed to be increased or at least normal. It therefore was unexpected when Beshyah and colleagues [13] observed an increased prevalence of abnormal glucose tolerance among GH-deficient adults as compared with healthy, age- and sex-matched controls. Determinants of abnormal glucose tolerance included age, female sex, and obesity. Moreover, the area under the curve (AUC) for insulin and the insulin AUC: glucose AUC ratio during oral glucose tolerance testing (OGTT) were elevated in the patients. Fasting plasma levels of glucose and insulin were normal. Johansson and colleagues [14] also observed distinctly impaired insulin sensitivity ($>50\%$) in 15 adult patients by means of the glucose clamp technique after correction for differences in body size. Again, fasting levels of plasma glucose and insulin were comparable between patients and controls. Similar results have been obtained by Hew and colleagues [15], who also documented decreased insulin-stimulated glycogen synthase activity. The mechanisms underlying the impairment of insulin sensitivity in untreated growth hormone deficiency in adults (GHDA) are unknown, but a plausible candidate is increased FFA flux from visceral fat because visceral adiposity is a hallmark of GHDA. It is noteworthy that a normal body mass index does not exclude visceral obesity [16]. The degree to which additional pituitary deficits or life style factors contribute to insulin resistance in GHDA remains elusive.

In a study of adolescent GH-deficient patients receiving GH replacement therapy, the present authors [17] evaluated the impact of replacing one daily (evening) injection with a 10-hour intravenous infusion of either saline or GH in a low dose (35 μg/h), starting the evening before the study. Continued GH infusion was associated with reduced basal rates of glucose as compared with saline-treated and healthy untreated control subjects. Reciprocal changes were observed regarding lipid oxidation. Insulin sensitivity was increased relative to control subjects during saline infusion and became reduced during GH infusion to a level comparable to the control group.

Fowelin and colleagues [18] studied insulin sensitivity and glucose metabolism in nine patients participating in a double-blind, placebo-controlled crossover trial including assessments at baseline and after 6 and 26 weeks of GH treatment. Fasting plasma levels of glucose and insulin increased after 6 weeks of GH but returned toward baseline values after 26 weeks. The glucose infusion rate during the clamp decreased significantly by 35% after 6 weeks of GH treatment. After 26 weeks the glucose infusion rate was decreased by 25% of baseline levels, a difference that was no longer significant. In an open design O'Neal and colleagues [19] studied 10 patients who had adult-onset disease after 1 week and after 3 months of GH replacement, respectively. Fasting plasma levels of glucose and insulin increased after 1 week but normalized after 3 months. Hemoglobin A_{1c} levels did not change during the study. In a 12-month study insulin sensitivity (Hemostatic Model Assessment) decreased, and first-phase insulin secretion (intravenous glucose) increased significantly, despite favorable alterations in body composition [20]. Moreover, fasting plasma glucose levels increased slightly within the normal range, whereas hemoglobin A_{1c} levels remained normal and unchanged. Christopher and colleagues [21] reported sustained insulin resistance in 11 patients treated with GH (0.22 IU/kg/wk) for 24 months. Beshyah and colleagues [22] performed OGTT after 6, 12, and 18 months of GH replacement in patients who had (predominantly) adult-onset disease

[22]. After 18 months elevated fasting plasma levels of both glucose and insulin were recorded together with elevated AUC for both glucose and insulin during the OGTTs. Impairment of glucose tolerance and moderate insulin resistance in combination with increased secretion and clearance of insulin also were recorded after 30 months of GH substitution (approximately 0.5 mg/d) in an open trial [23]. These findings contrast with data from two studies lasting 1 and 5 years, respectively, in which insulin sensitivity was either improved [24] or normal [25]. A subsequent, quasi-controlled study of 10 years of GH replacement in adult-onset patients did not detect changes in fasting levels of glucose, insulin, or C-peptide [26].

A double-blind, placebo-controlled, parallel study assessed the impact on body composition and insulin sensitivity of discontinuing GH replacement after completion of longitudinal growth [27]. The patients were assigned randomly to either continued GH replacement or placebo for 12 months followed by 12 months of open-labeled GH therapy in both groups. Body composition and insulin sensitivity were measured at baseline and after 12 and 24 months of therapy, respectively. In the group that continued GH therapy no significant changes were recorded in either body composition or insulin sensitivity. By contrast, placebo treatment was accompanied by an increase in insulin sensitivity despite a concomitant increase in fat mass. After the resumption of GH treatment in the placebo group, fat mass decreased together with insulin sensitivity. These data suggest that the direct insulin-antagonistic effects of GH are not balanced fully by the favorable long-term effects on body composition.

Effects of growth hormone on substrate metabolism and insulin sensitivity during fasting

Endogenous GH secretion in normal subjects is stimulated during fasting and is accompanied by low levels of insulin. Concomitantly, substrate metabolism is shifted toward lipid oxidation. When glycogen stores are depleted, glucose oxidation relies on gluconeogenesis from protein. In a series of studies Norrelund and colleagues [28] have demonstrated that in the fasting state (42 hours) GH is essential for the ongoing release and oxidation of FFA [28]. At the same time GH reduces insulin sensitivity as indicated by muscle uptake of glucose. These effects translate into reduced protein breakdown in muscle and at the whole-body level. The lipolytic effects of GH during fasting, which could be considered the natural domain of GH, thus seem to result in protein sparing.

Cellular mechanisms

Is there crosstalk between growth hormone and insulin signaling?

A central question to be answered was whether a GH bolus translates into GH receptor signaling in muscle and fat. To this end the authors [29] studied six healthy male subjects who, in a single-blinded design, received an intravenous bolus of either GH (0.5 mg) or saline in the postabsorptive phase on two separate occasions. Muscle and fat biopsies were taken 30 to 60 minutes after each bolus. Tyrosine phosphorylation of signal transducer and activator of transcription-5 (Stat5b), measured by Western blotting, was detected in both muscle and fat after GH (Fig. 2), whereas specific binding of phosphorylated Stat5b to DNA was demonstrated only in fat biopsies. Moreover, evidence of genomic effects in terms of increased expression of mRNA for insulin-like growth factor-1 and suppressor of cytokine signaling 3 was detectable in fat [29]. In the same study a GH-induced stimulation of lipolysis was evidenced by significantly elevated serum FFA levels. This study thus suggests that GH exerts acute and direct effects in muscle and fat in humans by means of the Janus kinase/Stat pathway as previously demonstrated in cell lines and in rodent models [30]. It thus is feasible that the acute and prolonged effects of GH on glucose metabolism in muscle could be mediated by inhibitory crosstalk between GH and insulin signaling in the target tissues in accordance with

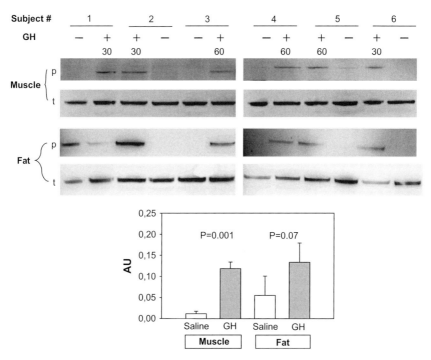

Fig. 2. Effect of GH (+) versus saline (−) on Stat5b tyrosine phosphorylation in muscle and fat biopsies (obtained after either 30 or 60 minutes as indicated) from the six participants (1–6). Western blots are shown for phosphorylated (p) and total (t) Stat5b. For bar graphs, content of phosphorylated Stat5b was normalized to the amount of total Stat5b in each issue. (*From* Jørgensen JOL, Jessen N, Pedersen SB, et al. Growth hormone receptor signaling in skeletal muscle and adipose tissue in human subjects following exposure to an intravenous GH bolus. Am J Physiol Endocrinol Metab 2006;291:E899–905; with permission.)

data obtained from other models [31]. It is, however, also possible that these effects of GH are secondary to the lipolytic effects, because data from normal subjects and from patients who have type 2 diabetes mellitus suggest a causal role of FFA for inhibiting insulin-stimulated glucose disposal in skeletal muscle [32]. In support of this possibility, Nielsen and colleagues [33] demonstrated that administration of acipimox, a nicotinic acid derivative that inhibits the hormone-sensitive lipase (HSL), was able to suppress GH-induced FFA release in GH-deficient patients. At the same time acipimox abrogated GH-induced insulin resistance during a glucose clamp. Studies by Lowell and Shulman [32] indicate that these putative effects of FFA involve inhibition of the activity of insulin-signaling proteins, especially insulin receptor substrate 1 (IRS1)–associated phosphoinositol-3 (PI3) kinase, important for the translocation and activation of glucose transporter 4 in skeletal muscle (Fig. 3). Somewhat surprisingly, Jessen and colleagues [34] were unable to demonstrate any significant suppressive effect of GH on PI3 kinase activity in normal subjects even though GH significantly stimulated lipolysis and induced insulin resistance. In accordance with this finding, the present authors [29] also observed that GH in the postabsorptive state did not influence basal (ie, non–insulin-stimulated) IRS1-associated PI3 kinase activity.

Lipolysis

The ability of exogenous GH to stimulate the release and oxidation of FFA has been known for many years [35], but the underlying mechanisms are not yet fully characterized. Using the microdialysis technique, Djurhuus and

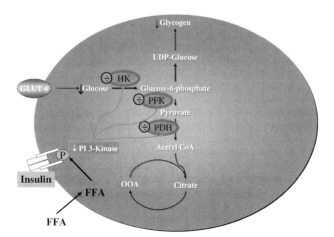

Fig. 3. The effects of free fatty acid on glucose metabolism in skeletal muscle as hypothesized by Shulman and colleagues [32]. Free fatty acit (FFA) inhibits insulin receptor substrate 1–associated phosphoinositol-3 (PI3) kinase activity, which is central for translocation of active glucose transporter 4 (GLUT4) to the cell surface. It remains unproven whether similar mechanisms are involved regarding the effects of GH. COA, coenzyme A; UDP, uridine diphosphate.

colleagues [36] have shown that an intravenous GH bolus is associated with release of glycerol from fat tissue compatible with a direct lipolytic effect on the adipocyte in vivo. Moreover, the data of Nielsen and colleagues [33] suggest that GH mediates its effects by activation of the HSL. HSL traditionally is activated by ligand binding to β-adrenergic receptors on the adipocytes, stimulating adenylate cyclase activity by activation of the guanine nucleotide-binding (Gs) protein or suppression of the inhibitory guanine nucleotide-binding (Gi) protein. This process generates cyclic AMP, which activates protein kinase A. Protein kinase A, in turn, phosphorylates HSL, which translocates to the surface of the lipid droplet and stimulates hydrolysis of triglyceride to glycerol and FFA [37]. In isolated adipocytes obtained from GH-deficient adults after 6 months of GH replacement, enhanced response to epinephrine-induced lipolysis was observed [38]. In a comparable study in obese women, Pedersen and colleagues [39] could not detect an effect of previous GH treatment on either basal or epinephrine-stimulated lipolysis ex vivo, whereas glucose oxidation was decreased. In general, a direct lipolytic effect of GH has been difficult to document in vitro and often requires pretreatment of the cells with dexamethasone [40]. This finding contrasts with the previously mentioned observation that GH directly stimulates glycerol release in human adipose tissue in vivo. In rodent models, however, evidence suggests that GH stimulates adenylate cyclase activity through a reduction in Gi protein expression in concomitance with enhanced expression of HSL and β-adrenergic receptors [41]. Based on an in vitro bioassay using human skin fibroblasts, Leung and Ho [42] reported a direct GH-induced stimulation of FFA oxidation, which indicates that the increase in lipid oxidation after GH exposure in vivo may not be solely secondary to increased substrate availability. It is assumed that GH exposure stimulates fat oxidation in muscle, because whole-body oxidation of lipid is increased together with increased uptake of FFA across the human forearm, and because skeletal muscle is the major determinant of energy metabolism. In addition, the present authors have observed recently that the intramyocellular content of triglyceride increases significantly after short-term GH administration in healthy human subjects (unpublished data). By contrast, it remains to be shown whether GH stimulates lipoprotein lipase activity in muscle, which theoretically would provide an additional mechanism whereby GH would promote lipid oxidation. Several groups, however, have shown that GH treatment inhibits lipoprotein lipase activity in vitro and ex vivo in fat tissue from human subjects, which supposedly contributes to the GH-induced reduction of adipose tissue [43,44]. The known sites of action of GH on FFA metabolism in human models are depicted in Fig. 4.

Discussion

Based on a continuing number of experiments since the studies of Houssay [1] and Rabinowitz and colleagues [2], there is little doubt that in a matter of few hours a GH bolus strongly antagonizes the effects of insulin on glucose disposal in skeletal muscle. In patients who have acromegaly, whose GH

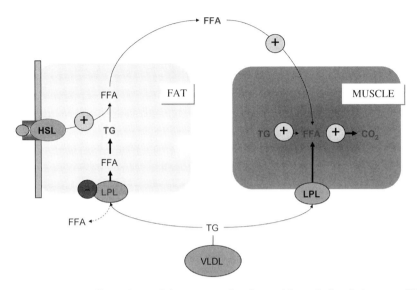

Fig. 4. The known effects of growth hormone on free fatty acid metabolism in humans. GH stimulates lipolysis in adipocytes by activation of hormone-sensitive lipase (HSL). The uptake of and oxidation of free fatty acid (FFA) is stimulated directly by growth hormone (based on in vitro data in human cells). Growth hormone administration is accompanied by accumulation of triglyceride (TG) in myocytes in conjunction with increased FFA oxidation. Growth hormone suppressed lipoprotein lipase activity in fat cells, but it remains to be shown if growth hormone stimulates lipoprotein lipase activity in muscle. LPL, lipoprotein lipase; VLDL, very low density lipoprotein.

levels also are elevated in the postprandial period, this antagonism may result in glucose intolerance and overt diabetes mellitus. Whether these effects of GH also have clinically important implications for GH replacement therapy in GH-deficient patients is another matter. GH usually is administered as a daily insulin-like subcutaneous injection in the evening, and the dose is tailored to maintain the level of growth factor-1 within the normal range [45]. With this regimen GH replacement therapy seems to be safe as regards glycemic control. Whether the favorable effects of GH on body composition and physical fitness fully balance the metabolic effects remains controversial.

During fasting the insulin-antagonistic effects of GH on glucose disposal may constitute an important and favorable adaptation by impeding the demand of gluconeogenesis from protein [28]. As such, GH is a unique counterregulatory hormone with combined lipolytic and protein-sparing actions (Fig. 5).

The underlying molecular mechanisms remain unclear and seem to differ from those observed in type 2 diabetes mellitus. Based on the study by Nielsen and colleagues [33] elevated FFA levels play a causal role, but there are no experimental in vivo data in humans to support the suggestion that insulin-signaling pathways are affected [29,34]. More studies in this field are needed. Attention should be paid to the dosing and timing of GH and

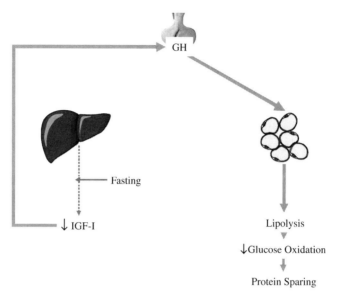

Fig. 5. A hypothetical and simplified model of the action of growth hormone during fasting on substrate metabolism. Hepatic insulin-like growth factor-I (IGF-I) production is suppressed by fasting, and the low circulating levels of IGF-I amplify pituitary release of growth hormone (GH). Subsequently, GH actions are partitioned toward lipolysis and lipid oxidation. This process ultimately translates into protein sparing by impeding the demand for glucose oxidation and thus gluconeogenesis from protein.

insulin and to the optimal time point of obtaining the muscle biopsy. It also is likely that newer methods such as gene arrays and proteomics may be important supplements to assays that measure only the activity of single proteins. Another important issue to address in the future is the relationship between GH and the so-called "adipokines" or "adipocytokines" (eg, leptin, adiponectin, visfatin, and resistin). Further research into the interaction of GH with glucose and FFA metabolism is clinically important for optimizing the treatment of GH-related disorders. It also may broaden the understanding of the role of cytokines in general in important conditions such as critical illness and type 2 diabetes mellitus.

Summary

Experimental data in human subjects demonstrate that GH acutely inhibits glucose disposal in skeletal muscle. The insulin-antagonistic effects are significant clinically because active acromegaly is accompanied by glucose intolerance, whereas children who have GH deficiency may develop fasting hypoglycemia. At the same time GH stimulates the turnover and oxidation of FFA, and experimental evidence suggests a causal link between elevated FFA levels and insulin resistance in skeletal muscle. During fasting, the induction of insulin resistance by GH is associated with enhanced lipid oxidation and protein conservation, which seems to constitute a favorable metabolic adaptation. Observational data in GH-deficient adults do not indicate that GH replacement is associated with significant impairment of glucose tolerance, but it is recommended that overdosing be avoided and glycemic control be monitored.

References

[1] Houssay BA. The hypophysis and metabolism. N Engl J Med 1936;214:961–85.
[2] Rabinowitz D, Klassen GA, Zierler KL. Effect of human growth hormone on muscle and adipose tissue metabolism in the forearm of man. J Clin Invest 1965;44:51–61.
[3] Bratusch-Marrain PR, Smith D, DeFronzo RA. The effect of growth hormone on glucose metabolism and insulin secretion in man. J Clin Endocrinol Metab 1982;55(5):973–82.
[4] Rizza RA, Mandarino LJ, Gerich JE. Effects of growth hormone on insulin action in man. Mechanisms of insulin resistance, impaired suppression of glucose production, and impaired stimulation of glucose utilization. Diabetes 1982;31(8 Pt 1):663–9.
[5] Møller N, Butler PC, Antsiferov MA, et al. Effects of growth hormone on insulin sensitivity and forearm metabolism in normal man. Diabetologia 1989;32(2):105–10.
[6] Møller N, Jørgensen JO, Schmitz O, et al. Effects of a growth hormone pulse on total and forearm substrate fluxes in humans. Am J Physiol 1990;258(1 Pt 1):E86–91.
[7] Randle PJ, Garland PB, Hales CN, et al. The glucose fatty-acid cycle. Its role in insulin sensitivity and the metabolic disturbances of diabetes mellitus. Lancet 1963;1:785–9.
[8] Møller N, Schmitz O, Jørgensen JO, et al. Basal- and insulin-stimulated substrate metabolism in patients with active acromegaly before and after adenomectomy. J Clin Endocrinol Metab 1992;74(5):1012–9.
[9] Puder JJ, Nilavar S, Post KD, et al. Relationship between disease-related morbidity and biochemical markers of activity in patients with acromegaly. J Clin Endocrinol Metab 2005;90(4):1972–8.
[10] Kopchick JJ, Parkinson C, Stevens EC, et al. Growth hormone receptor antagonists: discovery, development, and use in patients with acromegaly. Endocr Rev 2002;23(5): 623–46.
[11] Drake WM, Rowles SV, Roberts ME, et al. Insulin sensitivity and glucose tolerance improve in patients with acromegaly converted from depot octreotide to pegvisomant. Eur J Endocrinol 2003;149(6):521–7.
[12] Jørgensen JOL, Feldt-Rasmussen U, Frystyk J, et al. Cotreatment of acromegaly with a somatostatin analog and a growth hormone receptor antagonist. J Clin Endocrinol Metab 2005;90(10):5627–31.
[13] Beshyah SA, Gelding SV, Andres C, et al. Beta-cell function in hypopituitary adults before and during growth hormone treatment. Clin Sci (Lond) 1995;89(3):321–8.
[14] Johansson JO, Fowelin J, Landin K, et al. Growth hormone-deficient adults are insulin-resistant. Metabolism 1995;44(9):1126–9.
[15] Hew FL, Koschmann M, Christopher M, et al. Insulin resistance in growth hormone-deficient adults: defects in glucose utilization and glycogen synthase activity. J Clin Endocrinol Metab 1996;81(2):555–64.
[16] Vahl N, Jørgensen JO, Jurik AG, Christiansen JS. Abdominal adiposity and physical fitness are major determinants of the age associated decline in stimulated GH secretion in healthy adults. J Clin Endocrinol Metab 1996;81(6):2209–15.
[17] Jørgensen JO, Møller J, Alberti KG, et al. Marked effects of sustained low growth hormone (GH) levels on day-to-day fuel metabolism: studies in GH-deficient patients and healthy untreated subjects. J Clin Endocrinol Metab 1993;77(6):1589–96.

[18] Fowelin J, Attvall S, Lager I, et al. Effects of treatment with recombinant human growth hormone on insulin sensitivity and glucose metabolism in adults with growth hormone deficiency. Metabolism 1993;42(11):1443–7.
[19] O'Neal DN, Kalfas A, Dunning PL, et al. The effect of 3 months of recombinant human growth hormone (GH) therapy on insulin and glucose-mediated glucose disposal and insulin secretion in GH-deficient adults: a minimal model analysis. J Clin Endocrinol Metab 1994; 79(4):975–83.
[20] Weaver JU, Monson JP, Noonan K, et al. The effect of low dose recombinant human growth hormone replacement on regional fat distribution, insulin sensitivity, and cardiovascular risk factors in hypopituitary adults. J Clin Endocrinol Metab 1995;80(1):153–9.
[21] Christopher M, Hew FL, Oakley M, et al. Defects of insulin action and skeletal muscle glucose metabolism in growth hormone-deficient adults persist after 24 months of recombinant human growth hormone therapy. J Clin Endocrinol Metab 1998;83(5):1668–81.
[22] Beshyah SA, Henderson A, Niththyananthan R, et al. The effects of short and long-term growth hormone replacement therapy in hypopituitary adults on lipid metabolism and carbohydrate tolerance. J Clin Endocrinol Metab 1995;80(2):356–63.
[23] Rosenfalck AM, Maghsoudi S, Fisker S, et al. The effect of 30 months of low-dose replacement therapy with recombinant human growth hormone (rhGH) on insulin and C-peptide kinetics, insulin secretion, insulin sensitivity, glucose effectiveness, and body composition in GH-deficient adults. J Clin Endocrinol Metab 2000;85(11):4173–81.
[24] Hwu CM, Kwok CF, Lai TY, et al. Growth hormone (GH) replacement reduces total body fat and normalizes insulin sensitivity in GH-deficient adults: a report of one-year clinical experience. J Clin Endocrinol Metab 1997;82(10):3285–92.
[25] Jørgensen JOL, Vahl N, Nyholm B, et al. Substrate metabolism and insulin sensitivity following long-term growth hormone (GH) replacement therapy in GH-deficient adults. Endocrinol Metab 1996;3:281–6.
[26] Gibney J, Wallace JD, Spinks T, et al. The effects of 10 years of recombinant human growth hormone (GH) in adult GH-deficient patients. J Clin Endocrinol Metab 1999;84(8):2596–602.
[27] Norrelund H, Vahl N, Juul A, et al. Continuation of growth hormone (GH) therapy in GH-deficient patients during transition from childhood to adulthood: impact on insulin sensitivity and substrate metabolism. J Clin Endocrinol Metab 2000;85(5):1912–7.
[28] Norrelund H. The metabolic role of growth hormone in humans with particular reference to fasting. Growth Horm IGF Res 2005;15(2):95–122.
[29] Jørgensen JOL, Jessen N, Pedersen SB, et al. Growth hormone receptor signaling in skeletal muscle and adipose tissue in human subjects following exposure to an intravenous GH bolus. Am J Physiol Endocrinol Metab 2006;291:E899–905.
[30] Herrington J, Carter-Su C. Signaling pathways activated by the growth hormone receptor. Trends Endocrinol Metab 2001;12(6):252–7.
[31] Dominici FP, Argentino DP, Munoz MC, et al. Influence of the crosstalk between growth hormone and insulin signalling on the modulation of insulin sensitivity. Growth Horm IGF Res 2005;15(5):324–36.
[32] Lowell BB, Shulman GI. Mitochondrial dysfunction and type 2 diabetes. Science 2005; 307(5708):384–7.
[33] Nielsen S, Møller N, Christiansen JS, et al. Pharmacological antilipolysis restores insulin sensitivity during growth hormone exposure. Diabetes 2001;50(10):2301–8.
[34] Jessen N, Djurhuus CB, Jørgensen JO, et al. Evidence against a role for insulin-signaling proteins PI 3-kinase and Akt in insulin resistance in human skeletal muscle induced by short-term GH infusion. Am J Physiol Endocrinol Metab 2005;288(1):E194–9.
[35] Raben MS, Hollenberg CH. Effect of growth hormone on plasma fatty acids. J Clin Invest 1959;38(3):484–8.
[36] Djurhuus CB, Gravholt CH, Nielsen S, et al. Additive effects of cortisol and growth hormone on regional and systemic lipolysis in humans. Am J Physiol Endocrinol Metab 2004;286(3): E488–94.

[37] Gilman AG. G proteins: transducers of receptor-generated signals. Annu Rev Biochem 1987;56:615–49.
[38] Beauville M, Harant I, Crampes F, et al. Effect of long-term rhGH administration in GH-deficient adults on fat cell epinephrine response. Am J Physiol Endocrinol Metab 1992; 263(3):E467–72.
[39] Pedersen SB, Borglum J, Jørgensen JOL, et al. growth hormone treatment of obese premenopausal women: effects on isolated adipocyte metabolism. Endocrinol Metab 1995;2:251–8.
[40] Davidson MB. Effect of growth hormone on carbohydrate and lipid metabolism. Endocr Rev 1987;8(2):115–31.
[41] Yang S, Mulder H, Holm C, et al. Effects of growth hormone on the function of {beta}-adrenoceptor subtypes in rat adipocytes. Obes Res 2004;12(2):330–9.
[42] Leung KC, Ho KK. Stimulation of mitochondrial fatty acid oxidation by growth hormone in human fibroblasts. J Clin Endocrinol Metab 1997;82(12):4208–13.
[43] Richelsen B. Effect of growth hormone on adipose tissue and skeletal muscle lipoprotein lipase activity in humans. J Endocrinol Invest 1999;22(5 Suppl):10–5.
[44] Ottosson M, Vikman-Adolfsson K, Enerback S, et al. Growth hormone inhibits lipoprotein lipase activity in human adipose tissue. J Clin Endocrinol Metab 1995;80(3):936–41.
[45] Jørgensen JO. Human growth hormone replacement therapy: pharmacological and clinical aspects. Endocr Rev 1991;12(3):189–207.

Growth Hormone Effects on Protein Metabolism

Niels Møller, MD[a], Kenneth C. Copeland, MD[b], K. Sreekumaran Nair, MD, PhD[c],*

[a]*Medical Research Laboratories, Medical Dep. M, Århus University Hospital, DK-8000 Århus C, Denmark*
[b]*Department of Pediatrics, University of Oklahoma College of Medicine, 940 NE 13th Street, Oklahoma City, OK 73104, USA*
[c]*Division of Endocrinology, Mayo Clinic, 200 First Street SW, Joseph 5-194, Rochester, MN 55905, USA*

Growth hormone (GH) is generally accepted as an anabolic hormone because it promotes growth, results in abnormal growth (gigantism) when secreted in excess, and its deficiency results in stunted growth or dwarfism. The above effects on somatic growth are well described and justify the name GH; however, other biologic effects of GH, especially the anabolic role of GH in adult humans, remain to be defined fully. Excess GH is associated with abnormal bone growth, enlargement of certain/most internal organs, and many other metabolic features that suggest that excess GH may be anabolic to some, but not all, tissues. GH production declines by more than 50% in healthy older adults [1,2]. Administration of GH in older people increased fat-free muscle mass, but it may not have any positive impact on muscle functions and may cause complications (eg, carpal tunnel syndrome) [3–5]. It seems that the anabolic effect of GH is not a universal effect on all tissues or proteins, but is specific to certain tissue types. This article reviews the reported data on this topic.

The magnitude of GH effects on protein metabolism seems to be partially dependent on the circulating form of GH itself [6]; its effects on protein in vitro and in vivo seem to be dissimilar in various stores of the body (eg, muscle versus nonmuscle), and even vary according to the type of skeletal muscle examined [7]. Examination of muscle tissue from acromegalic patients (who are subject to a state of GH excess) reveals an increase in type 1 fibers

* Corresponding author.
E-mail address: nair.sree@mayo.edu (K.S. Nair).

and an atrophy of type 2 fibers [8]; however, the chronicity of GH exposure in this condition implies that these effects may have little, if anything, to do with GH acting directly, and may be related to stimulation of insulin-like growth factor (IGF)-1, insulin, or other effects.

GH is known to be secreted in vivo in a pulsatile fashion, and its effects on the elaboration/secretion of IGF-1 vary according to the dose of GH and the frequency or pulsatility of its administration [9–11]. Little is known about the time course or frequency of GH exposure on metabolic fuels, including carbohydrate, fat, or protein [12–14].

In vitro and animal studies

The metabolic effects of GH on amino acid and protein metabolism have been studied extensively over many years, initially using crude in vitro methods for examining amino acid uptake into protein. More recently, more sophisticated methods have been used to examine the fate of metabolic fuels in vitro and in vivo. Acute GH effects in stimulating amino acid incorporation into protein in vitro are observed within minutes, using crude [15] and pure [6] preparations of human GH. The rapid onset of these metabolic effects of GH in in vitro experiments suggests that these actions are a direct effect of GH itself, rather than through the elaboration of IGF-1 or other anabolic hormones [4]. Some effects of GH seem to be similar to those of an infusion of amino acids, although alterations in amino acid levels have not been reported. The exact mechanisms of these effects are unknown, however, and whether these actions are mediated through the elaboration of translation factors or phosphorylation of translation factors (eg, 4EBP-1 or 70-kD kinase effects on ribosomal protein S6), as observed during the process of direct amino acid stimulation of protein synthesis, is unclear [16].

Experiments in hypophysectomized rats provided evidence that GH may act on the liver to decrease urea synthesis, and, in parallel, to increase glutamine release, thereby diminishing hepatorenal clearance of the circulating nitrogen pool [17].

In vitro, GH seems to have rapid and profound effects on stimulating amino acid incorporation into protein. In vivo, in animals and humans, GH exerts its effects differentially in various protein stores in the body, and its effects are discordant, even in different types of skeletal muscle.

Studies in humans

Postabsorptive protein metabolism

Extensive evidence suggests that the earliest and most profound effect of GH on amino acid metabolism is a profound reduction in amino acid oxidation [18–21], followed by an increase in whole-body protein synthesis [22].

GH also was shown to counteract prednisone-induced protein catabolism by reducing protein breakdown and amino acid oxidation [19]. Perfusion of GH into the forearm demonstrated a rapid effect on increasing forearm blood flow and tissue uptake of phenylalanine and leucine into skeletal muscle tissue [23], although these apparent effects may be mediated by an elaboration of IGF-1 in peripheral tissues, because IGF-1 is known to have profound and rapid effects on forearm blood flow [24].

The effect of a combined, local infusion of GH and insulin also revealed stimulation of muscle protein synthesis while antagonizing insulin's inhibitory effect on protein breakdown [25]. In addition, it was reported that 6 weeks of high-dose GH treatment to malnourished patients who were undergoing hemodialysis stimulated muscle protein synthesis without any effects on muscle protein degradation [26]. Conversely, systemic infusion of GH failed to show any effect on muscle protein synthesis [18,27]. In one of these studies, however, it was noted that isotopically measured muscle protein breakdown across the leg tends to be lower after acute GH exposure with borderline significance ($P = .05$ for phenylalanine appearance rate and $P = .09$ for leucine appearance rate) [18]. The above studies were performed to determine the specific effect of GH, separate from its secondary effect that is due to insulin secretion by infusing somatostatin and replacing insulin and glucagon.

A study of protein turnover in GH-deficient adults demonstrated reduced rates of protein synthesis and breakdown and subsequent normal net protein loss compared with normal controls [28]; this was in line with earlier observations of the effect of chronic GH deficiency on protein metabolism [29]. An initial decline in lean body mass (LBM) may occur because of GH insufficiency; however, clinical experience suggests that LBM stabilizes at a reduced level, and that this adaptation may explain the development of stable, albeit reduced, protein and LBM in GH deficiency [28]. GH replacement for 6 weeks in GH-deficient adults revealed increased net protein synthesis and unaltered total protein turnover [30], consistent with a dose-response study [31].

In the postabsorptive state, Wolthers and colleagues [32,33] showed unchanged urea synthesis during short-term GH exposure and a decrease with more prolonged administration. This suggested that the established anabolic effect of GH on whole-body protein metabolism in normal subjects postabsorptively could be exerted through modulation of peripheral protein synthesis and degradation and liver urea nitrogen synthesis.

Protein metabolism during prolonged fasting

Studies have examined the potential role of GH on fasting longer than overnight that occur in the postabsorptive studies. GH secretion is increased during fasting, and it is likely that some of the most significant physiologic effects of GH are exerted during fasting. Administration of GH to normal subjects during dietary restriction [34] or a hyponitrogenous diet [35] decreases serum urea concentration and urinary urea excretion. Suppression of GH during fasting led to a 50% increase in urea-nitrogen excretion, muscle protein breakdown increased by 25%, and forearm phenylalanine release increased by 40% [36,37]. In GH-deficient adults, urea excretion, serum urea, and urea nitrogen synthesis rates increased by 50% during fasting and GH withdrawal [36]. The increase in whole-body protein loss was accounted for by a net reduction in protein synthesis. Furthermore, a significant decrease in branched-chain amino acid levels, consistent with decreased proteolysis, was seen during fasting with GH replacement [36,37]. When urea fluxes are measured with a steady-state isotope dilution technique ($[^{13}C]$urea), urea turnover decreases by 30% to 35% during fasting with GH replacement [38].

Obesity

Obesity is associated with suppressed levels of circulating GH, which is characterized by fewer secretory events and a shorter half-life compared with normal weight subjects [39]. Whether this abnormality is initiated by a primary neuroregulatory defect or by enhanced feedback inhibition from circulating hormones (eg, free IGF-I or substrate) is unsettled; however, the condition is partly reversible following weight loss [40]. Although controversial, this issue may pose potential problems in the treatment of obesity with fasting or a very low calorie diet, and future research may address the potential benefits versus risks in using GH administration in enhancing weight loss while maintaining lean tissue. The metabolic response to GH during prolonged fasting in obese subjects was studied 30 years ago by Felig and colleagues [41]. Suprphysiologic doses were used and a significant reduction in urinary urea could be demonstrated, implying that GH reduces protein catabolism. GH treatment in combination with a hypocaloric diet was studied by Clemmons and Snyder and colleagues [42,43]. Twenty obese subjects received 75 kJ/kg body weight for 13 weeks in combination with GH or saline. Nitrogen balance was significantly more positive in the group that received GH, but the effect vanished after 33 days. No significant increase of fat loss could be demonstrated, as estimated by underwater weighing. During dietary restriction in obese patients, Tagliaferri and colleagues [44] evaluated the effects of administration of GH; they found significant preservation of fat free mass in the GH-treated participants compared with those who received vehicle. In line with the observation of Tagliaferri and colleagues, Norrelund and colleagues [45] found reduced loss of fat-free mass in GH-treated obese patients, together with a decrease in urine urea excretion and serum urea. Protein breakdown decreased in both groups during the very low calorie diet. Phenylalanine degradation in relation to phenylalanine concentration—an index of phenylalanine hydroxylase activity independent of circulating phenylalanine—decreased by 9% in the GH group, whereas an increase of 8% was observed in the placebo group; this suggested that GH decreased protein breakdown.

Mechanisms underlying the protein metabolic effects of growth hormone

The role of lipids

It is well described that prolonged fasting induces insulin resistance [46], related to high circulating levels of GH and free fatty acid (FFA) [36]. Because GH promotes lipolysis in adipose tissue and elevation of GH is associated with increased plasma FFA level [11,47,48], increased lipid availability has been suggested to be responsible for the protein-retaining effects and the insulin resistance that are observed with elevated GH levels.

The interaction between lipid and protein metabolism was suggested first by Krebs (1935) as an aspect of substrate competition in cellular oxidative pathways. In particular, the increase of FFAs and their oxidative products, ketone bodies, during fasting may contribute to the protein-sparing mechanisms following reduced exogenous substrate supply [49]. The mechanism by which GH affects protein metabolism remains to be fully understood. Studies showed that beta-hydroxybutyrate enhances the synthesis rate of muscle protein in healthy humans [50]. During periods of fasting that are associated with increases in plasma FFA concentrations, decreased rates of whole-body leucine and lysine flux were observed [51,52]. Several lines of evidence support a protein-sparing effect of lipids. Infusion of ketone bodies was shown to decrease circulating levels of alanine [53], leucine oxidation, and leucine incorporation in protein [50]. Elevation of circulating levels by infusion of intralipid and heparin in humans was shown to have a hypoaminoacidemic action [54]. An inhibition of muscular amino acid release by systemic lipid infusion has been described in humans [55], with a parallel decrease in amino acids. The large increase of amino acid efflux from exercising leg muscle was shown to be reduced, in part, by a simultaneous intralipid infusion and the resulting systemic increase of FFA concentration [56]. In dogs, Tessari and colleagues [57] demonstrated that acute changes in plasma FFA concentrations within the physiologic range observed with feeding and fasting by way of the infusion of triglyceride, heparin, or nicotinic acid were related inversely to estimates of whole-body leucine carbon flux and oxidation. Reduction of FFA concentrations in healthy subjects by the administration of acipimox increased total urea-nitrogen production [58] and urea excretion during fasting [59]. Postabsorptively, pharmacologic suppression of FFA is accompanied by augmented whole-body protein degradation [60]. In normal subjects, inhibition of lipolysis increased urea-nitrogen production and muscle protein breakdown by nearly 50%, and neutralized the ability of GH to reduce ureagenesis and to restrict muscle protein breakdown during short-term fasting [61]. Restoration of high FFA levels by means of an infusion normalized urea and phenylalanine metabolism [61]. Thus, the finding that inhibition of lipolysis blocks the ability of GH to restrict protein loss suggests that stimulation of lipolysis is instrumental for the protein-conserving effects of GH under fasting conditions.

The role of insulin and insulin-like growth factor

GH counteracts the effects of insulin on lipid and glucose metabolism, and compensatory hyperinsulinemia is a frequent occurrence during GH treatment. In humans, the rates of whole-body [62–64], splanchnic [65], skeletal [65–67], and heart [68] muscle protein degradation are decreased in response to physiologic replacement or increase in plasma insulin. The effect of insulin on muscle protein metabolism has been studied after 60 hours of fasting [67]. During short-term fasting, muscle amino acid output accelerates s a result of increased proteolysis. Despite marked impairment in insulin-mediated glucose disposal, muscle remains sensitive to insulin antiproteolytic action.

Systemic infusion of GH not only increases systemic IGF-I concentrations but also stimulates local generation of IGF-I [69]. IGF-I inhibits whole-body protein breakdown and causes hypoaminoacidemia and hypoglycemia [70]. When given acutely in vivo at doses that have no stimulatory effect on glucose uptake, IGF-I stimulates muscle protein synthesis, whereas muscle protein breakdown seems to be inhibited by the higher doses of IGF-I that also promote glucose uptake [71]. IGF-I increases the phosphorylation of enzymes and leads to protein synthesis [72]. In contrast, there is a lack of stimulation of whole-body protein synthesis with systemic infusion of IGF-I [73], which probably results from the decrease in circulating insulin or amino acid concentration that accompanies systemic IGF-I infusion in vivo. When amino acids [74] or insulin [75] is replaced, IGF-I clearly stimulates whole-body protein synthesis.

The role of amino acids and glucose

Active acromegaly clearly reveals the diabetogenic properties of GH. In the basal state, plasma glucose is increased, despite compensatory hyperinsulinemia. Previous studies implied that the administration of small amounts of glucose during fasting [76] and hyperglycemia [77] may be protein sparing. One study failed to observe any effect of glucose on amino acid metabolism after a 10-day fast, despite increased insulin levels [78]. In a similar study protocol [79], whole-body protein breakdown, however, was decreased following parenteral nutrition in comparison with starvation. Some studies suggested that the decrease in whole-body proteolysis during the infusion of glucose is mediated by increased plasma insulin concentrations and not the ambient glucose concentration or the rate of glucose use [80]. Luzi and colleagues [81] demonstrated that hyperglycemia with maintenance of basal insulin and glucagon concentrations in normal subjects did not affect the plasma concentration of total amino acids, branched chain amino acids, leucine or ketoisocaproate, nor did it affect net leucine balance, basal leucine flux, basal leucine oxidation, or the rate of nonoxidative leucine disposal.

In general, GH administration leads to a decrease in muscle release of branched-chain amino acids [25,26,37]. Whether GH leads to overall changes in amino acid levels is controversial. No overall change has been reported during comparable insulin and IGF-I levels [18,82]. Despite an increase in IGF-I during GH treatment, some studies failed to show an overall change in amino acid level [26,37]. During fasting in growth hormone deficient adults and semifasting in obesity, a decrease in the concentrations of some amino acids is seen [36,83].

The role of thyroid function

GH increases peripheral deiodination of thyroxine (T_4) to triiodothyronine (T_3) [84,85]. Sato and colleagues [84] followed eight GH-deficient children and adolescents, of whom three had biochemical signs of central hypothyroidism, during 12 months of GH therapy. They noted a decrease in serum T_4 and an increase in T_3 during GH therapy in the euthyroid group and similar findings in the hypothyroid group after initiation of T_4 substitution. A GH-induced enhancement of peripheral deiodination of T_4 to T_3 was confirmed during 4 months of replacement therapy in GH-deficient patients [85], and increased peripheral T_3 generation was found in obese women who were treated with GH [86]. Thyroid hormones have anabolic and catabolic effects. Thyroid hormones stimulate lipolysis and increases in plasma levels of free fatty acids [87,88]. Patients who have hyperthyroidism have increased amino acid turnover and an increased muscle protein breakdown. The extent to which the modest increase in T_4 to T_3 conversion contributes to the effects of GH on insulin sensitivity and protein metabolism is uncertain.

Summary

GH has a pivotal role in regulating in vivo protein metabolism. GH was shown to enhance protein anabolism at the whole-body level, mainly by stimulating protein synthesis. It remains incompletely understood whether this important GH effect on protein synthesis occurs in all tissues. It is possible that GH's effect on protein metabolism is specific to certain proteins in a tissue rather than having a global effect on all proteins. This effect of GH may be different with acute versus chronic GH administration. These differences in the GH exposure may have different effects based not only on direct GH stimulation of protein synthesis but also the variable effects at the level of gene transcription that ultimately affect protein metabolism. Other GH effects are likely to be mediated by changes in various metabolites and hormones that probably also differ based on the duration of GH administration.

GH effects on protein synthesis and breakdown probably are influenced by age as well. It is likely that GH effect on different tissues is influenced by whether it is investigated in children, adults, or the elderly. Because many of the biologic effects of GH are mediated by insulin and IGF-I, the responsiveness of these hormones to GH also may determine GH effect on protein metabolism.

References

[1] Veldhuis JD, Iranmanesh A, Bowers CY. Joint mechanisms of impaired growth-hormone pulse renewal in aging men. J Clin Endocrinol Metab 2005;90(7):4177–83.
[2] Zadik Z, Chalew SA, McCarter RJ Jr, et al. The influence of age on the 24-hour integrated concentration of growth hormone in normal individuals. J Clin Endocrinol Metab 1985; 60(3):513–6.
[3] Yarasheski KE, Zachwieja JJ, Campbell JA, et al. Effect of growth hormone and resistance exercise on muscle growth and strength in older men. Am J Physiol 1995;268:E268–76.
[4] Copeland KC, Nair KS. Effects of rhGH on glucose, lipid, and amino acid metabolism. In: Blackman MR, Roth J, Harman SM, et al, editors. GHRH, GH, and IGF-I. Basic and clinical advances. New York: Springer-Verlag; 1995. p. 185–94.
[5] Schwartz RS. Trophic factors supplementation: effect of age-associated changes in body composition. J Gerontol 1995;50A:151–6.
[6] Cameron CM, Kostyo JL, Adamafio NA, et al. The acute effects of growth hormone on amino acid transport and protein synthesis are due to its insulin-like action. Endocrinology 1988;122(2):471–4.
[7] Ayling CM, Moreland BH, Zanelli JM, et al. Human growth hormone treatment of hypophysectomized rats increases the proportion of type-1 fibres in skeletal muscle. J Endocrinol 1989;123(3):429–35.
[8] Nagulesparen M, Trickey R, Davies MJ, et al. Muscle changes in acromegaly. BMJ 1976; 2(6041):914–5.
[9] Copeland KC, Underwood LE, Van Wyk JJ. Induction of immunoreactive somatomedin C human serum by growth hormone: dose-response relationships and effect on chromatographic profiles. J Clin Endocrinol Metab 1980;50(4):590–7.
[10] Jorgensen JO, Moller N, Lauritzen T, et al. Pulsatile versus continuous intravenous administration of growth hormone (GH) in GH-deficient patients: effects on circulating insulin-like growth factor-I and metabolic indices. J Clin Endocrinol Metab 1990;70(6):1616–23.
[11] Moller N, Jorgensen JO, Schmitz O, et al. Effects of a growth hormone pulse on total and forearm substrate fluxes in humans. Am J Physiol Endocrinol Metab 1990;258:E86–91.
[12] Laursen T, Jorgensen JOL, Christiansen JS. Metabolic response to growth hormone (GH) administered in a pulsatile, continuous or combined pattern. Endocrinol Metab 1994;1: 33–40.
[13] Laursen T, Gravholt CH, Heickendorff L, et al. Long-term effects of continuous subcutaneous infusion versus daily subcutaneous injections of growth hormone (GH) on the insulin-like growth factor system, insulin sensitivity, body composition, and bone and lipoprotein metabolism in GH-deficient adults. J Clin Endocrinol Metab 2001;86(3): 1222–8.
[14] Vahl N, Moller N, Lauritzen T, et al. Metabolic effects and pharmacokinetics of a growth hormone pulse in healthy adults: relation to age, sex, and body composition. J Clin Endocrinol Metab 1997;82(11):3612–8.
[15] Kostyo JL. Rapid effects of growth hormone on amino acid transport and protein synthesis. Ann N Y Acad Sci 1968;148(2):389–407.

[16] Liu Z, Long W, Fryburg DA, et al. The regulation of body and skeletal muscle protein metabolism by hormones and amino acids. J Nutr 2006;136(1 Suppl):212S–7S.
[17] Welbourne T, Joshi S, McVie R. Growth hormone effects on hepatic glutamate handling in vivo. Am J Physiol 1989;257:E959–62.
[18] Copeland KC, Nair KS. Acute growth hormone effects on amino acid and lipid metabolism. J Clin Endocrinol Metab 1994;78:1040–7.
[19] Horber FF, Haymond MW. Human growth hormone prevents the protein catabolic side effects of prednisone in humans. J Clin Invest 1990;86:265–72.
[20] Fryburg DA, Barrett EJ. Growth hormone acutely stimulates skeletal muscle but not whole-body protein synthesis in humans. Metabolism 1993;42(9):1223–7.
[21] Buijs MM, Romijn JA, Burggraaf J, et al. Growth hormone blunts protein oxidation and promotes protein turnover to a similar extent in abdominally obese and normal-weight women. J Clin Endocrinol Metab 2002;87(12):5668–74.
[22] Gibney J, Wolthers T, Johannsson G, et al. Growth hormone and testosterone interact positively to enhance protein and energy metabolism in hypopituitary men. Am J Physiol Endocrinol Metab 2005;289(2):266–71.
[23] Fryburg DA, Gelfand RA, Barrett EJ. Growth hormone acutely stimulates forearm muscle protein synthesis in normal humans. Am J Physiol Endocrinol Metab 1991;260(23): E499–504.
[24] Copeland KC, Nair KS. Recombinant human insulin-like growth factor-I increases forearm blood flow. J Clin Endocrinol Metab 1994;79:230–2.
[25] Fryburg DA, Louard RJ, Gerow KE, et al. Growth hormone stimulates skeletal muscle protein synthesis and antagonizes insulin's antiproteolytic action in humans. Diabetes 1992;41: 424–9.
[26] Garibotto G, Barreca A, Russo R, et al. Effects of recombinant human growth hormone on muscle protein turnover in malnourished hemodialysis patients. J Clin Invest 1997;99: 97–105.
[27] Zachwieja JJ, Bier DM, Yarasheski KE. Growth hormone administration in older adults: effects on albumin synthesis. Am J Physiol 1994;266:E840–4.
[28] Hoffman DM, Pallasser R, Duncan M, et al. How is whole body protein turnover perturbed in growth hormone-deficient adults? J Clin Endocrinol Metab 1998;83(12):4344–9.
[29] Beshyah SA, Sharp PS, Gelding SV, et al. Whole-body leucine turnover in adults on conventional treatment for hypopituitarism. Acta Endocrinol (Copenh) 1993;129(2): 158–64.
[30] Russell-Jones DL, Weissberger AJ, Bowes SB, et al. The effects of growth hormone on protein metabolism in adult growth hormone deficient patients. Clin Endocrinol (Oxf) 1993; 38(4):427–31.
[31] Lucidi P, Lauteri M, Laureti S, et al. A dose-response study of growth hormone (GH) replacement on whole body protein and lipid kinetics in GH-deficient adults. J Clin Endocrinol Metab 1998;83(2):353–7.
[32] Wolthers T. Effects of growth hormone and fasting on urea synthesis in humans [doctoral dissertation]. Aarhus, Denmark: Aarhus University; 1995.
[33] Katusic ZS. Vascular endothelial dysfunction: does tetrahydrobiopterin play a role? Am J Physiol Heart Circ Physiol 2001;281(3):H981–6.
[34] Manson JM, Wilmore DW. Positive nitrogen balance with human growth hormone and hypocaloric intravenous feeding. Surgery 1986;100(2):188–97.
[35] Lundeberg S, Belfrage M, Wernerman J, et al. Growth hormone improves muscle protein metabolism and whole body nitrogen economy in man during a hyponitrogenous diet. Metab Clin Exp 1991;40(3):315–22.
[36] Norrelund H, Moller N, Nair KS, et al. Continuation of growth hormone (GH) substitution during fasting in GH-deficient patients decreases urea excretion and conserves protein synthesis. J Clin Endocrinol Metab 2001;86(7):3120–9.

[37] Nørrelund H, Nair KS, Jorgensen JOL, et al. The protein-retaining effects of growth hormone during fasting involve inhibition of muscle-protein breakdown. Diabetes 2001;50: 96–104.
[38] Norrelund H, Djurhuus C, Jorgensen JO, et al. Effects of GH on urea, glucose and lipid metabolism, and insulin sensitivity during fasting in GH-deficient patients. Am J Physiol Endocrinol Metab 2003;285(4):E737–43.
[39] Veldhuis JD, Iranmanesh A, Ho KK, et al. Dual defects in pulsatile growth hormone secretion and clearance subserve the hyposomatotropism of obesity in man. J Clin Endocrinol Metab 1991;72(1):51–9.
[40] Williams T, Berelowitz M, Joffe SN, et al. Impaired growth hormone responses to growth hormone-releasing factor in obesity. A pituitary defect reversed with weight reduction. N Engl J Med 1984;311(22):1403–7.
[41] Felig P, Marliss EB, Cahill GF Jr. Metabolic response to human growth hormone during prolonged starvation. J Clin Invest 1971;50(2):411–21.
[42] Clemmons DR, Snyder DK, Williams R, et al. Growth hormone administration conserves lean body mass during dietary restriction in obese subjects. J Clin Endocrinol Metab 1987;64(5):878–83.
[43] Snyder DK, Clemmons DR, Underwood LE. Treatment of obese, diet-restricted subjects with growth hormone for 11 weeks: effects on anabolism, lipolysis, and body composition. J Clin Endocrinol Metab 1988;67(1):54–61.
[44] Tagliaferri M, Scacchi M, Pincelli AI, et al. Metabolic effects of biosynthetic growth hormone treatment in severely energy-restricted obese women. Int J Obes Relat Metab Disord 1998;22(9):836–41.
[45] Norrelund H, Borglum J, Jorgensen JO, et al. Effects of growth hormone administration on protein dynamics and substrate metabolism during 4 weeks of dietary restriction in obese women. Clin Endocrinol (Oxf) 2000;52(3):305–12.
[46] Bjorkman O, Eriksson LS. Influence of a 60 hour fast on insulin-mediated splanchnic and peripheral glucose metabolism in humans. J Clin Invest 1985;76:87–92.
[47] Fowelin J, Attvall S, von Schenck H, et al. Characterization of the insulin-antagonistic effect of growth hormone in man. Diabetologia 1991;34(7):500–6.
[48] Jorgensen JO, Moller J, Alberti KG, et al. Marked effects of sustained low growth hormone (GH) levels on day-to-day fuel metabolism: studies in GH-deficient patients and healthy untreated subjects. J Clin Endocrinol Metab 1993;77(6):1589–96.
[49] Felig P, Owen OE, Wahren J, et al. Amino acid metabolism during prolonged starvation. J Clin Invest 1969;48:584–94.
[50] Nair KS, Welle SL, Halliday D, et al. Effect of beta-hydroxybutyrate on whole-body leucine kinetics and fractional mixed skeletal muscle protein synthesis in humans. J Clin Invest 1988; 82:198–205.
[51] Adibi SA, Stanko RT, Morse EL. Modulation of leucine oxidation and turnover by graded amounts of carbohydrate intake in obese subjects. Metabolism 1982;31(5):578–88.
[52] Henson LC, Heber D. Whole body protein breakdown rates and hormonal adaptation in fasted obese subjects. J Clin Endocrinol Metab 1983;57(2):316–9.
[53] Sherwin RS, Hendler RG, Felig P. Effect of ketone infusion on amino acid and nitrogen metabolism in man. J Clin Invest 1975;55:1382–90.
[54] Ferrannini E, Barrett EJ, Bevilacqua S, et al. Effect of free fatty acids on blood amino acid levels in humans. Am J Physiol Endocrinol Metab 1986;250:E686–94.
[55] Wicklmayr M, Rett K, Schwiegelshohn B, et al. Inhibition of muscular amino acid release by lipid infusion in man. Eur J Clin Invest 1987;17(4):301–5.
[56] Graham TE, Kiens B, Hargreaves M, et al. Influence of fatty acids on ammonia and amino acid flux from active human muscle. Am J Physiol Endocrinol Metab 1991;261: E168–76.
[57] Tessari P, Nissen SL, Miles JM, et al. Inverse relationship of leucine flux and oxidation to free fatty acid availability in vivo. J Clin Invest 1986;77:575–81.

[58] Fery F, Plat L, Baleriaux M, et al. Inhibition of lipolysis stimulates whole body glucose production and disposal in normal postabsorptive subjects. J Clin Endocrinol Metab 1997;82(3):825–30.
[59] Fery F, Plat L, Melot C, et al. Role of fat-derived substrates in the regulation of gluconeogenesis during fasting. Am J Physiol Endocrinol Metab 1996;270:E822–30.
[60] Nielsen S, Jorgensen JOL, Hartmund T, et al. Effects of lowering circulating free fatty acid levels on protein metabolism in adult growth hormone deficient patients. Growth Horm IGF Res 2002;12:425–33.
[61] Norrelund H, Nielsen S, Christiansen JS, et al. Modulation of basal glucose metabolism and insulin sensitivity by growth hormone and free fatty acids during short-term fasting. Eur J Endocrinol 2004;150(6):779–87.
[62] Nair KS, Garrow JS, Ford C, et al. Effect of poor diabetic control and obesity on whole body protein metabolism in man. Diabetologia 1983;25:400–3.
[63] Brodsky IG, Suzara D, Furman M, et al. Proteasome production in human muscle during nutritional inhibition of myofibrillar protein degradation. Metab Clin Exp 2004; 53(3):340–7.
[64] Tessari P, Trevisan R, Inchiostro S, et al. Dose-response curves of effects of insulin on leucine kinetics in humans. Am J Physiol Endocrinol Metab 1986;251:E334–42.
[65] Nair KS, Ford GC, Ekberg K, et al. Protein dynamics in whole body and in splachnic and leg tissues in type I diabetic patients. J Clin Invest 1995;95:2926–37.
[66] Gelfand RA, Barrett EJ. Effect of physiologic hyperinsulinemia on skeletal muscle protein synthesis and breakdown in man. J Clin Invest 1987;80:1–6.
[67] Fryburg DA, Jahn LA, Hill SA, et al. Insulin and insulin-like growth factor-I enhance human skeletal muscle protein anabolism during hyperaminoacidemia by different mechanisms. J Clin Invest 1995;96:1722–9.
[68] McNulty PH, Louard RJ, Deckelbaum LI, et al. Hyperinsulinemia inhibits myocardial protein degradation in patients with cardiovascular disease and insulin resistance. Circulation 1995;92:2151–6.
[69] Butler AA, Le Roith D. Control of growth by the somatropic axis: growth hormone and the insulin-like growth factors have related and independent roles. Ann Rev Physiol 2001;63: 141–64.
[70] Le Roith D, Bondy C, Yakar S, et al. The somatomedin hypothesis: 2001. Endocr Rev 2001; 22(1):53–74.
[71] Fryburg DA. Insulin-like growth factor I exerts growth hormone- and insulin-like actions on human muscle protein metabolism. Am J Physiol Endocrinol Metab 1994;267: E331–6.
[72] Liu Z, Barrett EJ. Human protein metabolism: its measurement and regulation. Am J Physiol Endocrinol Metab 2002;283:E1105–12.
[73] Elahi D, McAloon-Dyke M, Fukagawa NK, et al. Effects of recombinant human IGF-I on glucose and leucine kinetics in men. Am J Physiol Endocrinol Metab 1993;265:E831–8.
[74] Russell-Jones DL, Umpleby AM, Hennessy TR, et al. Use of leucine clamp to demonstrate that IGF-I actively stimulates protein synthesis in normal humans. Am J Physiol 1994;267: E591–8.
[75] Jacob R, Hu X, Niederstock D, et al. IGF-I stimulation of muscle protein synthesis in the awake rat: permissive role of insulin and amino acids. Am J Physiol 1996;270:E60–6.
[76] Cahill GF. Starvation in man. N Engl J Med 1970;282(12):668–75.
[77] Felig P, Wahren J, Sherwin R, et al. Amino acid and protein metabolism in diabetes mellitus. Arch Intern Med 1977;137:507–13.
[78] Albert JD, Legaspi A, Horowitz GD, et al. Extremity amino acid metabolism during starvation and intravenous refeeding in humans. Am J Physiol Endocrinol Metab 1986;251: 604–10.
[79] Tracey KJ, Legaspi A, Albert JD, et al. Protein and substrate metabolism during starvation and parenteral feeding. Clin Sci 1988;74:123–32.

[80] Heiling V, Campbell PJ, Gottesman IS, et al. Differential effects of hyperglycemia and hyperinsulinemia on leucine rate of appearance in normal humans. J Clin Endocrinol Metab 1993;76:203–6.

[81] Luzi L, Castellino P, Simonson DC, et al. Leucine metabolism in IDDM. Role of insulin and substrate availability. Diabetes 1990;39:38–48.

[82] Nørrelund H. The decisive role of free fatty acids for protein conservation during fasting in humans with and without growth hormone. J Clin Endocrinol Metab 2003;88:4371–8.

[83] Norrelund H, Vahl N, Juul A, et al. Continuation of growth hormone (GH) therapy in GH-deficient patients during transition from childhood to adulthood: impact on insulin sensitivity and substrate metabolism. J Clin Endocrinol Metab 2000;85(5):1912–7.

[84] Sato T, Suzukui Y, Taketani T, et al. Enhanced peripheral conversion of thyroxine to triiodothyronine during hGH therapy in GH deficient children. J Clin Endocrinol Metab 1977;45(2):324–9.

[85] Jorgensen JO, Pedersen SA, Laurberg P, et al. Effects of growth hormone therapy on thyroid function of growth hormone-deficient adults with and without concomitant thyroxine-substituted central hypothyroidism. J Clin Endocrinol Metab 1989;69(6):1127–32.

[86] Jorgensen JOL, Pedersen SB, Borglum J, et al. Fuel metabolism, energy expenditure, and thyroid function in growth hormone-treated obese women: a double-blind placebo-controlled study. Metabolism 1994;43:872–7.

[87] Pucci E, Chiovato L, Pinchera A. Thyroid and lipid metabolism. Int J Obes Relat Metab Disord 2000;24(Suppl 2):S109–12.

[88] Riis AL, Gravholt CH, Djurhuus CB, et al. Elevated regional lipolysis in hyperthyroidism. J Clin Endocrinol Metab 2002;87(10):4747–53.

What Endocrinologists Should Know About Growth Hormone Measurements

Martin Bidlingmaier, MD[a],*,
Christian J. Strasburger, MD[b]

[a]*Endocrine Research Laboratories, Medizinische Klinik–Innenstadt, Ludwig-Maximilians University, Ziemssenstrasse 1, Munich 80336, Germany*
[b]*Division of Clinical Endocrinology, Charité Universitätsmedizin, Mitte Charitéplatz 1, Berlin 10117, Germany*

Diagnosis of growth hormone deficiency (GHD) and growth hormone excess (acromegaly) is based to a large extent on the measurement of serum growth hormone (GH) concentrations. Because of the pulsatile nature of pituitary GH secretion, dynamic tests assessing either stimulation or suppression of GH concentration are used rather than single baseline measurements. Several consensus guidelines have been published aiming to define diagnosis of GHD in children and adolescents [1], GHD in adults [2,3], and acromegaly [4,5]. Although of course such consensus guidelines help harmonize diagnostic standards and thereby contribute to improved patient care, several problems remain unsolved. On the one hand, there is a large variability in the amount of GH secreted during stimulation tests, depending on the stimulus used [6–10] and on individual factors such as age, sex [11], clinical situation [12,13], and body composition [14,15]. On the other hand, the heterogeneity of GH assay results obtained by different methods on the market remains a problem [16,17]. The consensus guidelines recommend using assay-specific cut-off values for diagnostic tests, but published data are lacking for almost all commercially available methods.

Growth hormone assay heterogeneity

GH concentration in circulation is highly variable and is regulated by several physiologic factors such as age, body composition, nutrition, stress,

* Corresponding author.
E-mail address: martin.bidlingmaier@med.uni-muenchen.de (M. Bidlingmaier).

and sleep. Once a sample is drawn, however, the GH concentration exhibits a remarkable stability in the preanalytic period. Several studies have shown that GH in vitro is stable for more than 24 hours at room temperature and for even longer periods at 4°C [18,19]. Therefore, the potential for problems in the preanalytic period affecting the results of GH measurements in the laboratory seems to be rather small. In striking contrast, however, is the great variability in GH concentrations measured by different laboratories in national external quality assessment schemes, as reported by several authors during the last 15 years [20–23]. The method-dependent variability is in excess of 100%, making it extremely difficult to compare results from one study with another or from one laboratory with another. Studies directly comparing results from different assays in the same clinical samples also showed considerable disagreement between methods [24–26]. Some authors recommend the central reassessment of samples by a defined GH assay method to improve the comparability of study results from several centers [27]. There is increasing awareness of the problem [28,29], and recently an international collaborative started to discuss options for improving the situation [30]. This article focuses on the molecular, structural, and methodologic background of the heterogeneity of assay results and on possible next steps toward standardization.

Heterogeneity of growth hormone molecules

In contrast to the recombinant GH preparations used therapeutically, which consist only of a defined 191–amino acid protein with a molecular weight of 22 kD, GH as secreted by the pituitary gland and circulating in the blood stream is a heterogeneous mixture of different isoforms and fragments [31]. Two different genes, *GH-N* and *GH-V*, exist in humans [32]. *GH-N* is expressed mainly in the somatotroph cells of the pituitary, whereas the variant *GH-V* is expressed in the syncytiotrophoblast layer of the placenta [33,34]. The respective gene products differ by only 13 amino acids, and therefore potential interference from GH-V in GH assays must be considered when GH measurements are made during pregnancy. A shorter 20-kD GH form results from alternative splicing of the *GH-N* gene [35], in which amino acid residues 32 through 46 are deleted. In addition, various other isoforms have been described [36]. Dimers of GH molecules occur, both as hetero- and homodimers, and multimers also are present in human serum [37]. For 20-kD GH, the process of dimerization has been shown to be reversible [38]. All these isoforms and fragments of GH are present in variable abundance, with differences occurring within a subject and even more so between subjects. In particular, specific regulation has been shown for the whole group of non–22-kD isoforms [39] and for the 20-kD isoform [40].

Assay design

The heterogeneity of GH molecules in circulation affects the results of GH measurements. The GH concentration in a blood sample is determined by immunoassay techniques that have changed considerably over the last 20 years: Until the early 1990s, classic radioimmunoassays (RIAs) were used most frequently for GH measurements. These assays usually use polyclonal antisera against GH together with radiolabeled GH as a tracer. GH molecules present in a sample compete with the radiolabeled GH for binding to the antiserum; the higher the GH concentration is in a sample, the lower is the radioactivity bound to the antiserum. During the last decade sandwich immunoassays have become more popular. In these assays an immobilized antibody captures GH molecules in a sample, and the bound molecules then are detected by a second detection antibody. The second detection antibody can be labeled radioactively, but nonisotopic techniques using colorimetric or chemiluminescence signals are used more frequently. This change in the assay technique has several implications for GH assay results.

Generally, the older RIAs were less sensitive than today's assays based on high-affinity monoclonal antibodies. The lower detection limit is especially important in the evaluation of acromegaly, when the nadir GH concentration after oral glucose load is used as a diagnostic criterion. Along with the improvement in assay sensitivity, the nadir GH levels after oral glucose indicating cure or control of acromegaly decreased from 2.5 µg/L to 1.0 µg/L. More recently, with very sensitive assays, it could be demonstrated that nadir levels in healthy subjects are usually below 0.14 µg/L [41,42]; therefore the cut-off levels recommended by guidelines should be decreased further.

Antibodies recognize their target molecules by binding to a distinct surface structure (epitope). Polyclonal antibodies represent a mixture of different antibodies raised by the host animal's immune system. As a consequence, assays involving polyclonal antisera tend to translate into a signal a broader spectrum of molecules and molecular isoforms than do assays using monoclonal antibodies. In case of GH, polyclonal assays therefore usually give higher results than monoclonal assays. In contrast, GH assays involving monoclonal antibodies can be designed to measure only, or at least preferentially, a specific isoform of GH [43–45]. It is obvious that GH assay results from the two types of assays will differ substantially; it also is important to see that the discrepancy between assay results is determined by the isoform composition in a certain sample. The importance of this source of variability also has been by highlighted external quality assessment schemes that demonstrated a worsening (from about 17% to about 30%) in the between-method agreement with the introduction of monoclonal antibody–based assays [21]. Apparently, the more specific assays that recognize only a specific fraction of the GH molecules present in a sample lead to higher variability than the more permissive, less specific polyclonal RIAs. It is a matter of debate whether a more specific or a more permissive recognition of GH isoforms or fragments by an immunoassay is preferable from a clinical point of view. Clearly, the clinical chemist tends to prefer measurement of a well-defined substance by a very specific method. In GH, however, different isoforms are biologically active [46]. Assays measuring only one isoform may miss important information [47].

Impact of growth hormone binding protein

Another important factor contributing to the GH assay heterogeneity is the presence of a high-affinity GH binding protein (GHBP) in human serum [48]. GHBP corresponds to the extracellular domain of the GH receptor, and its concentration varies with nutritional and metabolic conditions [49]. In circulation, up to 50% of GH is complexed with GHBP. Depending on the antibodies used in a given assay, complexed GH may not be recognized because the respective epitope of the antibodies used might not be accessible. Again, this problem was less pronounced in the past when assays with polyclonal antisera and long incubation times were used [50]. Today's assays use monoclonal antibodies with comparably short incubation times and therefore are more prone to interference from GHBP. It has been shown that the negative bias introduced by increasing concentrations of GHBP is considerable, approaching 50% for GHBP concentrations still in the physiologic range [51–53].

Standard preparations and units

Among the many problems contributing to GH assay variability, two can be resolved quite easily. The first is the use of different standard preparations to calibrate assay systems and the use of two units of measurement, mass units (mU/L) and μg/L). Some important journals in the field have decided to publish papers on GH data only if these data are expressed in mass units of the most recent International Standard 98/574 [30]. Historically, the first international reference preparations (IRP) for GH were of pituitary origin. These preparations (66/217, introduced in 1969, and 80/505, introduced in 1982) were poorly defined extracts, containing a variety of GH isoforms. They were assigned to contain 2.0 and 2.6 U/mg, respectively. With the introduction of recombinant human GH, the IRP 88/624 became available. Being of recombinant origin, this preparation consisted of pure 22-kD GH only and was assigned a biopotency of 3.0 U/mg. More recently, another recombinant IRP of 22-kD GH (98/674) became available, and this preparation should be the reference material of choice to ensure traceability of GH assay results to a uniform standard. In the past that the use of different conversion factors between international units and mass units frequently led to confusion among clinicians and among scientists. IRP 98/674 is of recombinant origin, pure, and well defined, and results should be reported in mass units. The use of international units should be abandoned. Of course, calibration against a uniform reference preparation will not solve all problems. For example, the amount of the reference preparation available for antibody binding will be different if dissolved in horse serum, which contains considerable amounts of GHBP, rather than sheep serum, which is almost free of GHBP. Such differences can be responsible for a twofold deviation in GH results [24]. In a nationwide study the Japanese Foundation for Growth Science showed convincingly that recalibration of all assays to a single, recombinant reference preparation leads to a significant reduction in between-laboratory variability in GH assay results, from more than 35% to less than 20% [23].

Summary

The heterogeneity of GH assay results from different laboratories is a serious problem in using published consensus criteria for diagnosis and monitoring treatment of GH-related diseases. The main reasons for the discrepancies between results are associated with the heterogeneity of GH molecules in circulation, the presence of a GHBP interfering with assays, the use of different assay designs involving monoclonal or polyclonal antibodies, and the availability of different standard preparations to calibrate GH assays. International efforts to harmonize GH assays have led to the recommendation that GH assay results be reported only in mass units of the new IRP 98/574. Consensus on isoform recognition and the implementation of measures to eliminate the interference from GHBP and other matrix components might be important issues for the future. From a clinical perspective, however, the most important recommendation is to apply only assay-specific cut-off values for diagnostic decisions. Using uniform cut-off values published in consensus statements while relying on different methods to determine the GH value makes it difficult to provide a consistently satisfactory quality of patient care.

References

[1] GH Research Society. Consensus guidelines for the diagnosis and treatment of growth hormone (GH) deficiency in childhood and adolescence: summary statement of the GH Research Society. J Clin Endocrinol Metab 2000;85(11):3990–3.
[2] Attanasio A, Attie K, Baxter R, et al. Consensus guidelines for the diagnosis and treatment of adults with growth hormone deficiency: Summary statement of the Growth Hormone Research Society workshop on adult growth hormone deficiency. J Clin Endocrinol Metab 1998;83:379–81.
[3] Molitch ME, Clemmons DR, Malozowski S, et al. Evaluation and treatment of adult growth hormone deficiency: an Endocrine Society Clinical Practice Guideline. J Clin Endocrinol Metab 2006;91(5):1621–34.
[4] Giustina A, Barkan A, Casanueva FF, et al. Criteria for cure of acromegaly: a consensus statement. J Clin Endocrinol Metab 2000;85(2):526–9.

[5] Biochemical assessment and long-term monitoring in patients with acromegaly: statement from a joint consensus conference of the Growth Hormone Research Society and the Pituitary Society. J Clin Endocrinol Metab 2004;89(7):3099–102.
[6] Biller BM, Samuels MH, Zagar A, et al. Sensitivity and specificity of six tests for the diagnosis of adult GH deficiency. J Clin Endocrinol Metab 2002;87(5):2067–79.
[7] Ghigo E, Bellone J, Aimaretti G, et al. Reliability of provocative tests to assess growth hormone secretory status. Study in 472 normally growing children. J Clin Endocrinol Metab 1996;81(9):3323–7.
[8] Aimaretti G, Corneli G, Razzore P, et al. Comparison between insulin-induced hypoglycemia and growth hormone (GH)-releasing hormone + arginine as provocative tests for the diagnosis of GH deficiency in adults. J Clin Endocrinol Metab 1998;83:1615–8.
[9] Hoeck HC, Jakobsen PE, Vestergaard P, et al. Differences in reproducibility and peak growth hormone responses to repeated testing with various stimulators in healthy adults. Growth Horm IGF Res 1999;9(1):18–24.
[10] Petersenn S, Jung R, Beil FU. Diagnosis of growth hormone deficiency in adults by testing with GHRP-6 alone or in combination with GHRH: comparison with the insulin tolerance test. Eur J Endocrinol 2002;146(5):667–72.
[11] Veldhuis JD, Patrie J, Wideman L, et al. Contrasting negative-feedback control of endogenously driven and exercise-stimulated pulsatile growth hormone secretion in women and men. J Clin Endocrinol Metab 2004;89(2):840–6.
[12] Pfeifer M, Kanc K, Verhovec R, et al. Reproducibility of the insulin tolerance test (ITT) for assessment of growth hormone and cortisol secretion in normal and hypopituitary adult men. Clin Endocrinol (Oxf) 2001;54(1):17–22.
[13] Darzy KH, Aimaretti G, Wieringa G, et al. The usefulness of the combined growth hormone (GH)-releasing hormone and arginine stimulation test in the diagnosis of radiation-induced GH deficiency is dependent on the post-irradiation time interval. J Clin Endocrinol Metab 2003;88(1):95–102.
[14] Bonert VS, Elashoff JD, Barnett P, et al. Body mass index determines evoked growth hormone (GH) responsiveness in normal healthy male subjects: diagnostic caveat for adult GH deficiency. J Clin Endocrinol Metab 2004;89(7):3397–401.
[15] Qu XD, Gaw Gonzalo IT, Al Sayed MY, et al. Influence of body mass index and gender on growth hormone (GH) responses to GH-releasing hormone plus arginine and insulin tolerance tests. J Clin Endocrinol Metab 2005;90(3):1563–9.
[16] Butler J. Role of biochemical test in assessing need for growth hormone therapy in children with short stature: Royal College of Pathologists Clinical Audit Project. Ann Clin Biochem 2001;38(1):i–xxiii.
[17] Rakover Y, Lavi I, Masalah R, et al. Comparison between four immunoassays for growth hormone (GH) measurement as guides to clinical decisions following GH provocative tests [in process citation]. J Pediatr Endocrinol Metab 2000;13(6):637–43.
[18] Derr RL, Cameron SJ, Golden SH. Pre-analytic considerations for the proper assessment of hormones of the hypothalamic-pituitary axis in epidemiological research. Eur J Epidemiol 2006;21(3):217–26.
[19] Evans MJ, Livesey JH, Ellis MJ, et al. Effect of anticoagulants and storage temperatures on stability of plasma and serum hormones. Clin Biochem 2001;34(2):107–12.
[20] Bidlingmaier F, Geilenkeuser WJ, Kruse R, et al. Our experience with quality control in current growth hormone assays. Horm Res 1991;36(Suppl 1):1–4.
[21] Seth J, Ellis A, Al-Sadie R. Serum growth hormone measurements in clinical practice: an audit of performance from the UK National External Quality Assessment scheme. Horm Res 1999;51(Suppl 1):13–9.
[22] Morsky P, Tiikkainen U, Ruokonen A, et al. Problematic determination of serum growth hormone: experience from external quality assurance surveys 1998–2003. Scand J Clin Lab Invest 2005;65(5):377–86.

[23] Tanaka T, Tachibana K, Shimatsu A, et al. A nationwide attempt to standardize growth hormone assays. Horm Res 2005;64(Suppl 2):6–11.
[24] Celniker AC, Chen AB, Wert RM Jr, et al. Variability in the quantitation of circulating growth hormone using commercial immunoassays. J Clin Endocrinol Metab 1989;68(2):469–76.
[25] Granada ML, Sanmarti A, Lucas A, et al. Assay-dependent results of immunoassayable spontaneous 24-hour growth hormone secretion in short children. Acta Paediatr Scand Suppl 1990;370:63–70 [discussion: 71].
[26] Markkanen H, Pekkarinen T, Valimaki MJ, et al. Effect of sex and assay method on serum concentrations of growth hormone in patients with acromegaly and in healthy controls. Clin Chem 2006;52(3):468–73.
[27] Hauffa BP, Lehmann N, Bettendorf M, et al. Central reassessment of GH concentrations measured at local treatment centers in children with impaired growth: consequences for patient management. Eur J Endocrinol 2004;150(3):291–7.
[28] Juul A, Bernasconi S, Clayton PE, et al. European audit of current practice in diagnosis and treatment of childhood growth hormone deficiency. Horm Res 2002;58(5):233–41.
[29] Wieringa GE, Barth JH, Trainer PJ. Growth hormone assay standardization: a biased view? Clin Endocrinol (Oxf) 2004;60(5):538–9.
[30] Trainer PJ, Barth J, Sturgeon C, et al. Consensus statement on the standardisation of GH assays. Eur J Endocrinol 2006;155(1):1–2.
[31] Baumann G. Growth hormone heterogeneity: genes, isohormones, variants, and binding proteins. Endocr Rev 1991;12(4):424–49.
[32] Cooke NE, Ray J, Watson MA, et al. Human growth hormone gene and the highly homologous growth hormone variant gene display different splicing patterns. J Clin Invest 1988;82(1):270–5.
[33] Igout A, Van Beeumen J, Frankenne F, et al. Purification and biochemical characterization of recombinant human placental growth hormone produced in Escherichia coli. Biochem J 1993;295:719–24.
[34] Hirt H, Kimelman J, Birnbaum MJ, et al. The human growth hormone gene locus: structure, evolution, and allelic variations. DNA 1987;6:59–70.
[35] Chapman GE, Rogers KM, Brittain T, et al. The 20,000 molecular weight variant of human growth hormone. Preparation and some physical and chemical properties. J Biol Chem 1981;256(5):2395–401.
[36] Baumann G. Growth hormone heterogeneity in human pituitary and plasma. Horm Res 1999;51(Suppl 1):2–6.
[37] Baumann G, Stolar MW, Buchanan TA. The metabolic clearance, distribution, and degradation of dimeric and monomeric growth hormone (GH): implications for the pattern of circulating GH forms. Endocrinology 1986;119(4):1497–501.
[38] Nagatomi Y, Ikeda M, Uchida H, et al. Reversible dimerization of 20 kilodalton human growth hormone (hGH). Growth Horm IGF Res 2000;10(4):207–14.
[39] Wallace JD, Cuneo RC, Bidlingmaier M, et al. The response of molecular isoforms of growth hormone to acute exercise in trained adult males. J Clin Endocrinol Metab 2001;86(1):200–6.
[40] Leung KC, Howe C, Gui LY, et al. Physiological and pharmacological regulation of 20-kDa growth hormone. Am J Physiol Endocrinol Metab 2002;283(4):E836–43.
[41] Freda PU, Nuruzzaman AT, Reyes CM, et al. Significance of ''abnormal'' nadir growth hormone levels after oral glucose in postoperative patients with acromegaly in remission with normal insulin-like growth factor-I levels. J Clin Endocrinol Metab 2004;89(2):495–500.
[42] Freda PU, Post KD, Powell JS, et al. Evaluation of disease status with sensitive measures of growth hormone secretion in 60 postoperative patients with acromegaly. J Clin Endocrinol Metab 1998;83(11):3808–16.
[43] Tsushima T, Katoh Y, Miyachi Y, et al. Serum concentration of 20K human growth hormone (20K hGH) measured by a specific enzyme-linked immunosorbent assay. Study Group of 20K hGH. J Clin Endocrinol Metab 1999;84(1):317–22.

[44] Wu Z, Bidlingmaier M, Dall R, et al. Detection of doping with human growth hormone. Lancet 1999;353(9156):895.
[45] Boguszewski CL, Boguszewski MC, de Zegher F, et al. Growth hormone isoforms in newborns and postpartum women. Eur J Endocrinol 2000;142(4):353–8.
[46] Hayakawa M, Shimazaki Y, Tsushima T, et al. Metabolic effects of 20-kilodalton human growth hormone (20K-hGH) for adults with growth hormone deficiency: results of an exploratory uncontrolled multicenter clinical trial of 20K-hGH. J Clin Endocrinol Metab 2004;89(4):1562–71.
[47] Popii V, Baumann G. Laboratory measurement of growth hormone. Clin Chim Acta 2004; 350(1–2):1–16.
[48] Baumann G. Growth hormone binding protein 2001. J Pediatr Endocrinol Metab 2001; 14(4):355–75.
[49] Amit T, Youdim MB, Hochberg Z. Clinical review 112: does serum growth hormone (GH) binding protein reflect human GH receptor function? J Clin Endocrinol Metab 2000;85(3): 927–32.
[50] Jan T, Shaw MA, Baumann G. Effects of growth hormone-binding proteins on serum growth hormone measurements. J Clin Endocrinol Metab 1991;72(2):387–91.
[51] Ebdrup L, Fisker S, Sorensen HH, et al. Variety in growth hormone determinations due to use of different immunoassays and to the interference of growth hormone-binding protein. Horm Res 1999;51(Suppl 1):20–6.
[52] Jansson C, Boguszewski C, Rosberg S, et al. Growth hormone (GH) assays: influence of standard preparations, GH isoforms, assay characteristics, and GH-binding protein. Clin Chem 1997;43(6 Pt 1):950–6.
[53] Hansen TK, Fisker S, Hansen B, et al. Impact of GHBP interference on estimates of GH and GH pharmacokinetics. Clin Endocrinol (Oxf) 2002;57(6):779–86.

Value of Insulin-like Growth Factor System Markers in the Assessment of Growth Hormone Status

David R. Clemmons, MD

Division of Endocrinology, University of North Carolina School of Medicine, University of North Carolina, CB #7170, 8024 Burnett-Womack, Chapel Hill, NC 27599, USA

Variables that regulate serum insulin-like growth factor-I concentrations

Because insulin-like growth factor-I (IGF-I) is the best-characterized growth hormone (GH)-dependent peptide, and because it has been measured extensively in a variety of clinical settings, it frequently is used to assess the clinical impact of disorders of GH secretion and to monitor the response to therapy in patients who have these disorders. Several points that are described more fully in other articles in this issue deserve emphasis to understand how IGF-I measurements are used as an index of GH secretion in a clinical context. The variables that regulate IGF-I are listed in Box 1.

The predominant determinant of a serum IGF-I is nutritional status. This observation derives from the fact that in primitive organisms (eg, *Caenorhabditis elegans*) there is no pituitary gland, and the link between nutrient intake and growth is provided by a hormone that is a precursor of both insulin and IGF-I [1]. This IGF-I/insulin precursor is stored in the olfactory apparatus, is secreted in response to food availability, is transported to distal tissues, and functions to signal to the organism that adequate nutrient intake has been ingested for the organism to respond with an anabolic response. In vertebrates, GH and insulin evolved, thus allowing organisms to function in a more complex environment wherein nutrient storage and breakdown were required to survive periods of food deprivation. Thus during fasting the organism synthesizes less IGF-I, and GH secretion increases [2,3]. The increased GH secretion serves to facilitate lipid breakdown to free fatty acid; this process provides substrate for energy use and thus conserves

---

This work was supported by Grant No. AG02331 from the National Institutes of Health.
E-mail address: endo@med.unc.edu

> **Box 1. Variables that regulate insulin-like growth factor-I secretion**
>
> 1. Nutritional status
> 2. Growth hormone secretion
> 3. Age
> 4. Genetic factors
> 5. IGF binding proteins
> 6. Insulin
> 7. Thyroxine
> 8. Cortisol
> 9. Testosterone
> 10. Estrogen
> 11. Obesity
> 12. Cytokines

glucose consumption by peripheral tissues. During periods of food abundance, mobilization of free fatty acids is suppressed, and insulin functions to store fat in adipose tissue. These responses are facilitated by absence of IGF-I receptors in liver and fat; therefore, the reduction in IGF-I helps augment the ability of GH to stimulate hepatic gluconeogenesis and to stimulate lipolysis. The clinical consequence is that during periods of extensive food deprivation IGF-I synthesis is relatively refractory to GH stimulation, and serum IGF-I concentrations decline [4]. Therefore disorders of nutrition such as anorexia, celiac disease, cystic fibrosis, or inflammatory bowel disease can result in substantial lowering of serum IGF-I.

Insulin is a major regulator of the liver's ability to synthesize IGF-I [5]. Low portal vein insulin concentrations that occur in type 1 diabetes result in refractoriness of IGF-I synthesis to GH stimulation [6]. Administration of insulin to insulin-dependent animals restores IGF-I synthesis to normal. Therefore maintenance of normal IGF-I concentrations requires adequate carbohydrate intake and insulin secretion. Frystyk and coworkers [7] have shown in experimental animals that starvation also inhibits the synthesis of acid-labile subunit (ALS) and insulin-like growth factor binding protein-3 (IGFBP-3). Thus insulin functions as a peripheral enhancer of IGF-I signaling and also regulates and coordinates the synthetic response of all three proteins to GH. A second major variable that regulates IGF-I is chronologic age [8]. During childhood mean IGF-I values rise fivefold from birth and peak during puberty between the ages of 13 and 15 years. IGF-I levels then decline fourfold between the ages of 15 and 70 years. These changes parallel the changes that occur in GH secretion. Thus the greatest increase in IGF-I values comes between the prepubertal and midpubertal years, and the greatest decrease is between midpuberty and age 20 years. The clinical consequence of these observations is that to assess IGF-I as an index of

GH secretion properly, large numbers of normal subjects in each age range are needed to establish precise normative values [9].

IGFBPs contribute to changes in total IGF-I concentrations and are the major determinant of free IGF-I concentrations. Although only a few gene polymorphisms have been described that regulate IGFBP secretion, some have been associated with significant changes in total serum IGF-I concentrations [10]. More importantly, certain disorders such as poorly controlled diabetes mellitus result in proteolysis of IGFBP-3, thus lowering total IGF-I [11].

There are six IGFBPs in serum, and their concentrations fluctuate in a variety of clinical conditions. In general IGFBP-3 and -5 are regulated by GH in a manner similar to IGF-I; however, the percent increase in IGFBP-3 or -5 in response to GH is substantially less than the percent increase in IGF-I because their ambient serum concentrations are also regulated by IGF-II [12]. Because serum IGF-II concentrations are threefold greater than IGF-I, and because IGF-II is minimally GH dependent, it serves to blunt GH-dependent changes that occur in IGFBP-3 and -5. IGFBP-2 is the second most abundant binding protein in serum. Unlike IGFBP-3 and -5, it is not saturated, and therefore it acts as a reservoir of excess IGF binding capacity [13]. Conditions that elevate IGFBP-2 (eg, severe chronic illness) can decrease free IGF-I. IGFBP-1 fluctuates widely throughout the day because its concentrations are suppressed by insulin secretion; thus after a meal there is a four- to fivefold decrease in IGFBP-1 [14]. This increase is associated with changes in free IGF-I although, because the absolute concentration of IGFBP-1 is substantially less than that of IGFBP-2 and –3, the changes in total IGF-I that occur in response to changes in IGFBP-1 are minimal. IGFBP-4 and -6 are not regulated by GH; therefore in disorders of GH secretion they do not contribute significantly to changes in total serum IGF-I or free IGF-I.

Total serum IGF-I is measured in most clinical conditions, and the factors that regulate serum IGFBP concentrations and their degree of saturation function to regulate total IGF-I concentrations indirectly. Although there are six forms of IGFBPs in serum, only two forms, IGFBP-3 and –5, bind to ALS. Therefore these two forms have the longest half-lives and constitute almost all of the IGF-binding activity in serum that is GH inducible. In assessing IGF-I concentrations, one of the primary variables that must be considered is ALS synthesis [15]. ALS is synthesized almost exclusively in liver and fat, and GH is the only known major hormonal determinant of ALS synthesis. ALS, IGFBP-3, and IGF-I function as a ternary complex in a coordinate manner. The combined molecular weight of the ternary complex is approximately 140,000 d. It is not filtered in the glomerulus; thus, its half-life is prolonged to 16 hours. Although binary complexes of IGF–I/IGFBP-3 or IGFBP–3/ALS form, they are much less stable. The stimulation of synthesis of all three components by GH allows IGF-I to be maintained in a stable ternary complex with a prolonged half-life [16].

The effects of androgens on the ability of GH to stimulate IGF-I secretion are complex. IGF-I measurements during puberty in boys show a strong correlation with the changes in testosterone. If testosterone is administered to healthy older man, there is an increase in GH release; therefore, at a low concentration androgen enhances GH and IGF-I; if administered in high doses that result in supraphysiologic concentrations, however, it is inhibitory [17].

Thyroxine has direct effects on IGF-I secretion and regulates GH responsiveness. Severely hypothyroid individuals respond poorly to GH in terms of increasing IGF-I [18]. In hyperthyroid individuals, serum IGF-I is increased by 20% to 40% above the normal range although it is not clear whether this increase results from to increased sensitivity to GH stimulation [19]. Because hyperthyroidism also results in increased IGFBP-3, this change accounts for part of the change in total IGF-I, but free IGF-I concentrations are also increased.

Genetic studies in identical twins have shown that there is a strong genetic component that regulates IGF-I [20]. These studies have shown that up to 50% of the variance in IGF-I levels can be accounted for by a genetic variable, and this variable also correlates with final adult height. This genetic determinant has not been identified, and it is possible that it regulates GH secretion or IGF-I responsiveness to GH.

Several gene polymorphisms contribute to variability in plasma IGF-I concentrations, but gene polymorphisms account for no more than of the 20% variability [21]. The polymorphism that has received the most in-depth analysis is a CA repeat that occurs approximately 500 base pairs proximal to the promoter [22]. This repeat is of variable length. The most common form, which contains 18 CA repeats, is present in 60% of white persons. The next most common contains 16 CA reagents. All other variations have been classified as noncarriers. In a Dutch study of white subjects, 11% were noncarriers. These individuals had lower serum IGF-I than the general population and were 2.3 cm shorter on average. When a group of individuals older the 60 years was studied, the noncarriers had a 2.2-fold increase in diabetes [23]. GH secretion in this group has not been analyzed, but increased secretion presumably would provide a mechanism to explain the effect of this gene polymorphism on insulin resistance. Other *IGF-I* gene polymorphisms have been described that result in decreased head circumference at birth and are associated with the small for gestational age phenotype.

Regulation of insulin-like growth factor-I synthesis by growth hormone

Most IGF-I in the circulation results from hepatic synthesis and secretion. Analysis of the liver has shown that the most abundant form of IGF-I mRNA is a 6-kb transcript and that its transcription is regulated by GH [24]. GH also increases the transcription of a 0.9-kb transcript. Both transcripts are believed to be the source of the secreted peptide. The hepatic IGF-I synthesis response to GH is mediated by signal transducer and activator of transcription-5b (Stat5b) [25]. After GH receptor activation, recruitment of Janus kinase-2 (JAK-2) results in phosphorylation of Stat5b, which is translocated to the nucleus and stimulates IGF-I transcription. Therefore variables that regulate the synthesis or activation of Stat5b can regulate IGF-I synthesis. IGF-I is secreted as one of two precursor forms, IGF-Ia or IGF-Ib. These forms have 15– and 27–amino acid C-terminal extensions, respectively, and their synthesis and secretion are stimulated by GH [26]. Neither form is a major contributor to serum IGF-I. After GH administration to GH-deficient individuals there is a major rise in serum IGF-I (mean, 5.5-fold) that is detectable after 4 hours; IGF-I values peak between 10 and 14 hours [27].

After Stat5b induction there is a counterregulation of the IGF-I synthesis by suppressor of cytokine signaling 1 (SOCS-1) or 2 (SOCS-2) or cytokine inducible SH2 protein. These proteins further negatively regulate the degree of increase in IGF-I by targeting Stat5b for degradation. Another variable that can affect serum IGF-I concentrations is stress-induced resistance to GH. Severe physical stress, such as that which occurs in patients who suffer burns or head trauma, results in a substantial decrease in IGF-I synthesis [28]. Cytokines such as tumor necrosis factor-α and interleukin-1β suppress IGF-I synthesis in response to GH [29]. In addition, IGFBP-3 proteolysis occurs in these clinical conditions, contributing to the overall decrease in total IGF-I [30]. In a manner similar to cytokine release, estrogens block GH signal transduction by induction of SOCS-2, leading to suppression of Stat-5b and thus lowering IGF-I [31]. Glucocorticoids also function through the SOCS proteins to attenuate GH stimulation of IGF-I. If administered over the long term, glucocorticoids can lead to significant skeletal muscle wasting, and IGF-I synthesis in skeletal muscle is notably attenuated [30].

Regulation of serum concentrations of insulin-like growth factor-I by growth hormone

The multiple variables that regulate serum IGF-I concentrations function either directly by regulating IGF-I synthesis or indirectly by altering sensitivity to GH. They also can function indirectly by regulating the concentrations of IGFBPs, which secondarily alters total serum IGF-I. Therefore, if any attempt is made to assess the role of GH in regulating IGF-I concentrations, the potential impact of changes in these other variables on GH secretion or GH sensitivity must be ascertained. There is a hierarchy among these variables, with nutritional status being predominant [32]. The age-related changes that occur in IGF-I concentrations clearly are related to changes in GH secretion. These are accompanied by parallel changes in IGFBP-3 and ALS [33]. Thus in midpuberty, when GH concentrations peak, there is the greatest increase in IGF-I. In contrast, during the newborn period, when there is minimal GH secretion, IGF-I levels are quite low [34]. Normal tall children seem to have higher levels of IGF-I and IGFBP-3 than children of normal stature [35]. The extent to which this increase in IGF-I and IGFBP-3 correlates with increased GH secretion in tall children has not been established. Estrogen increases resistance to GH actions by Socs-3, inhibiting the ability of STAT5b to increase IGF-I synthesis response to GH [31]. Thyroxine also modulates GH responsiveness. When GH-deficient children who are also hypothyroid have been analyzed, the IGF-I response to GH is attenuated, and it is enhanced after thyroxine replacement.

Regulation of acid-labile subunit and insulin-like growth factor binding protein-3

Several of the variables that regulate IGF-I concentrations also can regulate IGFBP-3 and ALS synthesis. Partial starvation in humans results in reduced IGFBP-3 and ALS levels, and these levels increase with regain of weight [36]. Acute illness can result in reduction in both ALS and IGFBP-3, and part of the decrease in IGFBP-3 results from enhanced cytokine-induced suppression of Stat5b actions [37]. Acute endotoxemia results in decreased ALS synthesis, and ALS is reduced in patients who have liver disease such as cirrhosis [38]. Similar to IGF-I, the induction of cytokines such as interleukin-1β can result in decreased ALS synthesis [39]. Patients who have poorly controlled diabetes mellitus or severe burns have decreased ALS levels, and insulin treatment results in stabilization of ALS [40]. During GH receptor blockade in monkeys, reinfusion of IGF-I resulted in increased ALS and IGFBP-3, suggesting that IGF-I can have direct effects on the concentration of these two proteins, probably by facilitating complex formation and thus prolonging their half-lives [41]. In contrast, when IGF-I is given to children who have GH receptor defects, there is no IGFBP-3 response to IGF-I [42]. GH-deficient children who are administered GH show major increases in both ALS and IGFBP-3 [43].

IGFBP-3 regulation is more complex than ALS. IGFBP-3 is not synthesized by hepatocytes but rather by connective tissue cells within the liver and other organs [44]. Endothelium also represents an important source of IGFBP-3. IGFBP-3 concentrations are low in patients who have GH deficiency (GHD) and increase after GH administration. There is a correlation between IGFBP-3 and GH secretion [45]. In healthy children IGFBP-3 levels correlate with spontaneous GH secretion. In adults this correlation has been more difficult to demonstrate. Both IGF-I and II have been shown to cause modest increases in IGFBP-3 [16,41], but they are not major regulatory variables. Testosterone increases IGFBP-3 synthesis in prepubertal and hypogonadal males. Hypothyroidism results in a low IGFBP-3 levels that increase after thyroid replacement [18]. Insulin is required for normal IGFBP-3 synthesis, and insulin-deficient diabetics have low IGFBP-3 [46,47]. Increases in insulin also lead to a decrease in IGFBP-3 proteolysis [11], an important variable in regulating IGFBP-3 concentrations in pregnancy [48]. Although the major fragment of IGFBP-3 retains some IGF-I binding capacity, its affinity it reduced 20-fold, and it has a reduced affinity for ALS.

Prolonged fasting results in minimal changes in IGFBP-3 that are clearly much less than the changes that occur in IGF-I [49]. Age is a major regulator of IGFBP-3; the levels are low at birth, increase during childhood, and peak at adolescence in a manner similar to IGF-I [8]. Genetic changes also regulate IGFBP-3. A polymorphism in the promoter region of the *IGFBP-3* gene has been associated with lower levels of IGFBP-3 [10]. Two additional polymorphisms have been shown to be associated with significant decreases in IGFBP-3 [50]. In severe liver disease in which both IGF-I and ALS mRNA levels are suppressed, IGFBP-3 mRNA levels are normal, but serum levels are decreased, presumably because of decreased IGF-I and ALS [51]. Glucocorticoids regulate IGFBP-3. High doses suppress its synthesis and increase its proteolytic degradation [52,53]. IGFBP-3 variations that occur in childhood seem to correlate with growth. There is a statistically significant correlation between IGFBP-3 levels and head circumference at birth. Similarly, birth length correlates with IGFBP-3 levels [34]. Estrogen does not seem to be required for postmenopausal women to respond to GH with an increase in IGFBP-3, but women in this age range receiving oral or cutaneous estrogen replacement have a blunted response to GH [54].

Efficacy of insulin-like growth factor-I, insulin-like growth factor binding protein-3, and acid-labile subunit in the diagnosis of growth hormone deficiency in children

The largest studies (ones that include at least 50 patients) show that a low IGF-I value (below -2 SD) has a sensitivity for predicting GHD that varies from 62% to 96% [33,55–62]. All children in these studies had undergone at least two provocative tests for GH secretion. The specificity of an IGF-I value for differentiating between other causes of short stature and GHD varied from 52% to 81%. Lowering the cut-off value to -2.5 SD improved the ability of low IGF-I to differentiate between GHD and idiopathic short stature [57]. When IGF-I values are used to decide which children should receive a provocative test, studies have shown that a cut-off of -0.83 SD results in a sensitivity of 92% and a specificity of 47% [61]. If values lower than -1 SD were used, 68% of the children who had ISS would not have to undergo a stimulation test [62]. Because age-adjusted normative data indicating what constitutes a normal GH response to provocative stimuli are sparse, it is difficult to ascertain why some studies show a discrepancy between the results of GH provocative testing and IGF-I values and others do not. One variable is the quality of GH assays. Analysis across several different laboratories has shown there is as much as a fourfold difference in GH values. Therefore, depending on the particular assay that is used, the cut-off values that are used to define a normal response may vary greatly. This variation increases the problems encountered in trying to compare the clinical usefulness of IGF-I and GH measurements. In summary, at present IGF-I is used in short children to select individuals who require a GH stimulation test. It is not used as a stand-alone test to diagnose GHD definitively because a value within the normal range does not exclude the diagnosis.

When IGFBP-3 was used to diagnose GHD in children, the sensitivity varied between 61% and 97%, and specificity ranged from 72 to 98% [55,56,60,63,64]. Blum and colleagues [55] reported the best results, showing a sensitivity of 95% and a specificity of 98%. Other studies have not found this excellent discrimination, and in most studies the sensitivity of IGFBP-3 has been lower than that of IGF-I, probably because most of the IGFBP-3 is bound to IGF-II, which is less GH dependent. For this reason IGFBP-3 often is used to confirm that GHD is not present in short children. In children younger than 6 years, however, IGFBP-3 may be of equal or greater discriminative value than IGF-I [57]. Only a few studies have assessed the usefulness of ALS, and they have had minimal numbers of subjects. At present the measurement of ALS seems to have no added value over the measurement of IGF-I and IGFBP-3. Whether increases in IGFBP-2, which has increased GH deficiency, contribute to the lack of more consistent suppression of IGF-I in children has not been studied. One study has suggested that inclusion of IGFBP-2 measurements improves diagnostic accuracy [65].

Diagnosis of growth hormone deficiency in adults

Most studies in adults have confirmed that low IGF-I values (eg, below the first percentile) predict the presence of GHD with 95% accuracy [66]. A normal value does not exclude the diagnosis, however. The degree to which a normal value will lead to a false-negative result depends on the age of the patient at diagnosis and the chronologic age at the time of testing. In adults older than 40 years false-negative values can be present in up to 50% of cases [67]. The published studies are best interpreted if the analysis is compartmentalized into adults who have childhood-onset GHD (CO-GHD) and those who have adult-onset GHD (AO-GHD). Even when adjusted for age, IGF-I levels are lower in adults who have CO-GHD than in those who have AO-GHD [68,69]. The exact reason for this difference is not clear but does not seem to be related to the severity of GHD or to the presence of other hormonal deficits. When adults who have CO-GHD are analyzed, they almost invariably have low IGF-I values [68,70]. In a large population of patients who had CO-GHD assessed before GH treatment, the mean IGF-I value was -3.5 SD, and IGF-I was within the normal range in only 11% of the patients [70]. Debour and colleagues [70] demonstrated that 96% of adults who had CO-GHD had values that were subnormal, and others have confirmed this finding [69–71]. Pooled analysis of the studies with more than 50 subjects shows that the sensitivity for the diagnosis of severe GHD for IGF-I varies between 76% and 96% in patients who have

CO-GHD [70–72]. Most of these studies include younger adults, and adults younger than 40 years who have either AO- or CO-GHD tend to have values that are much lower than the age-adjusted mean [68]. When young patients who have AO-GHD (ie, patients in their 20s and 30s) were evaluated, between 17% and 30% had IGF-I within the normal range in spite of severe GHD [68]. When AO-GHD patients of all ages were evaluated, the sensitivity of IGF-I for predicting GHD varied between 55% and 93% [66–68,73–80]. One study that analyzed 817 adults showed that an IGF-I measurement has most value when it is below the normal range. When a cut-off of 88 ng/mL (1% percentile) was used, 96% of the patients who had values below the cut-off had severe GHD [66]. This study showed that if the IGF-I measurement is combined with the presence of three or four additional hormonal deficits, the positive predictive value is 100%. Therefore a low IGF-I often reflects presence of GHD. It is not true, however, that an IGF-I value within the normal range can be used to exclude the diagnosis of GHD. In subjects who had IGF-I levels above 88 ng/mL, 32% had GHD as determined by the results of two provocative GH-stimulation tests. Amaretti and colleagues [68] showed that age stratification of GHD patients also improves the ability of IGF-I to diagnose GHD. In subjects younger than 40 years, IGF-I values were below normal in 78% of GHD subjects. In the 50- to 60-year range IGF-I was decreased significantly in only 56% of GHD subjects. In spite of these limitations IGF-I is widely used by physicians to predict the presence of GHD in adults, and many physicians have found it a useful tool in selecting which patients should undergo stimulation testing.

When IGFBP-3 has been analyzed, the sensitivity has varied between 61% and 97%, and specificity has varied between 72% and 98% [71,81–83]. One study evaluating adults showed that serum IGFBP-3 levels were 1.96 SD below the mean in 91% of patients who had AO-GHD and in 98% of adults who had CO-GHD [84]. Several studies, however, have reported IGFBP-3 values in the normal range in adults who had severe GHD [71,85–87]. ALS values were below normal in 71% of patients who had AO-GHD [83] and in 74% of those who had CO-GHD [82]. In summary, determination of IGFBP-3 or ALS in adults does not add accuracy to the diagnosis [78–80]. These measurements are less sensitive and less specific than are with total IGF-I values [71,73,84–88].

Prediction of growth response after growth hormone therapy

Several studies in children have attempted to determine if there is a correlation between the change in height velocity and the change in IGF-I in response to GH. Initial studies showed that, as a stand-alone test, an IGF-I value obtained before therapy did not predict changes in growth that occur in response to GH [89,90]. Subsequent studies with larger numbers of children have shown that both IGF-I and IGFBP-3 correlate with the subsequent growth response [91,92]. Baseline changes in IGF-I and IGFBP-3 also have been used successfully in multifactorial models to predict the growth response to GH. Some studies have reported a positive correlation between the changes in growth rate and the changes in IGF-I. In some larger studies both IGF-I and IGFBP-3 have been shown to correlate significantly with growth rate.

Monitoring growth hormone therapy in adults

IGF-I measurements have found their greatest usefulness in monitoring therapy in adults to avoid side effects. Because adults require lower doses of GH than children to achieve a maximum response and because of concern about maintaining high IGF-I levels in adults for prolonged periods, IGF-I concentrations have been evaluated extensively for their ability to predict GH excess [93]. Detectible increases in IGF-I can occur at doses as low as 2 µg/kg/d. de Boer and colleagues [94] reported that IGF-I was more sensitive and useful than IGFBP-3 and ALS for monitoring GH dosage in adults. They determined that weight-based dosing up to maximum of 11 µg/kg/d resulted in supraphysiologic concentrations of IGF-I in 20% of patients. In general, if side effects develop, the IGF-I levels are usually substantially higher than the upper limit of normal [95,96]. Subject age, sex, and route of estrogen administration are the major determinants of the amount of GH that will be required to normalize IGF-I [97–100]. Failure to adjust appropriately for these variables can result in significantly increased serum concentrations and side effects.

Changes in insulin-like growth factor binding proteins that result in a change in free insulin-like growth factor-I

In contrast to IGFBP-3 and -5, other binding proteins may be increased with GHD. IGFBP-2 often is increased in states of low GH secretion or impaired GH action such as hypopituitarism or severe, poorly controlled type 1 diabetes or in elderly subjects experiencing a severe, prolonged illness [65]. Severe protein restriction results in increased IGFBP-2 levels [101]. Because IGFBP-2 is much less abundant than IGFBP-3, these changes have minimal effects on total IGF-I; however, increases in IGFBP-2 may alter free IGF-I concentrations significantly. For similar reasons, changes in IGFBP-1 have a greater effect on free IGF-I concentrations. IGFBP-1 may increase as much as fivefold during an overnight fast and decrease by three- to fourfold after a meal [102]. These wide fluctuations in IGFBP-1 result in minimal changes in total IGFBP concentration in serum, but they may result in significant changes in the amount of available IGF binding capacity and thus lead to greater changes in free IGF-I. Although IGFBP-5 has a very stable half-life, similar to that of IGFBP-3, it is cleaved proteolytically to a great extent in serum. Therefore changes in the concentration of intact IGFBP-5, although GH dependent, have minimal influence on total IGF-I [103].

Free insulin-like growth factor-I and the diagnosis of growth hormone deficiency

The degree of change in free IGF-I in GHD is not influenced by changes in IGFBPs. Therefore changes in free IGF-I can provide a reliable index of GH secretion when changes in specific forms of IGFBPs such as IGFBP-1 and -2 and changes in IGF-I are discordant (eg, in fasting). An assay that measured the amount of IGF-I complexed IGFBP-1 showed that, when this complex increased, there was a decrease in measurable free IGF-I [104]. Like total IGF-I, free IGF-I varies across childhood, peaks a puberty, and declines in an age-dependent manner [105,106]. Direct comparisons of the usefulness of free IGF-I with that of total IGF-I, IGFBP-3, or ALS in discriminating between GH-deficient and normal subjects have been disappointing. In most studies free IGF-I has not provided any better discrimination in diagnosing GHD [106,107]. The exception is in obesity. Because free IGF-I levels are relatively increased compared with total IGF-I in obesity, free IGF-I measurements may be useful in discriminating between older adults who are normal and simply have low GH secretion because of obesity and those who are truly GH deficient [108]. Free IGF-I levels increase 4.5- to 6.1-fold after administration of GH to GH-deficient subjects [109]. The use of free IGF-I to monitor treatment of children or adults who have GHD and who are receiving GH treatment has not proven more useful than total IGF-I in predicting efficacy or avoiding toxicity, however [110]. Because the variability of the degree of increase in free IGF-I is less than that of total IGF-I, studies are still ongoing to determine whether free IGF-I may be a more useful parameter to monitor efficacy. The experience with using total IGF-I to monitor toxicity is extensive, however, and therefore it is the preferred measurement at present.

There are three clinical situations in which measurement of free IGF-I may be superior to total IGF-I. The first is in evaluating the presence of GH disorders in patients who are obese or have type 2 diabetes. Free IGF-I measurements consistently are substantially higher than total IGF-I in obese persons [111]. The change in free IGF-I correlates inversely with a change in IGFBP-1, and IGFBP-1 tends to be suppressed in obesity, thus explaining the relatively higher level of free IGF-I [111]. This finding also has been documented in children; therefore, free IGF-I may more closely reflect bioavailable IGF-I in obese children [112]. In type 1 diabetes there are wide swings in IGFBP-1; therefore, free IGF-I levels are significantly lower than normal as compared with changes in total IGF-I. In poorly controlled diabetes, free IGF-I levels are more severely suppressed than total IGF-I levels, reflecting the lack of GH responsiveness in the liver caused by inadequate insulinization [113]. In chronic renal failure there are multiple IGFBP abnormalities. When free IGF-I is measured using an ultrafiltration method, the levels generally are significantly lower than total IGF-I as compared with normal subjects [114,115]. This finding reflects the increase in available IGF binding capacity. In conditions where there is extensive IGFBP-3 proteolysis (ie, pregnant and postoperative patients), free IGF-I is relatively higher than total IGF-I as compared with normal persons [116,117]. Thus measurement of free IGF-I in these conditions may be superior to total IGF-I. For routine diagnosis of GHD, however, free IGF-I seems to offer no significant advantage.

Acromegaly

Adults who have excessive GH consistently have IGF-I concentrations that are greater than normal [118]. Unlike GHD, in acromegaly GH is the predominant determinant of total serum IGF-I. If one assumes that daily fluctuations in GH in normal subjects account for no more than 50% of the total stimulatory input for IGF-I, in GHD they probably account for less than 10%. In contrast, in acromegaly GH may account for more than 90% of the ambient IGF-I value. Therefore variables such as gene polymorphisms, the unknown genetic factor linked to adult height, changes in sex steroid secretion, thyroxine, and cortisol that account for most of the variability in normal subjects are relatively unimportant in acromegaly. Several studies have confirmed the diagnostic precision of IGF-I. Mean IGF-I levels are 8- to 10-fold greater than age-adjusted norms [118–121]. The sensitivity and specificity of IGF-I is at least equal and at times is superior to that of glucose suppression testing [119,122]. The only diagnostic difficulty occurs in adolescence: normal persons between ages of 13 and 18 years have IGF-I values that are substantially higher than those in persons older than 18 years. Because the upper limit of the normal range in adolescents is quite high, it can be difficult to tell the difference between a child who has mild giantism and a normal, rapidly growing, tall adolescent [123]. In this setting, GH suppression testing is mandated. In persons older than 18 years of age, this distinction is rarely an issue. The other clinical difficulty that is encountered results from the statistical definition of a normal IGF-I level; thus if the 95% confidence interval is used to define "normal," 2.5% of all normal adults will have values that are in the acromegalic range. Because acromegaly is such a rare disease, the incidence of high IGF-I values in the normal population will be greater than the incidence of acromegaly. When this situation arises, GH-suppression testing should be performed, if clinically indicated.

Other GH-dependent peptides may be altered in acromegaly. IGFBP-3 is elevated in 75% to 80% of patients who present with a GH secretory disorder [120,121,123,124]. IGFBP-3 is a less accurate measurement for diagnosing acromegaly because of the effect of IGF-II, which is not increased in acromegaly and restrains the degree of increase in IGFBP-3 as compared with IGF-I. Therefore IGFBP-3 is not used as a routine diagnostic test, although values are abnormal in at least 80% of subjects [123,124]. ALS also has diagnostic value in acromegaly. Its sensitivity and specificity have been reported to be greater than 90% [82]. Because ALS concentrations are partly determined by total IGFBP-3 concentrations, however, the absence of IGF-II elevation in acromegaly acts as a restraining influence on ALS, and ALS changes less than IGF-I in acromegaly (ie, mean ALS values are increased 3.5-fold) [125].

Total IGF-I is a superior to free IGF-I in predicting 24- hour GH secretion [126]. Free IGF-I is an equally precise diagnostic test, but total IGF-I seems to be superior in estimating the degree of disease activity. Because somatostatin directly increases IGFBP-1 levels, free IGF-I levels may predict improvement more reliably in persons who are treated with long-acting somatostatin analogues [127], but rigorous proof that free IGF-I levels is more useful than total IGF-I in this circumstance has not been reported.

A clinically useful feature of IGF-I is its correlation with disease severity; presumably because IGF-I is one step closer to the growth-stimulatory event and because it has a long half-life in serum. Plasma concentrations have been shown to correlate well with the degree of abnormality at the time of diagnosis in acromegaly and to correlate well with improvement in symptoms and signs after successful treatment [121,128–132]. When IGF-I values were compared with heelpad thicknesses, a measure of soft tissue thickness, the correlation coefficient was 0.73 [118]. After treatment, when symptom index scores or changes in ring size are correlated with changes in IGF-I, the R value is 0.4 [129].

The choice of GH values or IGF-I to determine long-term treatment outcomes is more controversial. Most epidemiologic studies that compare biochemical control and premature mortality in acromegaly have used GH measurements; therefore more long-term data are available for GH. One study by Holdaway and colleagues [133] followed 200 acromegalics for 13 years. In that study the standardized mortality ratio (SMR) was 2.6 for subjects who failed to suppress GH after glucose to less than 5 µg/L and was 1.6 for those who suppressed GH to between 5 µg/L and 2.5 µg/L. For those who suppressed GH to less than 1 µg/L, it was 1.1. In contrast, the SMR for an IGF-I value greater than 2 SD above normal was 3.5. A study by Swearingen and colleagues [134] also found that elevated IGF-I values were associated with an SMR of 1.7. This finding was confirmed in a recent study, which found an SMR value of 4.78 in persons who had an elevated IGF-I level, whereas those who failed to suppress GH to less than 2 µg/L had an SMR of only 1.6 [135]. In contrast, a British study found an SMR of 1.55, which was almost significant ($P < .07$), for patients who failed to suppress GH hormone to less than 2 µg/L, but they found no significant difference with IGF-I (SMR = 1.2) [136]. Several of the subjects who died prematurely had GH measurements but did not have IGF-I measurements;

therefore the comparison was not made in the identical population. The current recommendation is to use either failure to suppress GH or elevated IGF-I as an indicator of the need for further treatment.

Summary

Pituitary GH secretion is the predominant hormonal variable regulating serum IGF-I concentrations. In disorders of GH secretion, when large populations are examined, there is substantial increase or decrease in mean IGF-I values. Although IGF-I is the most widely used test to detect changes in GH secretion, it does not have sufficient precision to be used as a stand-alone test in the diagnosis of GHD. IGF-I does seem to be a useful screening test in adults under 40 years of age and in children, because a low value almost always predicts the presence of GHD. A normal value does not exclude this diagnosis, however, and stimulation testing is required. In patients who have acromegaly, IGF-I values are almost always elevated, and a normal value usually is sufficient to exclude the diagnosis. Although measurement of free IGF-I, IGFBP-3, or ALS may provide useful information regarding GH secretion in specific conditions, in general these tests have not proven superior to IGF-I for making the diagnosis of GHD or acromegaly. Future studies no doubt will define better the indications for more precise use of these measurements in combination with total IGF-I to determine the need for therapeutic intervention.

Acknowledgments

The author thanks Laura Lindsey for her help in preparing the manuscript.

References

[1] Kenyon C. The plasticity of aging: insights from long-lived mutants. Cell 2005;120:449–60.
[2] Clemmons DR, Klibanski A, Underwood LE, et al. Reduction of plasma immunoreactive somatomedin C during fasting in humans. J Clin Endocrinol Metab 1981;53:1247–50.
[3] Ho KY, Veldhuis JD, Johnson ML, et al. Fasting enhances growth hormone secretion and amplifies the complex rhythms of growth hormone secretion in man. J Clin Invest 1988;81: 968–75.
[4] Underwood LE, Thissen JP, Lemozy S, et al. Hormonal and nutritional regulation of IGF-I and its binding proteins. Horm Res 1994;42:145–51.
[5] Kaytor EN, Zhu JL, Pao CI, et al. Insulin-responsive nuclear proteins facilitate Sp1 interactions with the insulin-like growth factor-I gene. J Biol Chem 2001;276:36896–901.
[6] Hanaire-Broutin H, Sallerin-Caute B, Poncet MF, et al. Effect of intraperitoneal insulin delivery on growth hormone binding protein, insulin-like growth factor (IGF)-I, and IGF-binding protein-3 in IDDM. Diabetologia 1996;39:1498–504.
[7] Frystyk J, Delhanty PJ, Skjaerbaek C, et al. Changes in the circulating IGF system during short-term fasting and refeeding in rats. Am J Physiol 1999;277:E245–52.

[8] Juul A. Serum levels of insulin-like growth factor I and its binding proteins in health and disease. Growth Horm IGF Res 2003;13:113–70.
[9] Brabant G, von zur Muhlen A, Wuster C, et al. Serum insulin-like growth factor I reference values for an automated chemiluminescence immunoassay system: results from a multicenter study. Horm Res 2003;60:53–60.
[10] Deal C, Ma J, Wilkin F, et al. Novel promoter polymorphism in insulin-like growth factor-binding protein-3: correlation with serum levels and interaction with known regulators. J Clin Endocrinol Metab 2001;86:1274–80.
[11] Bereket A, Lang CH, Blethen SL, et al. Insulin-like growth factor binding protein-3 proteolysis in children with insulin-dependent diabetes mellitus: a possible role for insulin in the regulation of IGFBP-3 protease activity. J Clin Endocrinol Metab 1995;80:2282–8.
[12] Baxter RC. Circulating levels and molecular distribution of the acid-labile (alpha) subunit of the high molecular weight insulin-like growth factor-binding protein complex. J Clin Endocrinol Metab 1990;70:1347–53.
[13] McCusker RH, Campion DR, Jones WK, et al. The insulin-like growth factor-binding proteins of porcine serum: endocrine and nutritional regulation. Endocrinology 1989;125:501–9.
[14] Lewitt MS. Role of the insulin-like growth factors in the endocrine control of glucose homeostasis. Diabetes Res Clin Pract 1994;23:3–15.
[15] Boisclair YR, Hurst KR, Ueki I, et al. Regulation and role of the acid-labile subunit of the 150-kilodalton insulin-like growth factor complex in the mouse. Pediatr Nephrol 2000;14:562–6.
[16] Kupfer SR, Underwood LE, Baxter RC, et al. Enhancement of the anabolic effects of growth hormone and insulin-like growth factor I by use of both agents simultaneously. J Clin Invest 1993;91:391–6.
[17] Veldhuis JD, Anderson SM, Iranmanesh A, et al. Testosterone blunts feedback inhibition of growth hormone secretion by experimentally elevated insulin-like growth factor-I concentrations. J Clin Endocrinol Metab 2005;90:1613–7.
[18] Miell JP, Taylor AM, Zini M, et al. Effects of hypothyroidism and hyperthyroidism on insulin-like growth factors (IGFs) and growth hormone- and IGF-binding proteins. J Clin Endocrinol Metab 1993;76:950–5.
[19] Westermark K, Alm J, Skottner A, et al. Growth factors and the thyroid: effects of treatment for hyper- and hypothyroidism on serum IGF-I and urinary epidermal growth factor concentrations. Acta Endocrinol (Copenh) 1988;118:415–21.
[20] Hong Y, Pedersen NL, Brismar K, et al. Quantitative genetic analyses of insulin-like growth factor I (IGF-I), IGF-binding protein-1, and insulin levels in middle-aged and elderly twins. J Clin Endocrinol Metab 1996;81:1791–7.
[21] Harrela M, Koistinen H, Kaprio J, et al. Genetic and environmental components of interindividual variation in circulating levels of IGF-I, IGF-II, IGFBP-1, and IGFBP-3. J Clin Invest 1996;98:2612–5.
[22] Rosen CJ, Kurland ES, Vereault D, et al. Association between serum insulin growth factor-I (IGF-I) and a simple sequence repeat in IGF-I gene: implications for genetic studies of bone mineral density. J Clin Endocrinol Metab 1998;83:2286–90.
[23] Vaessen N, Heutink P, Janssen JA, et al. A polymorphism in the gene for IGF-I: functional properties and risk for type 2 diabetes and myocardial infarction. Diabetes 2001;50:637–42.
[24] Roberts CT Jr, Lasky SR, Lowe WL Jr, et al. Molecular cloning of rat insulin-like growth factor I complementary deoxyribonucleic acids: differential messenger ribonucleic acid processing and regulation by growth hormone in extrahepatic tissues. Mol Endocrinol 1987;1:243–8.
[25] Greenhalgh CJ, Alexander WS. Suppressors of cytokine signalling and regulation of growth hormone action. Growth Horm IGF Res 2004;14:200–6.

[26] Hiney JK, Srivastava V, Nyberg CL, et al. Insulin-like growth factor I of peripheral origin acts centrally to accelerate the initiation of female puberty. Endocrinology 1996;137: 3717–28.
[27] Jorgensen JO, Flyvbjerg A, Lauritzen T, et al. Dose-response studies with biosynthetic human growth hormone (GH) in GH-deficient patients. J Clin Endocrinol Metab 1988;67: 36–40.
[28] Frost RA, Nystrom GJ, Lang CH. Tumor necrosis factor-alpha decreases insulin-like growth factor-I messenger ribonucleic acid expression in C2C12 myoblasts via a Jun N-terminal kinase pathway. Endocrinology 2003;144:1770–9.
[29] Wolf M, Bohm S, Brand M, et al. Proinflammatory cytokines interleukin 1 beta and tumor necrosis factor alpha inhibit growth hormone stimulation of insulin-like growth factor I synthesis and growth hormone receptor mRNA levels in cultured rat liver cells. Eur J Endocrinol 1996;135:729–37.
[30] Lang CH, Nystrom GJ, Frost RA. Burn-induced changes in IGF-I and IGF-binding proteins are partially glucocorticoid dependent. Am J Physiol Regul Integr Comp Physiol 2002; 282:R207–15.
[31] Leung KC, Johannsson G, Leong GM, et al. Estrogen regulation of growth hormone action. Endocr Rev 2004;25:693–721.
[32] Merimee TJ, Zapf J, Froesch ER. Insulin-like growth factors in the fed and fasted states. J Clin Endocrinol Metab 1982;55:999–1002.
[33] Hasegawa Y, Hasegawa T, Takada M, et al. Plasma free insulin-like growth factor I concentrations in growth hormone deficiency in children and adolescents. Eur J Endocrinol 1996;134:184–9.
[34] Ong K, Kratzsch J, Kiess W, et al. Size at birth and cord blood levels of insulin, insulin-like growth factor I (IGF-I), IGF-II, IGF-binding protein-1 (IGFBP-1), IGFBP-3, and the soluble IGF-II/mannose-6-phosphate receptor in term human infants. The ALSPAC Study Team. Avon Longitudinal Study of Pregnancy and Childhood. J Clin Endocrinol Metab 2000;85:4266–9.
[35] Garrone S, Radetti G, Sidoti M, et al. Increased insulin-like growth factor (IGF)-II and IGF/IGF-binding protein ratio in prepubertal constitutionally tall children. J Clin Endocrinol Metab 2002;87:5455–60.
[36] Stoving RK, Hangaard J, Hagen C, et al. Low levels of the 150-kD insulin-like growth factor binding protein 3 ternary complex in patients with anorexia nervosa: effect of partial weight recovery. Horm Res 2003;60:43–8.
[37] Woelfle J, Rotwein P. In vivo regulation of growth hormone-stimulated gene transcription by STAT5b. Am J Physiol Endocrinol Metab 2004;286:E393–401.
[38] Kong SE, Firth SM, Baxter RC, et al. Regulation of the acid-labile subunit in sustained endotoxemia. Am J Physiol Endocrinol Metab 2002;283:E692–701.
[39] Barreca A, Ketelslegers JM, Arvigo M, et al. Decreased acid-labile subunit (ALS) levels by endotoxin in vivo and by interleukin-1beta in vitro. Growth Horm IGF Res 1998;8: 217–23.
[40] Bereket A, Wilson TA, Blethen SL, et al. Regulation of the acid-labile subunit of the insulin-like growth factor ternary complex in patients with insulin-dependent diabetes mellitus and severe burns. Clin Endocrinol (Oxf) 1996;44:525–32.
[41] Wilson ME. Insulin-like growth factor I (IGF-I) replacement during growth hormone receptor antagonism normalizes serum IGF-binding protein-3 and markers of bone formation in ovariectomized rhesus monkeys. J Clin Endocrinol Metab 2000;85:1557–62.
[42] Burren CP, Wanek D, Mohan S, et al. Serum levels of insulin-like growth factor binding proteins in Ecuadorean children with growth hormone insensitivity. Acta Paediatr Suppl 1999;88:185–91 [discussion: 192].
[43] Laursen T, Flyvbjerg A, Jorgensen JO, et al. Stimulation of the 150-kilodalton insulin-like growth factor-binding protein-3 ternary complex by continuous and pulsatile patterns of

growth hormone (GH) administration in GH-deficient patients. J Clin Endocrinol Metab 2000;85:4310–4.
[44] Zimmermann EM, Li L, Hoyt EC, et al. Cell-specific localization of insulin-like growth factor binding protein mRNAs in rat liver. Am J Physiol Gastrointest Liver Physiol 2000;278: G447–57.
[45] Blum WF, Albertsson-Wikland K, Rosberg S, et al. Serum levels of insulin-like growth factor I (IGF-I) and IGF binding protein 3 reflect spontaneous growth hormone secretion. J Clin Endocrinol Metab 1993;76:1610–6.
[46] Phillips LS, Pao CI, Villafuerte BC. Molecular regulation of insulin-like growth factor-I and its principal binding protein, IGFBP-3. Prog Nucleic Acid Res Mol Biol 1998;60: 195–265.
[47] Bideci A, Camurdan MO, Cinaz P, et al. Serum zinc, insulin-like growth factor-I and insulin-like growth factor binding protein-3 levels in children with type 1 diabetes mellitus. J Pediatr Endocrinol Metab 2005;18:1007–11.
[48] Hossenlopp P, Segovia B, Lassarre C, et al. Evidence of enzymatic degradation of insulin-like growth factor-binding proteins in the 150K complex during pregnancy. J Clin Endocrinol Metab 1990;71:797–805.
[49] Bang P, Brismar K, Rosenfeld RG, et al. Fasting affects serum insulin-like growth factors (IGFs) and IGF-binding proteins differently in patients with noninsulin-dependent diabetes mellitus versus healthy nonobese and obese subjects. J Clin Endocrinol Metab 1994;78: 960–7.
[50] Le Marchand L, Kolonel LN, Henderson BE, et al. Association of an exon 1 polymorphism in the IGFBP3 gene with circulating IGFBP-3 levels and colorectal cancer risk: the multiethnic cohort study. Cancer Epidemiol Biomarkers Prev 2005;14:1319–21.
[51] Shaarawy M, Fikry MA, Massoud BA, et al. Insulin-like growth factor binding protein-3: a novel biomarker for the assessment of the synthetic capacity of hepatocytes in liver cirrhosis. J Clin Endocrinol Metab 1998;83:3316–9.
[52] Chevalley T, Strong DD, Mohan S, et al. Evidence for a role for insulin-like growth factor binding proteins in glucocorticoid inhibition of normal human osteoblast-like cell proliferation. Eur J Endocrinol 1996;134:591–601.
[53] Bang P, Degerblad M, Thoren M, et al. Insulin-like growth factor (IGF) I and II and IGF binding protein (IGFBP) 1, 2 and 3 in serum from patients with Cushing's syndrome. Acta Endocrinol (Copenh) 1993;128:397–404.
[54] Lissett CA, Shalet SM. The impact of dose and route of estrogen administration on the somatotropic axis in normal women. J Clin Endocrinol Metab 2003;88:4668–72.
[55] Blum WF, Ranke MB, Kietzmann K, et al. A specific radioimmunoassay for the growth hormone (GH)-dependent somatomedin-binding protein: its use for diagnosis of GH deficiency. J Clin Endocrinol Metab 1990;70:1292–8.
[56] Smith WJ, Nam TJ, Underwood LE, et al. Use of insulin-like growth factor-binding protein-2 (IGFBP-2), IGFBP-3, and IGF-I for assessing growth hormone status in short children. J Clin Endocrinol Metab 1993;77:1294–9.
[57] Juul A, Skakkebaek NE. Prediction of the outcome of growth hormone provocative testing in short children by measurement of serum levels of insulin-like growth factor I and insulin-like growth factor binding protein 3. J Pediatr 1997;130:197–204.
[58] Rikken B, van Doorn J, Ringeling A, et al. Plasma levels of insulin-like growth factor (IGF)-I, IGF-II and IGF-binding protein-3 in the evaluation of childhood growth hormone deficiency. Horm Res 1998;50:166–76.
[59] Mitchell H, Dattani MT, Nanduri V, et al. Failure of IGF-I and IGFBP-3 to diagnose growth hormone insufficiency. Arch Dis Child 1999;80:443–7.
[60] Ranke MB, Schweizer R, Elmlinger MW, et al. Significance of basal IGF-I, IGFBP-3 and IGFBP-2 measurements in the diagnostics of short stature in children. Horm Res 2000;54: 60–8.

[61] Nunez SB, Municchi G, Barnes KM, et al. Insulin-like growth factor I (IGF-I) and IGF-binding protein-3 concentrations compared to stimulated and night growth hormone in the evaluation of short children—a clinical research center study. J Clin Endocrinol Metab 1996;81:1927–32.

[62] Rosenfeld RG, Wilson DM, Lee PD, et al. Insulin-like growth factors I and II in evaluation of growth retardation. J Pediatr 1986;109:428–33.

[63] Hasegawa Y, Hasegawa T, Aso T, et al. Clinical utility of insulin-like growth factor binding protein-3 in the evaluation and treatment of short children with suspected growth hormone deficiency. Eur J Endocrinol 1994;131:27–32.

[64] Tillmann V, Buckler JM, Kibirige MS, et al. Biochemical tests in the diagnosis of childhood growth hormone deficiency. J Clin Endocrinol Metab 1997;82:531–5.

[65] Ranke MB, Schweizer R, Elmlinger MW, et al. Relevance of IGF-I, IGFBP-3, and IGFBP-2 measurements during GH treatment of GH-deficient and non-GH-deficient children and adolescents. Horm Res 2001;55:115–24.

[66] Hartman ML, Crowe BJ, Biller BM, et al. Which patients do not require a GH stimulation test for the diagnosis of adult GH deficiency? J Clin Endocrinol Metab 2002;87:477–85.

[67] Span JP, Pieters GF, Sweep CG, et al. Plasma IGF-I is a useful marker of growth hormone deficiency in adults. J Endocrinol Invest 1999;22:446–50.

[68] Aimaretti G, Corneli G, Razzore P, et al. Usefulness of IGF-I assay for the diagnosis of GH deficiency in adults. J Endocrinol Invest 1998;21:506–11.

[69] Janssen YJ, Frolich M, Roelfsema F. A low starting dose of genotropin in growth hormone-deficient adults. J Clin Endocrinol Metab 1997;82:129–35.

[70] de Boer H, Blok GJ, Popp-Snijders C, et al. Diagnosis of growth hormone deficiency in adults. Lancet 1994;343:1645–6.

[71] Juul A, Kastrup KW, Pedersen SA, et al. Growth hormone (GH) provocative retesting of 108 young adults with childhood-onset GH deficiency and the diagnostic value of insulin-like growth factor I (IGF-I) and IGF-binding protein-3. J Clin Endocrinol Metab 1997;82:1195–201.

[72] Attanasio AF, Howell S, Bates PC, et al. Confirmation of severe GH deficiency after final height in patients diagnosed as GH deficient during childhood. Clin Endocrinol (Oxf) 2002;56:503–7.

[73] Svensson J, Johannsson G, Bengtsson BA. Insulin-like growth factor-I in growth hormone-deficient adults: relationship to population-based normal values, body composition and insulin tolerance test. Clin Endocrinol (Oxf) 1997;46:579–86.

[74] Bates AS, Evans AJ, Jones P, et al. Assessment of GH status in adults with GH deficiency using serum growth hormone, serum insulin-like growth factor-I and urinary growth hormone excretion. Clin Endocrinol (Oxf) 1995;42:425–30.

[75] Cuneo RC, Judd S, Wallace JD, et al. The Australian Multicenter Trial of Growth Hormone (GH) Treatment in GH-Deficient Adults. J Clin Endocrinol Metab 1998;83:107–16.

[76] Gillberg P, Bramnert M, Thoren M, et al. Commencing growth hormone replacement in adults with a fixed low dose. Effects on serum lipoproteins, glucose metabolism, body composition, and cardiovascular function. Growth Horm IGF Res 2001;11:273–81.

[77] Kehely A, Bates PC, Frewer P, et al. Short-term safety and efficacy of human GH replacement therapy in 595 adults with GH deficiency: a comparison of two dosage algorithms. J Clin Endocrinol Metab 2002;87:1974–9.

[78] Sassolas G, Chazot FB, Jaquet P, et al. GH deficiency in adults: an epidemiological approach. Eur J Endocrinol 1999;141:595–600.

[79] Colao A, Cerbone G, Pivonello R, et al. The growth hormone (GH) response to the arginine plus GH-releasing hormone test is correlated to the severity of lipid profile abnormalities in adult patients with GH deficiency. J Clin Endocrinol Metab 1999;84:1277–82.

[80] Hilding A, Hall K, Wivall-Hellleryd IL, et al. Serum levels of insulin-like growth factor I in 152 patients with growth hormone deficiency, aged 19–82 years, in relation to those in healthy subjects. J Clin Endocrinol Metab 1999;84:2013–9.

[81] Aimaretti G, Corneli G, Rovere S, et al. Insulin-like growth factor I levels and the diagnosis of adult growth hormone deficiency. Horm Res 2004;62(Suppl 1):26–33.
[82] Fukuda I, Hizuka N, Itoh E, et al. Acid-labile subunit in growth hormone excess and deficiency in adults: evaluation of its diagnostic value in comparison with insulin-like growth factor (IGF)-I and IGF-binding protein-3. Endocr J 2002;49:379–86.
[83] Marzullo P, Di Somma C, Pratt KL, et al. Usefulness of different biochemical markers of the insulin-like growth factor (IGF) family in diagnosing growth hormone excess and deficiency in adults. J Clin Endocrinol Metab 2001;86:3001–8.
[84] Kim HJ, Kwon SH, Kim SW, et al. Diagnostic value of serum IGF-I and IGFBP-3 in growth hormone disorders in adults. Horm Res 2001;56:117–23.
[85] Hoffman DM, O'Sullivan AJ, Baxter RC, et al. Diagnosis of growth-hormone deficiency in adults. Lancet 1994;343:1064–8.
[86] Baum HB, Biller BM, Katznelson L, et al. Assessment of growth hormone (GH) secretion in men with adult-onset GH deficiency compared with that in normal men—a clinical research center study. J Clin Endocrinol Metab 1996;81:84–92.
[87] Juul A, Moller S, Mosfeldt-Laursen E, et al. The acid-labile subunit of human ternary insulin-like growth factor binding protein complex in serum: hepatosplanchnic release, diurnal variation, circulating concentrations in healthy subjects, and diagnostic use in patients with growth hormone deficiency. J Clin Endocrinol Metab 1998;83:4408–15.
[88] Granada ML, Murillo J, Lucas A, et al. Diagnostic efficiency of serum IGF-I, IGF-binding protein-3 (IGFBP-3), IGF-I/IGFBP-3 molar ratio and urinary GH measurements in the diagnosis of adult GH deficiency: importance of an appropriate reference population. Eur J Endocrinol 2000;142(3):243–53.
[89] Rosenfeld RG, Kemp SF, Hintz RL. Constancy of somatomedin response to growth hormone treatment of hypopituitary dwarfism, and lack of correlation with growth rate. J Clin Endocrinol Metab 1981;53:611–7.
[90] de Muinck Keizer-Schrama SM, Rikken B, Wynne HJ, et al. Dose-response study of biosynthetic human growth hormone (GH) in GH-deficient children: effects on auxological and biochemical parameters. Dutch Growth Hormone Working Group. J Clin Endocrinol Metab 1992;74:898–905.
[91] Kristrom B, Jansson C, Rosberg S, et al. Growth response to growth hormone (GH) treatment relates to serum insulin-like growth factor I (IGF-I) and IGF-binding protein-3 in short children with various GH secretion capacities. Swedish Study Group for Growth Hormone Treatment. J Clin Endocrinol Metab 1997;82:2889–98.
[92] Wikland KA, Kristrom B, Rosberg S, et al. Validated multivariate models predicting the growth response to GH treatment in individual short children with a broad range in GH secretion capacities. Pediatr Res 2000;48:475–84.
[93] Juul A, Andersson AM, Pedersen SA, et al. Effects of growth hormone replacement therapy on IGF-related parameters and on the pituitary-gonadal axis in GH-deficient males. A double-blind, placebo-controlled crossover study. Horm Res 1998;49:269–78.
[94] de Boer H, Blok GJ, Popp-Snijders C, et al. Monitoring of growth hormone replacement therapy in adults, based on measurement of serum markers. J Clin Endocrinol Metab 1996;81:1371–7.
[95] Abs R, Bengtsson BA, Hernberg-Stahl E, et al. GH replacement in 1034 growth hormone deficient hypopituitary adults: demographic and clinical characteristics, dosing and safety. Clin Endocrinol (Oxf) 1999;50:703–13.
[96] Chipman JJ, Attanasio AF, Birkett MA, et al. The safety profile of GH replacement therapy in adults. Clin Endocrinol (Oxf) 1997;46:473–81.
[97] Toogood AA, Shalet SM. Growth hormone replacement therapy in the elderly with hypothalamic-pituitary disease: a dose-finding study. J Clin Endocrinol Metab 1999;84:131–6.
[98] Burman P, Johansson AG, Siegbahn A, et al. Growth hormone (GH)-deficient men are more responsive to GH replacement therapy than women. J Clin Endocrinol Metab 1997;82:550–5.

[99] Drake WM, Coyte D, Camacho-Hubner C, et al. Optimizing growth hormone replacement therapy by dose titration in hypopituitary adults. J Clin Endocrinol Metab 1998;83:3913–9.
[100] Cook DM, Ludlam WH, Cook MB. Route of estrogen administration helps to determine growth hormone (GH) replacement dose in GH-deficient adults. J Clin Endocrinol Metab 1999;84:3956–60.
[101] Smith WJ, Underwood LE, Clemmons DR. Effects of caloric or protein restriction on insulin-like growth factor-I (IGF-I) and IGF-binding proteins in children and adults. J Clin Endocrinol Metab 1995;80:443–9.
[102] Busby WH, Snyder DK, Clemmons DR. Radioimmunoassay of a 26,000-dalton plasma insulin-like growth factor-binding protein: control by nutritional variables. J Clin Endocrinol Metab 1988;67:1225–30.
[103] Ono T, Kanzaki S, Seino Y, et al. Growth hormone (GH) treatment of GH-deficient children increases serum levels of insulin-like growth factors (IGFs), IGF-binding protein-3 and -5, and bone alkaline phosphatase isoenzyme. J Clin Endocrinol Metab 1996;81:2111–6.
[104] Frystyk J, Hojlund K, Rasmussen KN, et al. Development and clinical evaluation of a novel immunoassay for the binary complex of IGF-I and IGF-binding protein-1 in human serum. J Clin Endocrinol Metab 2002;87:260–6.
[105] Frystyk J, Skjaerbaek C, Dinesen B, et al. Free insulin-like growth factors (IGF-I and IGF-II) in human serum. FEBS Lett 1994;348:185–91.
[106] Juul A, Holm K, Kastrup KW, et al. Free insulin-like growth factor I serum levels in 1430 healthy children and adults, and its diagnostic value in patients suspected of growth hormone deficiency. J Clin Endocrinol Metab 1997;82:2497–502.
[107] Frystyk J. Free insulin-like growth factors–measurements and relationships to growth hormone secretion and glucose homeostasis. Growth Horm IGF Res 2004;14:337–75.
[108] Frystyk J, Vestbo E, Skjaerbaek C, et al. Free insulin-like growth factors in human obesity. Metabolism 1995;44:37–44.
[109] Skjaerbaek C, Frystyk J, Moller J, et al. Free and total insulin-like growth factors and insulin-like growth factor binding proteins during 14 days of growth hormone administration in healthy adults. Eur J Endocrinol 1996;135:672–7.
[110] Lee PD, Durham SK, Martinez V, et al. Kinetics of insulin-like growth factor (IGF) and IGF-binding protein responses to a single dose of growth hormone. J Clin Endocrinol Metab 1997;82:2266–74.
[111] Nam SY, Lee EJ, Kim KR, et al. Effect of obesity on total and free insulin-like growth factor (IGF)-1, and their relationship to IGF-binding protein (BP)-1, IGFBP-2, IGFBP-3, insulin, and growth hormone. Int J Obes Relat Metab Disord 1997;21:355–9.
[112] Argente J, Caballo N, Barrios V, et al. Multiple endocrine abnormalities of the growth hormone and insulin-like growth factor axis in prepubertal children with exogenous obesity: effect of short- and long-term weight reduction. J Clin Endocrinol Metab 1997;82:2076–83.
[113] Janssen JA, Jacobs ML, Derkx FH, et al. Free and total insulin-like growth factor I (IGF-I), IGF-binding protein-1 (IGFBP-1), and IGFBP-3 and their relationships to the presence of diabetic retinopathy and glomerular hyperfiltration in insulin-dependent diabetes mellitus. J Clin Endocrinol Metab 1997;82:2809–15.
[114] Frystyk J, Ivarsen P, Skjaerbaek C, et al. Serum-free insulin-like growth factor I correlates with clearance in patients with chronic renal failure. Kidney Int 1999;56:2076–84.
[115] Jehle PM, Ostertag A, Schulten K, et al. Insulin-like growth factor system components in hyperparathyroidism and renal osteodystrophy. Kidney Int 2000;57:423–36.
[116] Lassarre C, Binoux M. Insulin-like growth factor binding protein-3 is functionally altered in pregnancy plasma. Endocrinology 1994;134:1254–62.
[117] Skjaerbaek C, Frystyk J, Orskov H, et al. Differential changes in free and total insulin-like growth factor I after major, elective abdominal surgery: the possible role of insulin-like growth factor-binding protein-3 proteolysis. J Clin Endocrinol Metab 1998;83:2445–9.

[118] Clemmons DR, Van Wyk JJ, Ridgway EC, et al. Evaluation of acromegaly by radioimmunoassay of somatomedin-C. N Engl J Med 1979;301:1138–42.
[119] Barkan AL, Beitins IZ, Kelch RP. Plasma insulin-like growth factor-I/somatomedin-C in acromegaly: correlation with the degree of growth hormone hypersecretion. J Clin Endocrinol Metab 1988;67:69–73.
[120] Thissen JP, Ketelslegers JM, Maiter D. Use of insulin-like growth factor-I (IGF-I) and IGF-binding protein-3 in the diagnosis of acromegaly and growth hormone deficiency in adults. Growth Regul 1996;6:222–9.
[121] Paramo C, Andrade OM, Fluiters E, et al. Comparative study of insulin-like growth factor-I (IGF-I) and IGF-binding protein-3 (IGFBP-3) level and IGF-I/IGFBP-3 ratio measurements and their relationship with an index of clinical activity in the management of patients with acromegaly. Metabolism 1997;46:494–8.
[122] Stoffel-Wagner B, Springer W, Bidlingmaier F, et al. A comparison of different methods for diagnosing acromegaly. Clin Endocrinol (Oxf) 1997;46:531–7.
[123] Jorgensen JO, Moller N, Moller J, et al. Insulin-like growth factors (IGF)-I and -II and IGF binding protein-1, -2, and -3 in patients with acromegaly before and after adenomectomy. Metabolism 1994;43:579–83.
[124] Grinspoon S, Clemmons D, Swearingen B, et al. Serum insulin-like growth factor-binding protein-3 levels in the diagnosis of acromegaly. J Clin Endocrinol Metab 1995;80:927–32.
[125] Feelders RA, Bidlingmaier M, Strasburger CJ, et al. Postoperative evaluation of patients with acromegaly: clinical significance and timing of oral glucose tolerance testing and measurement of (free) insulin-like growth factor I, acid-labile subunit, and growth hormone-binding protein levels. J Clin Endocrinol Metab 2005;90:6480–9.
[126] van der Lely AJ, de Herder WW, Janssen JA, et al. Acromegaly: the significance of serum total and free IGF-I and IGF-binding protein-3 in diagnosis. J Endocrinol 1997;155(Suppl 1):S9–13 [discussion: S15–6].
[127] Kaal A, Frystyk J, Skjaerbaek C, et al. Effects of intramuscular microsphere-encapsulated octreotide on serum growth hormone, insulin-like growth factors (IGFs), free IGF's and IGF-binding proteins in acromegalic patients. Metab 1995;44(Suppl 1):6–19.
[128] Lindholm J, Giwercman B, Giwercman A, et al. Investigation of the criteria for assessing the outcome of treatment in acromegaly. Clin Endocrinol 1987;27:553–62.
[129] Trainer PJ, Drake WM, Katznelson L, et al. Treatment of acromegaly with the growth hormone-receptor antagonist pegvisomant. N Engl J Med 2000;342:1171–7.
[130] Wass JA, Clemmons DR, Underwood LE, et al. Changes in circulating somatomedin-C levels in bromocriptine-treated acromegaly. Clin Endocrinol (Oxf) 1982;17:369–77.
[131] Jasper H, Pennisi P, Vitale M, et al. Evaluation of disease activity by IGF-I and IGF binding protein-3 (IGFBP3) in acromegaly patients distributed according to a clinical score. J Endocrinol Invest 1999;22:29–34.
[132] Arafah BM, Rosenzweig JL, Fenstermaker R, et al. Value of growth hormone dynamics and somatomedin C (insulin-like growth factor I) levels in predicting the long-term benefit after transsphenoidal surgery for acromegaly. J Lab Clin Med 1987;109:346–54.
[133] Holdaway IM, Rajasoorya RC, Gamble GD. Factors influencing mortality in acromegaly. J Clin Endocrinol Metab 2004;89:667–74.
[134] Swearingen B, Barker FG II, Katznelson L, et al. Long-term mortality after transsphenoidal surgery and adjunctive therapy for acromegaly. J Clin Endocrinol Metab 1998;83:3419–26.
[135] Biermasz NR, Dekker FW, Pereira AM, et al. Determinants of survival in treated acromegaly in a single center: predictive value of serial insulin-like growth factor I measurements. J Clin Endocrinol Metab 2004;89:2789–96.
[136] Ayuk J, Clayton RN, Holder G, et al. Growth hormone and pituitary radiotherapy, but not serum insulin-like growth factor-I concentrations, predict excess mortality in patients with acromegaly. J Clin Endocrinol Metab 2004;89:1613–7.

Growth Hormone Treatment of Non–Growth Hormone-Deficient Growth Disorders

Charmian A. Quigley, MBBS, FRACP

Lilly Research Laboratories, Drop Code 5015, Lilly Corporate Center, Indianapolis, IN 46285, USA

When human growth hormone (GH) was first introduced as a therapeutic agent in 1958 [1], its use was restricted to children who had the most severe forms of GH deficiency associated with hypopituitarism. Availability was limited by the supply of the hormone, because of its human source and the complex and time-consuming extraction and purification procedures required. Nevertheless, the 1983 International Conference on Uses and Abuses of Growth Hormone noted a need for studies in "short children who do not have growth hormone deficiency" [2]. After the withdrawal of pituitary-derived human GH in the wake of the Jacob-Creutzfelt disease tragedy, and the subsequent introduction of recombinant DNA-derived forms of GH in the mid-1980s, treatment became available for children who had less severe forms of GH deficiency. In 1985, recombinant methionyl GH (ie, GH with an additional methionine residue at the amino-terminal end of the molecule, with the generic name "somatrem") was approved for treatment of pediatric patients whose growth failure resulted from "inadequate secretion of endogenous GH." Natural-sequence recombinant DNA–derived GH (somatropin, the exact 191–amino acid sequence of human pituitary somatotropin) was first approved for treatment of pediatric GH deficiency in 1987.

With greater availability of GH products, the potential effectiveness of GH treatment for impaired growth caused by various nonhypopituitary conditions began to be investigated. Spurred by the recommendation from

---

Based on the conversion factors 2.7 IU = 1 mg (the standard IU/mg conversion until 1999, the period during which most of the studies discussed herein were performed) and 1 m^2 = 27.25 kg (based on fiftieth percentile height and weight values for an 8-year-old male), the following conversions from IU/m^2 to mg/kg have been used for consistency: 1 $IU/m^2/d$ = 0.0136 mg/kg/d = 13.6 µg/kg/d = 0.0952 mg/kg/wk.

E-mail address: qac@lilly.com

the 1983 international conference, numerous studies were conducted in a variety of non–GH-deficient (GHD) conditions in the first few years after the introduction of recombinant GH. Such initial efforts paved the way for the subsequent approval in the United States (and, variably, elsewhere) for GH treatment of five additional pediatric disorders generally considered not to result from deficiency of GH secretion: chronic renal insufficiency (CRI), Turner syndrome (TS), *short stature homeobox-containing* (*SHOX*) gene deficiency, persistent short stature in children born small for gestational age (SGA), and so-called "idiopathic short stature" (ISS) (Table 1). Although Prader-Willi syndrome (PWS) also is an approved indication for GH treatment, irrespective of GH secretion status, it is not discussed in this article because it generally is thought that patients with PWS have evidence of disturbed hypothalamic control of GH secretion. The five conditions mentioned include three for which there is a recognized medical cause of the short stature or growth failure (CRI, TS, and SHOX deficiency) and two defined solely on the basis of stature or size—SGA and ISS. Acknowledging the varying etiology of the growth problem among non-GHD conditions, the primary goal of treatment of children who have non-GHD growth disorders is to increase height velocity and normalize height relative to peers, not to treat the underlying condition. Therefore, for these non-GHD conditions, GH is used in a pharmacologic manner rather than as physiologic replacement. The approved GH dosage range in the United States across all pediatric indications (including GH deficiency) is 26 to 100 μg/kg/d (0.18–0.70 mg/kg/wk). In all the conditions listed, GH improves height velocity, leading to progressive normalization of height standard deviation (SD) score (SDS) during childhood. In TS and ISS there also is definitive evidence of improvement in adult height.

Although there is much debate in the pediatric endocrine literature regarding the appropriateness of providing GH treatment to children who

Table 1
Non–growth hormone-deficient growth disorders: year of approval of growth hormone treatment and approved dosages in the United States and Europe

Growth disorder	First approved in United States	First approved in Europe	GH dosage range (μg/kg/d)
Chronic renal insufficiency	1993	1995	50
Turner syndrome	1996	1993	50 (Eu)–54 (US)
Small for gestational age	2001	1995 (France) 2003 (EMEA)	35 (Eu)–70 (US)
Idiopathic short stature	2003	NA	Up to 53
SHOX deficiency	2006	Pending	50

Abbreviations: EMEA, European Medicines Evaluation Agency; Eu, Europe; NA, not approved; SHOX, *short stature homeobox-containing* gene; US, United States.

do not have GH deficiency, this complex area of discussion is beyond the scope of this article, which focuses on the published efficacy and safety data.

Definitions

Growth velocity and height velocity

As typically used, the terms "growth velocity" and "height velocity" refer to the same parameter—the annualized rate of linear growth. Growth velocity is probably the more commonly used term, but because it may refer to the growth rate of any physical characteristic (eg, weight, head circumference, height, girth, arm span), this article uses the term "height velocity" to refer to the rate of linear growth.

Growth failure and short stature

Growth failure and short stature are among the most common reasons for which children are evaluated by pediatric endocrinologists. The term "growth failure" refers to a decline in the rate of linear growth (height velocity) that, if persistent, results in short stature. Because height velocity varies considerably with age and sex, there is no exact numerical criterion defining the lower limit of the normal range. Average height velocities at representative ages are ~8 cm/y at age 3 years; ~5 cm/y at age 10 years; and ~9.5 cm/y at pubertal peak. Any child whose height velocity is consistently below average (ie, below the fiftieth percentile or 0 SD for age and sex) will fall progressively away from his or her previously established height percentile. A normal height velocity, however, may be incorrectly assumed to indicate that overall growth is normal. The following hypothetical examples demonstrate that this is not always the case. An 11-year-old boy who presents for evaluation with height at −3.0 SDS and subsequently maintains a fiftieth percentile height velocity will attain an adult height of −2.7 SDS; a similar 11-year-old boy whose height is at −3.0 SDS but who grows with a twenty-fifth percentile height velocity (deemed by some to be the lower limit of normal) will attain an adult height of only −3.2 SDS.

Like the definition of growth failure, the definition of short stature is variable and changes with time and experience; therefore various criteria exist in the literature. Older studies tended to use the fifth or third percentile as the lower limit of the normal height range (equivalent to −1.64 and −1.88 SDS, respectively). More recent studies and guidelines generally use a more stringent definition: the American Association of Clinical Endocrinologists and the American Academy of Pediatrics define short stature as height more than 2.0 SD below the gender-specific population mean [3,4], approximately equivalent to the second percentile. Based on adult height standards for the United States population [5], this height is equivalent to 161.5 cm (63.6 inches) for a man and 150.1 cm (59.1 inches) for a woman.

Final height, near-final height, adult height, final adult height

The use of the terms "final height," "near-final height," "adult height," and "final adult height" in the growth literature is inconsistent and confusing, but, in essence, all terms refer to the same developmental point, the completion of linear growth. There may be some subtle differences, in that "final height" often is used when referring to a specific endpoint achieved during a clinical trial (ie, the final height measurement obtained at the end of the subject's study participation), which may not represent the ultimate adult stature. "Near-final" height acknowledges that growth may be not quite complete. "Adult" height typically refers to an unequivocal endpoint when the epiphyses have fused and no additional height gain is possible. It has been suggested facetiously that the term "final adult height" refers to the last height of an individual while still living, so this term is perhaps best avoided.

Growth hormone deficiency

A discussion of GH treatment of non-GHD growth disorders should begin by defining the term "GH deficiency," but this task would represent the conclusion of an unresolved 20-year debate among pediatric endocrinologists. Unlike simple blood chemistry parameters, there is no clear normal range for serum GH concentration. The elegant but complex positive and negative feedback mechanisms controlling physiologic GH secretion, and the resultant endogenous pulsatility, make it impossible to interpret random, single blood levels. The problem of defining normality is compounded by the variation in GH secretion with time of day, nutrition, physical activity, age, sex steroid hormone status, and a variety of other factors. Consequently, evaluation of pituitary GH reserve traditionally has been based on response to stimulation testing using a variety of pharmacologic agents. The inappropriateness of defining an individual as "GH deficient" or "not GH deficient" on the basis of a single numeric threshold on one or two nonphysiologic tests of GH reserve has been well recognized and much discussed [6–12].

When GH was scarce, there was little doubt or controversy about the diagnosis of childhood GH deficiency. Affected patients had profound growth failure associated with hypopituitarism and essentially imperceptible GH secretion. Since the introduction of recombinant human GH, however, the diagnosis has broadened to include children who have much milder forms of reduced GH secretion, so-called "partial" GH deficiency. The response to provocative stimulation used as the criterion for diagnosis of GH deficiency slowly crept up from a threshold of 3 µg/L to 5 µg/L, then to 7 µg/L, and eventually to 10 µg/L over the years, for no apparent scientific reason. Indeed, one school of thought argues that GH stimulation testing should be abandoned altogether [13]. Perhaps most logical and pragmatic is the Australian approach, in place since 1988 [14]. Under Australian regulations, auxologic criteria are used exclusively to define treatment eligibility and to mandate discontinuation of therapy. Australian pediatric endocrinologists nevertheless continue to perform GH stimulation tests to understand an individual patient's GH secretion status and compare outcomes among patients relative to their GH response to provocative testing [14,15]. Reopening the debate, much less attempting to resolve it, is beyond the scope of this article. For discussion regarding the outcomes of GH treatment in non-GHD growth disorders, this article assumes that the absence of GH deficiency has been confirmed by standard methods, applying current, locally accepted diagnostic criteria, whatever they may be.

Mechanism of exogenous growth hormone action in non–growth hormone-deficient growth disorders

GH has both direct and indirect actions on linear bone growth [16,17]. The direct effects are those that occur through GH interactions with GH receptors (GHR) at the epiphysis itself. The indirect effects are those that occur through the production of GH-regulated peptides, particularly insulin-like growth factor-I (IGF-I), either locally or by the liver (Fig. 1). More specifically, GH has a direct stimulatory effect that makes prechondrocytes (largely undifferentiated stem cells) responsive to IGF-I, whereas IGF-I causes chondrocytes to replicate and multiply in the proliferative zone of the epiphysis [17].

Children who have non-GHD growth disorders may have subtle defects at one of many points along the GH–IGF growth plate axis (Fig. 1). When

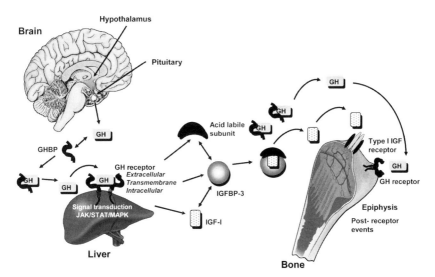

Fig. 1. The growth hormone–insulin-like growth factor–growth plate axis. GH, growth hormone; GHBP, growth hormone binding protein; IGF, insulin-like growth factor; IGFBP, insulin-like growth factor binding protein; JAK, Janus kinase; MAPK, mitogen-activated protein kinase; STAT, signal transducer and activator of transcription.

used to improve linear growth or to correct short stature in children who do not have classic GH deficiency, however, exogenous GH is used as a pharmacologic agent rather than as a physiologic replacement. The rationale for GH treatment in non-GHD patients relates to the strong dose-dependent growth-promoting effect of GH [18–21].

Exogenous GH is believed to act through the same GHR as endogenous GH and to induce the same intracellular signaling response, irrespective of the underlying disease-specific defects. The GH treatment effect in most cases is associated with increased serum concentrations of the GH-dependent peptides, IGF-I, and insulin-like growth factor binding protein (IGFBP)-3 [18,20,21]. In both mice [22] and humans, however, some of the growth-promoting effects of GH are clearly independent of hepatic IGF-I production, because the correlations between height gain in various GH-treated growth disorders and increases in circulating IGF-I (standardized for age and sex) explain only a portion of the variations in height in most studies [23–26]; no correlation at all has been reported in some studies of TS and SGA [27,28].

In addition to the changes in IGF-I, factors suggested to influence GH treatment outcome include differential responses mediated by the polymorphic short form of the GHR encoded by an isoform of the *GHR* gene that lacks exon 3 (d3GHR) [29] and other postreceptor factors. Apart from intrinsic individual variations in responsiveness, the magnitude of the growth response to GH depends on other external factors including frequency of administration, concomitant replacement of other hormones, nutritional status, age at treatment initiation, GH dose, and duration of GH treatment.

Growth hormone treatment of non–growth hormone-deficient short stature

Chronic renal insufficiency

Renal dysfunction is categorized as CRI when the glomerular filtration rate (GFR) is between 25 and 75 mL/min/1.73 m^2 and as the more severe chronic renal failure (CRF) when the GFR is between 5 and 25 mL/min/1.73 m^2 [30]. With respect to the growth disorder associated with renal disease, however, the term "CRI" generally is used to encompass all severities of renal dysfunction and is used in that sense in this article. CRI and CRF in childhood are rare, with an estimated annual incidence in individuals under 20 years old of approximately 12 new cases per million of the age-related population [31]. Growth failure is common in CRI, which was the first non-GHD growth disorder for which GH was approved for use by the US Food and Drug Administration (US FDA), in 1993. This milestone opened the way for the subsequent US FDA approval of the use of GH in four additional non-GHD growth disorders over the next 13 years. GH treatment for CRI was approved by European regulatory authorities in 1995.

Growth pattern and untreated adult height

In 2005 the North American Pediatric Renal Transplant Cooperative Study (NAPRTCS) reported that height fell below the third percentile (−1.88 SDS) in 36% of children who had CRI [32]. Notably, growth failure in CRI is associated with increased morbidity and mortality [32,33], probably reflecting greater severity of renal disease. The growth pattern in children who have CRI varies, depending on the age of onset of renal insufficiency. When the disease develops in infancy, growth failure occurs during the first 2 years of life, mainly because of nutritional deficiency, and is then followed by a period of relatively normal height velocity (ie, height increasing parallel to the standard percentile channels) but without evidence of catch-up growth [34,35]. Subsequently, abnormal pubertal growth compounds the height deficit: pubertal growth has been reported to be delayed by 2.5 years and to be of lower magnitude (ie, reduced total height gain); because of reduced peak height velocity and shorter duration of the pubertal growth period, total pubertal height gain is reduced to only about 50% of normal [36,37]. Adult height is below the expected genetic potential in the majority and is below −2.0 SDS in about half of affected individuals [36,38].

Etiology of the growth disturbance

Many factors contribute to the subnormal linear growth that accompanies CRI in childhood, including the specific etiology of the renal disease, protein-calorie malnutrition, acid-base disturbances, hyperparathyroidism, and derangements in the GH–IGF axis [30,32,35,39,40]. There is an apparent alteration in sensitivity to endogenous GH, the concentrations of which are elevated, primarily because of reduced renal GH clearance [41–45], accompanied in one study by an increase in GH secretion rate [44]. In addition, hepatic GHR expression seems to be decreased, as evidenced by reduced concentrations of circulating GH-binding protein (GHBP) [46–48], which derives from proteolytic cleavage of the extracellular domain of the GHR. Furthermore, there is evidence of impaired function of the GHR intracellular signaling mechanism (the Janus kinase/signal transducer and activator of transcription system) [49]. This state of GH insensitivity is manifest biochemically by reduced IGF-I:IGFBP ratios, resulting in decreased free IGF-I concentrations. Specifically, there have been reports of increased concentrations of IGFBP-1, -2, -3, and -6 in CRI, which, like the increased GH concentrations, result from reduced renal clearance [35,50–52]. A summary of the proposed pathophysiology of the growth disturbance in CRI is provided in Fig. 2.

Efficacy of growth hormone treatment

Because of the significant early-onset growth failure of children who have CRI, studies of GH treatment to improve linear growth in this condition began soon after the availability of recombinant GH. The first data reported

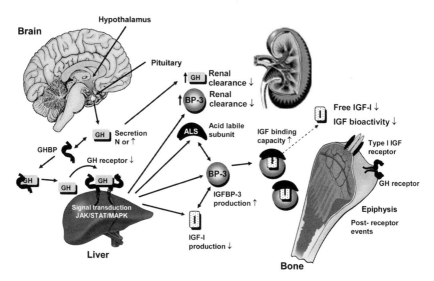

Fig. 2. The growth hormone–insulin-like growth factor–growth plate axis in renal insufficiency. GH, growth hormone; GHBP, growth hormone binding protein; IGF, insulin-like growth factor; IGFBP, insulin-like growth factor binding protein; JAK, Janus kinase; MAPK, mitogen-activated protein kinase; STAT, signal transducer and activator of transcription.

significant short term gains (ie, height velocity increase) in a small, open-label study using GH doses greater than those used for treatment of GH deficiency at the time, reflecting the partial GH resistance that characterizes CRI [53]. Initial small-scale proof-of-concept studies in Europe and the United States [54–56], including a Dutch double-blind, placebo-controlled, crossover study [57], generally demonstrated a doubling of pretreatment height velocity in the first year of treatment. These findings were supported by subsequent larger uncontrolled [58] and controlled trials ([59,60], reviewed in [38]).

The US FDA approval for GH treatment of children who have CRI was based on one multicenter, randomized, double-blind, placebo-controlled trial [59] and one multicenter, open-label, randomized trial [60] designed to determine the effect of GH on growth rate and height deficit. The specific indication for GH treatment in CRI is, in fact, not short stature (ie, height below a particular threshold) but rather growth failure (ie, subnormal height velocity), the exact wording being, "the treatment of growth failure associated with chronic renal insufficiency up to the time of renal transplantation" [61]. Both studies used a GH dose of 50 µg/kg/d (0.35 mg/kg/wk) administered daily by subcutaneous injection. Across the two studies 62 patients received 2 years of GH, and 28 received either placebo or no injections. Overall there was a highly significant difference in mean first-year height velocity between the GH-treated and untreated patients (GH, 10.8 cm/y; non-GH, 6.5 cm/y; $P < 0.00005$). In the second year the GH-treated growth rate waned, but the GH-treated patients still grew significantly faster than the untreated patients. Similarly, there was a significant increase in mean height SDS in the GH group (from -2.9 at baseline to -1.5 at 2 years) but no significant change in the controls (from -2.8 at baseline to -2.9 at 2 years).

Although treatment approval was granted on the basis of only short-term data, subsequent follow-up of a small number of GH-treated patients in the NAPRTCS database suggested that GH improves adult height by up to 0.7 SDS (approximately 5 cm) [62]. Similarly, data from a long-term Dutch study demonstrated progressive improvement of height SDS in a small group of children, with average height SDS approaching target height (ie, mid-parental height) SDS by 6 years of treatment [63]. A thorough meta-analysis of randomized, controlled trials of GH treatment of children with CRI concluded that all the catch-up growth occurred in the first year of treatment and that there was no further catch-up thereafter [38]. Continued GH treatment probably serves primarily to prevent the progressive growth failure that otherwise would occur and to avoid the catch-down growth pattern typically seen when GH is discontinued [64]. Review of the limited data available indicates that adult height may be improved by an average of 1.0 to 1.5 SDS (approximately 7 to 11 cm) [62,65,66]. Definitive data on the effect of GH on the adult height of patients who have CRI during childhood are absent and are unlikely to be forthcoming, because no randomized, controlled study to adult height has been undertaken in this condition.

The beneficial effect of exogenous GH treatment in the face of elevated endogenous GH concentrations is believed to result at least in part from the relatively greater GH-stimulated increase in IGF-I versus IGFBP-3 secretion, thereby enhancing free IGF-I concentrations and IGF-I bioactivity [55,60]. There also is a positive effect of the reversal of the catabolic state associated with uremia [18,54]. The GH treatment effect will be suboptimal, however, unless other aspects of the uremic state, such as nutritional deficiencies and acid-base disturbances, are addressed.

Safety of growth hormone treatment

There are two primary concerns regarding the safety of GH treatment in children who have CRI: deterioration of renal function and disturbances of carbohydrate metabolism. After about 20 years of experience GH treatment is considered safe in children who have CRI. A study from the NAPRTCS revealed no increase in the rate of adverse events in GH-treated children versus non–GH-treated children [67]. In one small study, plasma creatinine increased in a few children, so they were withdrawn from the study [68]. The primary registration studies, however, reported no deterioration of renal function in GH-treated children [64]. Although GH treatment is approved for use in CRI only before renal transplantation, a number of studies also have treated children after transplantation [62,69–71]. This fact is mentioned because some reports suggest an increase in the rate of graft rejection in GH-treated children, especially those who have a previous history of rejection episodes [69,71,72]. Overall, despite the general impression of the safety of GH treatment in children who have CRI, careful monitoring of renal function during such treatment is mandatory.

The second major concern regarding GH treatment of children with CRI is the potential for induction of glucose intolerance or type 2 diabetes. GH acts as a physiologic insulin antagonist with respect to carbohydrate metabolism through mechanisms such as stimulation of hepatic gluconeogenesis and suppression of insulin-stimulated glucose uptake by peripheral tissues. As an extension of this physiologic action, supraphysiologic GH concentrations, such as those induced by exogenous GH administration, may increase glucose production significantly enough to stimulate insulin secretion to maintain normoglycemia. In addition, because children with CRI have relative glucose intolerance even in the absence of exogenous GH, carbohydrate metabolism in GH-treated children has been scrutinized closely. Although no cases of overt diabetes were reported in the initial clinical trials [64], or in a review of GH-treated versus non–GH-treated children in the NAPRTCS database [67], analysis of a large multicenter GH-postmarketing surveillance study reported diabetes or impaired glucose tolerance in approximately 1% of GH-treated patients with CRI [73]. Furthermore, because GH treatment in children with CRI does increase fasting insulin concentrations [64,74], careful monitoring is recommended, especially in children who have a family history of diabetes or those receiving concomitant glucocorticoids.

Turner syndrome

Described in 1938 by Henry Turner, the syndrome that bears his name comprises a variable constellation of physical and functional anomalies that result from partial or complete deficiency of the second sex chromosome [75]. Although Turner was the first to report the clinical picture in a group of unrelated patients, 8 years earlier a German physician, Otto Ullrich, had reported a patient with the classical clinical features of TS and made the prescient speculation that the anomalies might be caused by generalized dilatation of the lymphatics [76]. Thus the condition is commonly known as "Ullrich-Turner syndrome" in Europe. TS occurs in approximately 1 of 2000 live female births, making it one of the most common chromosomal disorders [77]. Short stature relative to the height of family members (mid-parental height) or to that of the general female population of the same ethnicity is a consistent feature of individuals with TS. GH treatment of the short stature associated with TS was approved in France in 1990, soon after it was first launched for GH deficiency (1987); approval followed in 1993 in the rest of Europe and in 1996 in the United States.

Growth pattern and untreated adult height

In the decades following Turner's original paper, many studies evaluated growth disturbance of TS in detail, determining that the typical growth pattern in TS is characterized by mild intrauterine growth retardation, with birth weight and length about 1 SD below the mean for healthy females [78–80], slow growth during infancy and childhood [79–81], and absence of the normal pubertal growth spurt [78,82]. Untreated adult women, on average, are 20 cm shorter than the average height of the general female population in the same country [78,83–85] and are about 20 cm shorter than their midparental (target) heights [84,86]. However, there remains a strong correlation between untreated adult height and target height [84,86,87], indicating that despite the impact of X chromosome deficiency, a significant component of linear growth in TS is modulated by non–X chromosomal genes.

In addition to short stature, girls and women with TS have disproportionate growth, with greater shortening of the limbs than trunk, relative widening of the trunk, and relatively large hands and feet [88]. Additional skeletal anomalies such as short neck, scoliosis and kyphosis, cubitus valgus, short metacarpals, Madelung deformity, congenital hip dislocation, and genu valgum are reported with variable frequency [89–91].

Etiology of the growth disturbance

In his 1938 report of the syndrome, Turner described seven patients between the ages of 15 and 23 years who were referred to him for dwarfism and lack of sexual development [75]. Assuming that the short stature and lack of secondary sexual development resulted from pituitary dysfunction, he treated a number of his patients with various pituitary extracts, without significant clinical benefit. Subsequently, a number of studies suggested the existence of deficits in GH secretion in TS [92–96]; however, other work indicated that GH secretion was essentially normal in TS when sex steroid deficiency and relative obesity were accounted for [97,98]. Although GH secretion is probably normal, there is evidence of reduced free IGF-I and increased proteolysis of IGFBP-3 in TS [99], and one study postulated an element of IGF-I resistance [100]. Despite some evidence for subtle alterations in IGF-I physiology, the growth deficit of individuals with TS is believed to result primarily from haploinsufficiency of one copy of the *SHOX* gene located within the pseudoautosomal region on the distal short arm of the X (and Y) chromosomes [101–103], as discussed in further detail in the section on SHOX deficiency. Although SHOX deficiency accounts for a significant portion of the growth failure typical of TS, it probably is not the whole explanation. Other factors, such as haploinsufficiency of other X-chromosomal genes, aneuploidy or chromosomal imbalance [104,105], and absence of the normal estrogen-induced increase in endogenous GH secretion that drives the pubertal growth spurt in healthy girls, may contribute.

Efficacy of growth hormone treatment

As mentioned previously, GH treatment to improve the height of girls with TS began with Turner himself [75]. His attempts at treatment, using the best knowledge and technology available at the time, were unsuccessful, probably because of a combination of factors including the use of bovine GH (the only form then available) administered in subtherapeutic doses during the teenage years to girls who had limited growth potential. More than 40 years later, a small, placebo-controlled study, initiated in 1984 but aborted the next year when pituitary GH was withdrawn, formally demonstrated the proof-of-concept that GH could improve height velocity in TS significantly [106]. Subsequently, once recombinant GH became available in the mid-1980s, small-group, uncontrolled studies and large, randomized clinical trials of GH treatment for TS were initiated in the United States, Europe, and Australia [107–109]. Early results demonstrating improvements in height velocity in affected girls fueled enthusiasm for the use of GH in TS, and such treatment was used widely even before formal approval was obtained. In the 20 years after the initial launch of recombinant GH, dozens of papers reported results of GH treatment in TS, at various doses and frequencies of administration, either alone or in varying combinations with the nonaromatizable anabolic steroid oxandrolone or with sex steroids. Despite the plethora of studies, a recent Cochrane Center review identified only four studies in which GH treatment was compared in a randomized fashion with a concurrent nontreatment or placebo control [110]. The US FDA approval for GH treatment of "short stature associated with Turner syndrome" was based on interim data from four long-term, multicenter, clinical trials to adult or near-adult height. Two studies were open-label trials using historical controls [111,112]; one dose-response study included a placebo-control arm for the first 18 months [113]; one open-label single-dose study included an untreated control group to adult height [114].

In the two historically controlled studies [111,112], the long-term effect of GH (0.375 mg/kg/wk [equivalent to 54 μg/kg/d] given in divided doses as either three times per week or daily injections) was determined by comparing adult heights of treated patients with those of age-matched historical controls with TS who had received no growth-promoting therapy, or by comparison with the patients' own projected untreated adult heights (ie, extrapolation of the starting height along the TS-specific height percentile to adult height). The greatest improvement in adult height was seen in patients who received early GH treatment and in whom estrogen replacement to induce feminization was postponed until at least 14 years of age [111]. This improvement resulted in a mean difference in adult height of treated versus historical untreated patients of 7.4 cm after an average 7.6 years of treatment. In the second historically controlled study, patients treated with GH (dosage equivalent to 54 μg/kg/d) starting before 11 years of age were randomly assigned to receive estrogen replacement therapy, beginning at age 12 or 15 years. The adult height of girls whose estrogen replacement began at age 12 years was on average 5.1 cm greater than their pretreatment projected adult height, whereas that of the girls whose estrogen therapy was delayed to 15 years gained an average of 8.4 cm compared with their projected height [112].

In a large United States multicenter, randomized, long-term, dose-response study, 232 subjects with TS received either 0.27 or 0.36 mg/kg/wk in divided doses three or six times per week (dosages equivalent to 39 or 51 µg/kg/d) of GH with either low-dose ethinyl estradiol or oral placebo [113]. The study was placebo-controlled for both GH and estrogen for the first 18 months and remained placebo-controlled for estrogen for its duration. Near-adult height was available for 99 subjects whose bone age was at least 14 years, after an average of 5.5 years of therapy. The average height for the whole group, including some subjects who had received an initial 18 months of placebo injections, was 148.7 cm, representing at least a 5-cm gain over expected untreated adult height. The group that received the higher GH dose (51 µg/kg/d) attained a greater average near-final height of 150.4 cm.

Only one randomized, controlled study of GH treatment in TS has maintained an untreated, parallel control group to adult height [114]. In this landmark Canadian study, treated subjects received a GH dose of 43 µg/kg/d (0.3 mg/kg/wk) from a mean age of 11.7 years. Mean adult height of the GH-treated group was 7.2 cm greater than that of the untreated control group after an average 5.7 years (Fig. 3).

Overall, patients in the registration studies treated to adult or near-adult height achieved average height gains ranging from about 5 to 8 cm over

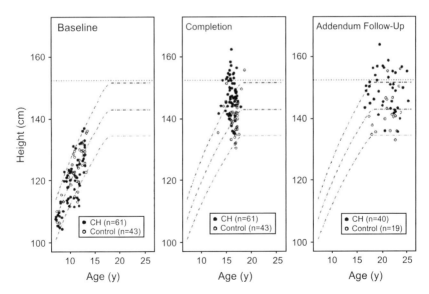

Fig. 3. Growth hormone treatment in Turner syndrome. Results from the Canadian randomized, controlled trial to adult height. The dashed lines represent the 10th, 50th, and 90th percentiles of the standard Turner syndrome growth curve. For reference, the horizontal dotted line represents the height of 5 ft (152.4 cm). GH, growth hormone. (*From:* Stephure DK on behalf or the Canadian Pediatric Endocrine Group. Impact of growth hormone supplementation on adult height in Turner syndrome: results of the Canadian randomized controlled trial. J Clin Endocrinol Metab 2005;90:3363; with permission.)

periods of treatment ranging from 5.5 to 7.6 years. This result may be interpreted as indicating that a height gain of about 1 cm per year of treatment is a reasonable expectation for the long-term outcome of GH treatment in patients who have TS.

Beyond the data from the registration trials, GH treatment results in TS are highly variable. Compared with the height projected at the beginning of treatment or with the height attained by historical control groups of patients with TS, GH treatment has been estimated to have essentially no effect on adult height in some studies [115,116], a modest effect in most, such as the registration studies described previously [111–114], and dramatic effects in a few, with height gains of 15 to 17 cm and normalization of adult height [19,20,27]. The wide variation in apparent treatment effect is probably the combined result of methodologic differences with respect to patient selection, the type of controls used (patients' own projected or predicted adult heights [117] versus historical controls [118] versus non-treatment control [114]), differences in treatment protocols (GH dose and frequency; concomitant treatment with oxandrolone or estrogen), and analytic methods.

On the basis of initial studies demonstrating greater height gains with delayed induction of puberty [111,112], the concept arose that height gain correlates with the duration of estrogen-free GH exposure. This assumption led to the general practice of postponing estrogen substitution well beyond the usual physiologic age of thelarche, with the intent of delaying the onset of estrogen-mediated epiphyseal maturation, thereby maximizing the duration of GH treatment. However, a recent small study using ultra-low-dose estradiol injections initiated at around 12 years of age demonstrated a synergistic effect of physiologic estrogen administration on adult height gain [119]. Furthermore, there is increasing concern about the potential negative effects of delayed estrogen replacement on psychosocial development, bone mineral accrual, uterine development, sexual maturation, self-esteem, and possibly cardiovascular risk [120–122]. Other approaches that avoid delaying feminization, such as earlier initiation of GH treatment or, for those whose GH treatment is unavoidably delayed, concomitant treatment with oxandrolone, may offer better overall benefit when aspects of well-being other than height are considered [121]. A recent randomized, controlled, 2-year study of 88 infants and toddlers with TS demonstrated that early initiation of GH treatment as young as 9 months of age prevented the growth failure that typically occurs in the first few years of life and restored height almost to average for age by as early as 4 to 6 years of age [123]. Although the data from this study so far represent only short-term outcomes, they do provide a rationale for considering early initiation of GH treatment in TS. Thus the updated guidelines for management of TS now state, "Treatment with GH should be considered as soon as growth failure (decreasing height percentiles on the normal curve) is demonstrated, and its potential risks and benefits have been discussed with the family" [121].

In summary, the large body of clinical data demonstrates that GH is effective in increasing short- and long-term height in TS, but there is significant variability in treatment response; patience is necessary, as height gains accumulate slowly over a long duration of treatment.

Safety of growth hormone treatment
Patients who have TS are inherently at increased risk of disorders of carbohydrate metabolism [124–127] and have a specific defect in glucose-stimulated insulin secretion [126]. Although most long-term data regarding exacerbation of such problems by GH in patients with TS have been reassuring [74,128–130], one report based on retrospective analysis of observational data from a large GH-postmarketing study reported a sixfold increase in type 2 diabetes mellitus relative to population rates in GH-treated children and adolescents who had various growth disorders, including TS, suggesting that GH may accelerate development of type 2 diabetes in predisposed individuals [131]. The appropriateness of the literature-derived general population control data used to generate these comparative diabetes rates, as well as other methodologic differences between the studies, have been questioned [132–135]. Furthermore, six of the seven girls with TS who developed diabetes also had received oxandrolone, potentially increasing the risk of insulin resistance [128]. Although there is no firm guidance on prospective monitoring for evidence of glucose intolerance during GH treatment in TS, such girls should have regular, careful review for any signs or symptoms of type 2 diabetes, and a fasting glucose measurement, at least, should be performed if there are any additional risk factors such as obesity or a family history of diabetes [121,136].

A second potential concern regarding GH treatment in TS relates to the increased prevalence of otitis media and hearing deficits, both conductive and sensorineural [137], in TS. In two of the four primary registration studies, a significantly greater rate of otitis media was reported for GH-treated subjects than for control subjects [113,114]. Because the data in both studies were gleaned from spontaneous, retrospective adverse event reports, rather than from prospective data collection, the biologic and clinical significance of the finding was unclear. Nevertheless, because the primary learning modality for girls who have TS is auditory, and otologic problems correlate with reduced self-esteem [122], any potential increase in the risk for hearing deficits poses a significant concern. Reassuringly, a more recent prospective, randomized study that included targeted collection of ear-related data in infants and toddlers with TS found similar rates of otitis media in GH-treated girls and untreated controls [138]. Data from postmarketing registries shed little light on this issue because they lack untreated controls.

Girls with TS have higher risks for scoliosis and kyphosis than the general population; these disorders occur in about 10% to 20% of affected girls and women [90,91]. Both problems may be exacerbated by the rapid increase in linear growth rate stimulated by GH [139]. Other problems seen with increased in frequency in TS include slipped capital femoral epiphysis [140] and an increase in the numbers of acquired benign cutaneous melanocytic nevi [141]. Although there is a hypothetical concern that GH treatment may increase the size or number of such nevi, no evidence has supported this possibility to date [142].

A major concern in caring for patients with TS is the rare but often fatal occurrence of aortic aneurysm, dissection, or rupture in relatively young individuals, usually associated with risk factors such as bicuspid aortic valve, coarctation or dilatation of the aorta, and systemic hypertension [143,144]. Fortunately, no adverse effects of GH treatment on aortic diameter or compliance have been detected by magnetic resonance imaging to date [121,145].

Short stature homeobox-containing gene deficiency

Discovered in 1997 during the search for genes underlying the short stature of TS, the *SHOX* gene, located in the pseudoautosomal regions at the distal ends of the X and Y chromosomes, encodes a homeodomain transcription factor responsible for a significant proportion of long bone growth [101]. In addition to its role underlying the growth deficit of TS, *SHOX* haploinsufficiency also is the primary cause of short stature in 50% to 70% of individuals who have Leri-Weill dyschondrosteosis (LWD) (also known as Leri-Weill syndrome) [146,147] and in about 2% to 3% of patients diagnosed clinically as having ISS [101,103,148–150]. Based on the screening studies published to date, the overall prevalence of SHOX deficiency is estimated to be approximately 1 in 2500 individuals, almost twice the prevalence of TS (the TS prevalence of approximately 1 in 2000 female births represents a population prevalence of 1 in 4000); some studies have reported even higher frequencies of SHOX deficiency in short-statured cohorts [151–153]. Although only recently approved in the United States (2006), GH treatment in SHOX deficiency is discussed here, directly after the discussion of TS, because of the etiologic relationship between these conditions.

Growth pattern and untreated adult height

SHOX deficiency is associated with a broad spectrum of phenotypic effects, ranging from short stature without dysmorphic signs to profound mesomelic skeletal dysplasia, a form of short stature characterized by disproportionate shortening of the middle (mesial) segments of the upper and lower limbs (ie, the forearms and lower legs) [102,149,153–155].

Longitudinal growth data in patients who have various forms of SHOX deficiency are limited and probably are subject to ascertainment bias. One Japanese study of 14 patients ascertained on the basis of short stature reported that skeletal lesions were more severe in females and became more obvious with age [102]. Patients who had the more severe LWD phenotype grew along the standard growth curves before puberty and had a relatively early but attenuated pubertal growth spurt and resultant significantly short stature. In contrast, two patients who had a milder non-LWD ("ISS-like") phenotype grew along a −2 SD growth curve before puberty and had a normal pubertal growth spurt, resulting in adult heights at the lower limits of the normal range. In a study by Ross and colleagues [155] height SDS of subjects who had clinically diagnosed LWD ranged from −5.5 to +0.1. Although the average height of females (−2.3 SDS) was somewhat lower than that of males (−1.8 SDS), the differences were not significant, and there was no significant correlation between height SDS and age.

Etiology of the growth disturbance

The *SHOX* gene encodes a homeodomain transcription factor whose target genes remain to be elucidated. Nevertheless, there is significant evidence for an important role for SHOX as mediator of linear growth. First, the gene is expressed in the developing skeleton during fetal life [156], specifically in bone marrow fibroblasts and proliferating and hypertrophic chondrocytes [101,157–160]. Second, deficiency of SHOX at the growth plate is associated with marked disorganization of chondrocyte proliferation [161]. Third, there is a dose-dependent association between the number of normal copies of the *SHOX* gene and height: *SHOX* gene haploinsufficiency is associated with short stature, whereas *SHOX* gene overdosage, as seen in sex chromosome polyploidy, is associated with tall stature [159,160,162].

At least 50 different mutations in the *SHOX* gene have been reported and are collated in the *SHOX* mutation database [163,164]. Although most reported *SHOX* gene defects are large-scale deletions, there also are heterogeneous intragenic mutations. The effects of SHOX deficiency on height vary markedly even among individuals who have the same molecular defect [165], probably because of epigenetic influences or variable penetrance of the genetically dominant *SHOX* defect.

Efficacy of growth hormone treatment

On the basis of the established effectiveness of GH treatment in TS, GH has been used on an empiric basis in patients with LWD and those with non-syndromic SHOX deficiency since soon after the 1997 discovery of the *SHOX* gene. Experience with GH treatment was limited to case reports and small, uncontrolled studies [162,166,167] until the recent completion of a 2-year, randomized, controlled, multicenter registration trial (Fig. 4) [150]. This study randomly assigned 52 patients who had molecularly proven *SHOX* defects to either GH treatment (n = 27) or no treatment (n = 25). In addition, to compare the response in isolated SHOX deficiency with that in TS, a group of 26 patients with TS also received GH. Mean baseline height SDS was −3.3 (average age, ∼7.4 years), and mean pretreatment height velocity was ∼4.9 cm/y. The GH-treated SHOX deficiency group had a significantly greater first-year height velocity than the untreated control group (mean ± SE: 8.7 ± 0.3 versus 5.2 ± 0.2 cm/y; $P < 0.001$), and first-year height velocity was similar to that in the TS group (8.9 ± 0.4 cm/y; $P = 0.592$). GH-treated subjects also had significantly greater second-year

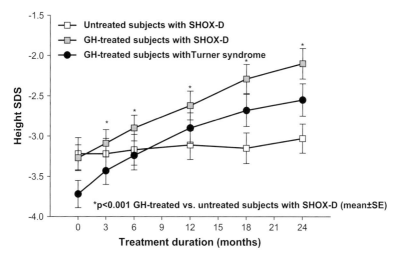

Fig. 4. Effect of 2 years of GH treatment in children with *short stature homeobox-containing gene (SHOX) deficiency* (*square symbols*) and with Turner syndrome (*filled circles*). (*From:* Blum WF, Crowe BJ, Quigley CA, et al. Growth hormone is effective in treatment of short stature associated with short stature homeobox-containing gene deficiency: two-year results of a randomized, controlled, multicenter trial. J Clin Endocrinol Metab 2007;92:222; with permission.)

height velocity (7.3 ± 0.2 versus 5.4 ± 0.2 cm/y; $P < 0.001$), second-year height SDS (-2.1 ± 0.2 versus -3.0 ± 0.2; $P < 0.001$), and second-year height gain (16.4 ± 0.4 cm versus 10.5 ± 0.4 cm; $P < 0.001$) than untreated subjects. This study formed the basis for the 2006 US FDA approval of GH treatment for SHOX deficiency. No data are available yet regarding GH effects on adult height in patients with SHOX deficiency.

Safety of growth hormone treatment

In the registration study discussed previously, no serious adverse events were reported for patients with SHOX deficiency, and treatment-emergent adverse events were consistent with typical childhood illnesses [150]. There was no evidence of any new safety concerns in this patient population, which, apart from short stature and, in some cases, mild-to-moderate skeletal dysplasia, represents an essentially healthy group of children. Because patient numbers were small, the study was only of 2 years' duration, and no postmarketing data have been reported, further follow-up is required.

Children born small for gestational age

The definition of SGA varies in the pediatric literature, from a birth weight below the tenth percentile for gestational age [168] to a birth weight or length at least 2 SD below the mean [169,170]. A recent consensus statement recommends the latter definition, with subclassifications of SGA for weight (SGA$_W$), SGA for length (SGA$_L$) or SGA for both weight and length (SGA$_{WL}$) [171]. Approximately 85% to 90% of infants born SGA undergo catch-up growth to return to their genetic potential within the first or second year of life. The remaining 10% to 15% who do not catch up are eligible for GH treatment. Based on these proportions, the prevalence of "short SGA" is about 2 to 3 in 1000 children or about 1 in 300 to 1 in 500 children. GH treatment for this patient population was approved in the United States in 2001 and in Europe in 2003.

Growth pattern and untreated adult height

The majority of infants born SGA undergo spontaneous catch-up growth in the first few months and years of life [172–176], with catch-up for weight occurring sooner than catch-up for height [177]. Notably, length at 6 months of age is the strongest predictor of catch-up growth and adult height in untreated individuals born SGA [172,176]. There is evidence that hypersecretion of endogenous GH may play a role in inducing the postnatal catch-up [178]. The 10% to 15% of children born SGA who do not catch up have persistent short stature during childhood and, if untreated, in adulthood [172,173,179,180]. These children grow with a low-normal height velocity during childhood [181]; puberty tends to occur at a normal to somewhat early age and to progress rapidly, particularly in girls. The rapid tempo of puberty is accompanied by diminished pubertal height gain [182–186], resulting in adult height about 1 SD (approximately 7 cm) below the mean for the general population [173,175,182,184–187] and ~4 cm below mid-parental (genetic target) height (which nevertheless remains a strong predictor of adult height) [180]. The largest long-term follow-up study of untreated children born SGA determined that about half remain relatively short in adulthood [173]. Notably, about 20% of adults who have short stature were born SGA [172,173].

Etiology of the growth disturbance

Poor fetal growth has a heterogeneous basis. The etiology ranges from defined syndromes, to maternal toxic exposures such as nicotine and alcohol, to undefined "idiopathic" SGA (reviewed in [169]). Potential influences include maternal sociodemographic factors such as age, income, race, overall body size, and uterine space; maternal health and habits such as hypertension, smoking, alcohol consumption, and drug use; fetal factors, such as birth order (firstborn infants typically weigh less than later-born infants) and multiple gestation; and a number of well-recognized syndromes (eg, Russell-Silver syndrome, Dubowitz syndrome) and chromosomal anomalies.

The impaired postnatal growth of short children born SGA is probably multifactorial. Contributing factors may include cellular hypoplasia [188], alterations of the GH–IGF axis manifest by changes in diurnal GH secretion patterns [181,189–192], and presence or absence of d3GHR, which is reported by some to affect GH sensitivity, although this finding remains controversial [193,194]. Significant deletions and mutations and of the *IGF-I* gene have been reported in rare patients who have profound intrauterine growth retardation, microcephaly, deafness, and postnatal growth failure [195–197], highlighting the critical role of IGF-I action in both pre- and postnatal growth (Table 2). Various more subtle alterations in the *IGF-I* gene (eg, polymorphisms, milder missense mutations) and reduced serum concentrations of IGF-I (and to a lesser extent IGFBP-3) have been reported in SGA cohorts [175,198–202]. In addition, one study reported reduced IGF-II expression in a cohort of SGA subjects who had Silver-Russell syndrome but not in the subjects did not have the Silver-Russell syndrome phenotype [203]. This reduced IGF-II expression seems to result from hypomethylation of the telomeric domain of chromosome 11p15 resulting in downregulation of *IGF-II* gene expression. Downstream of the IGFs themselves, there is some evidence of disturbed IGF-I action in children born SGA. Two patients who had intrauterine growth retardation and poor postnatal growth (one with elevated IGF-I) have been reported with mutations (one compound heterozygous, one nonsense) in the gene encoding the type 1 IGF receptor [204]. Because these two cases were detected from screening of fewer than 100 subjects, it seems likely that larger screening studies of children who have otherwise unexplained growth disturbance might detect either heterozygous or polymorphic changes in the *IGF-IR* gene. Genetic defects reported in subjects who have SGA phenotypes are summarized in Table 2.

Efficacy of growth hormone treatment

Because of persistent short stature in a subgroup of children born SGA, initial studies were performed in the late 1960s and early 1970s to investigate the effects of pituitary-derived GH on children who failed to catch up in height [205,206]. The recombinant GH era was characterized by numerous studies, mainly in Europe, that attempted to optimize treatment outcomes using various GH dosing and administration strategies (Fig. 5) [28,207–210] (reviewed in [169] and [211]).

The 2001 US FDA approval of GH for the "long-term treatment of growth failure in children born small for gestational age (SGA) who fail to manifest catch-up growth by age 2" was based on data from four randomized, controlled, open-label, clinical trials that enrolled 209 patients between the ages of 2 and 8 years. Height velocity was followed for 1 year, after which patients were assigned randomly to either GH (34 or 69 μg/kg/d [0.24 or 0.48 mg/kg/wk]) or no treatment for 2 years; thereafter, the untreated patients crossed over into treatment. Children who received GH at either dosage had significant increases in height velocity compared with untreated children; those who received the 69 μg/kg/d GH dose had gained ~ 0.5 SDS more in height after 2 years than children who received 34 μg/kg/d [212].

GH treatment for short children born SGA was approved initially in France in 1995 on the basis of short-term improvements in growth rate

[213] but because of lack of final height data was not approved for general use in Europe until 2003. The criteria for treatment are more stringent than in the United States: height SDS below −2.5, height velocity SDS below 0 during the previous year and a starting age of at least 4 years. Overall, efficacy data from dozens of diverse clinical studies conducted over many decades are available in the literature for thousands of patients born SGA; data for at least 2000 patients have been reported since 1994 (for review see Refs. [169,209,211,214]). Adult height data are far more limited, however. A meta-analysis of three randomized studies comprising 28 patients [209] reported that long-term GH treatment for 7 to 10 years, initiated at around 5 to 8 years of age at doses of 34 to 69 µg/kg/d, can be expected to increase adult height by approximately 1.0 to 1.4 SD. Another meta-analysis analysis of final height data for 56 children born SGA, performed on the basis of a request by European regulatory authorities, demonstrated a mean increase in height of 1.9 SDS for the 34 µg/kg/d dose and 2.2 SDS for the 69 µg/kg/d dose [215]. A larger, randomized, controlled French study reporting adult height data for 91 GH-treated versus 33 untreated adolescents born SGA found a lower between-group difference of 0.6 SDS [216]. GH treatment was started much later in this study, at around 12 years of age, and had a much shorter treatment duration of about 2.7 years. Although results vary among studies and among individual patients, on average GH treatment can be expected to stimulate approximately 1 cm of incremental height for each year of treatment. Factors that may increase overall height gain include greater magnitude of the child's height deficit relative to target height (ie, baseline height SDS minus mid-parental height SDS), greater GH dose (perhaps less important in the maintenance phase of treatment [209]), younger age at start of treatment, longer duration of treatment, and perhaps the presence of the d3GHR allele [169,211,217,218], although this latter point remains controversial. Some data suggest that efficacy of intermittent GH treatment regimens is similar to that of continuous treatment [207,219]; however, continuous treatment, avoiding the potential "catch-down" effect of interrupted treatment [220], is now considered standard of care.

Safety of growth hormone treatment

Like some other pediatric populations with growth disturbances (eg, CRI, TS), children born SGA, particularly those who have rapid postnatal weight gain, have reduced insulin-mediated glucose uptake and a greater likelihood of insulin resistance [177,221–226]. Nevertheless, there is no clear evidence of an increased risk of type 2 diabetes mellitus during childhood [225]. In view of the baseline increase in insulin resistance, there is appropriate concern regarding the effects of exogenous GH on glucose metabolism in this population. Indeed, mild, transient hyperglycemia was reported in a few children in the GH registration studies [212]. In addition, a number of studies have reported increased insulin resistance during GH treatment in children born SGA [227–229] that typically resolves after discontinuation of

Table 2
Molecular defects reported in individuals with phenotype of idiopathic short stature or small for gestational age

Level of defect	Locus	Change	Heterozygous or homozygous	Effect(s)	Phenotype	References
Pituitary	GHSR	Ala204Glu	Heterozygous	Not reported	Obesity	Wang 2004 [320]
	GHSR	Ala204Glu	a. Heterozygous b. Homozygous	Decreased cell surface expression of receptor; decreased constitutive activity	a. IGHD b. Familial SS	Pantel 2006 [259]
	GHSR	Phe279Leu	Heterozygous	Not reported	Obesity and SS	Wang 2004 [320]
	GH1	Polymorphism 5' UTR	Heterozygous	Reduced expression	SS of varying etiology	Counts 2001 [250]
	GH1	Promoter polymorphisms	Heterozygous	Altered promoter activity and GH1 gene expression	a. Normal-stature adult males b. Short stature	Horan 2003 [251]
	GH1	15 missense mutations and polymorphisms	Heterozygous	Reduced GH binding to GHR Reduced JAK/STAT activation	a. IGHD b. ISS	Millar 2003 [252]
	GH1	Ile79Met	Heterozygous	Reduced activation of ERK	Familial SS	Lewis 2004 [321]
	GH1	Cys53Ser	Homozygous	Disruption of disulphide bridge → bioinactive GH Reduced GH binding to GHR Reduced JAK/STAT activation Low IGF-I	ISS	Besson 2005 [258]
	GH1	Asp112Gly	Heterozygous	Bioinactive GH Prevents dimerization of GHR Low IGF-I	Severe ISS	Takahashi 1997 [322]

GHR and its intracellular signaling	GHR	Exon 3-deleted vs full-length	Heterozygous Homozygous	Not reported	SGA	Jensen 2006 [194] Audi 2006 [193]
	GHR	Polymorphisms in coding sequence	Heterozygous Homozygous	Not reported	ISS	Sjoberg 2001 [323]
	ECD	Cys122Ter	Heterozygous	Undetectable GHBP, mildly reduced IGF-I	Moderate ISS in proband, mild SS in carrier mother	Goddard 1995 [262]
		Arg161Cys	Heterozygous	Mildly reduced GHBP and IGF-I	Mild to moderate ISS	Goddard 1997 [263]
		Arg211His	Heterozygous	Undetectable GHBP, severely reduced GHR expression, mildly reduced IGF-I	Severe ISS	Goddard 1995 [262]
		Arg211His	Heterozygous	Elevated GHBP, normal IGF-I	Moderate ISS	Goddard 1997 [263]
		Glu224Asp	Heterozygous	Low GHBP, slight reduction in GHR affinity for GH, low IGF-I	Moderate ISS	Goddard 1995 [262]
		Glu44Lys and Arg161Cys	Compound Heterozygous	Low GHBP, severely reduced IGF-I Glu44Lys: severely reduced affinity for GH; Arg161Cys: minor effect on GH binding	Moderate ISS in proband, mild SS in carrier parents	Goddard 1995 [262]

(continued on next page)

Table 2 (continued)

Level of defect	Locus	Change	Heterozygous or homozygous	Effect(s)	Phenotype	References
GHR and its intracellular signaling (continued)	ECD	Arg161Cys	Heterozygous	Normal GHBP, mildly reduced IGF-I	ISS	Sjoberg 2001 [323]
	ECD	Val144Ile	Heterozygous	Variable GHBP levels	Familial ISS	Sanchez 1988 [324]
		Val144Ala	Heterozygous	Not reported	ISS	Bonioli 2005 [266][a]
	TMD	Splice junction mutation	Heterozygous	Truncated GHR forms heterodimers with full-length GHR → dominant -ve effect	Familial ISS	Ayling 1997 [325]
	ICD	Ala478Thr	Heterozygous	Normal GHBP Truncated GHR cytoplasmic domain	ISS	Goddard 1997 [263]
		3' splice acceptor mutation	Heterozygous	Failure of signal transduction	"Atypical" GHIS	Ayling 1999 [326]
	Stat5b SH2 domain	Ala630Pro	Homozygous	Signal transduction defect	Severe SS, immune dysfunction	Kofoed 2003 [327]

IGF-I and associated proteins	IGF-I	Deletion exons 4, 5 of *IGF-I* gene	Homozygous	Undetectable IGF-I	Severe SS, deafness, mental retardation	Woods 1996 [195]
	IGF-I	T→A transversion in polyadenylation signal (3′ UTR)	a. Homozygous b. Heterozygous	Altered IGF-I mRNA maturation; very low IGF-I (normal IGFBP3)	a. Severe IUGR, SS, microcephaly, deafness, mental retardation b. Mild SS	Bonapace 2003 [196]
	IGF-I	Val44Met	a. Homozygous b. Heterozygous	Bioinactive IGF-I; reduced affinity of IGF-I for receptor	a. SGA and severe SS, deafness, mental retardation; b. Mild SS in heterozygous carriers	Walenkamp 2005 [197]
	ALS	Nucleotide deletion → Glu35Lys and 120ter	Probably homozygous	Increased GH levels; normal GHBP levels; reduced IGF-I, IGFBP-3; undetectable ALS	Mild ISS and delayed puberty	Domene 2004 [271]
	ALS	Asp440Asn	Homozygous	Normal GH and GHBP levels; markedly reduced IGF-I and IGFBP-3; undetectable ALS	ISS; mild learning and speech disorders	Hwa 2006 [273]
	IGF-II	Hypomethylation telomeric 11p15	Not reported	Down-regulation of IGF-II	Silver-Russell syndrome	Netchine 2006 [203]

(continued on next page)

Table 2 (continued)

Level of defect	Locus	Change	Heterozygous or homozygous	Effect(s)	Phenotype	References
Growth plate	Type 1 IGF-R	Arg108Gly Lys115Asn	Compound heterozygous	Reduced IGF-R number; reduced fibroblast IGF-R function	SGA with persistent severe SS	Abuzzahab 2003 [204]
	Type 1 IGF-R	Arg709Gln	Heterozygous	Alters cleavage site → defective processing of proreceptor; reduced IGF-I binding capacity	SGA	Kawashima 2005 [328]
	NPR2	1092delT	Homozygous Heterozygous	Not reported	AMDM Mild SS	Olney 2006 [274]
	SHOX	Various deletions, missense and nonsense mutations	Heterozygous	Deficiency of SHOX transcription factor	LWD ISS	Rappold 2002 [103] Rappold 2007 [103]

Abbreviations: 3′, 3 prime end of complementary DNA; 5′, 5 prime end of complementary DNA; ALS, acid labile subunit (of the ternary complex); AMDM, acromesomelic dysplasia, Maroteaux type; d3 variant, exon 3 deleted form of GHR; ECD, extracellular domain; ERK, extracellular signal-regulated kinase; GH, growth hormone; GHBP, growth hormone binding protein; GHIS, GH insensitivity syndrome; GHR, GH receptor; GHSR, GH secretagogue receptor; ICD, intracellular domain; IGF-I, insulin-like growth factor-I; IGF-II, insulin-like growth factor-II; IGFBP, insulin-like growth factor binding protein; IGHD, isolated GH deficiency; IGF-R, type 1 IGF receptor; ISS, idiopathic short stature; IUGR, intrauterine growth retardation; JAK, Janus kinase; LWD, Leri-Weill dyschondrosteosis; NA, not applicable; NPR2, natriuretic peptide receptor B; SGA, small for gestational age; SH2, src homology 2 domain; *SHOX*, short stature homeobox-containing gene; SS, short stature; STAT5b, signal transducer and activator of transcription 5b; TMD, transmembrane domain; UTR, untranslated region.

[a] For summaries of GHR mutations estimated to be present in ∼5% of patients with apparent ISS, see Bonioli and colleagues [266] and Savage and colleagues [288].

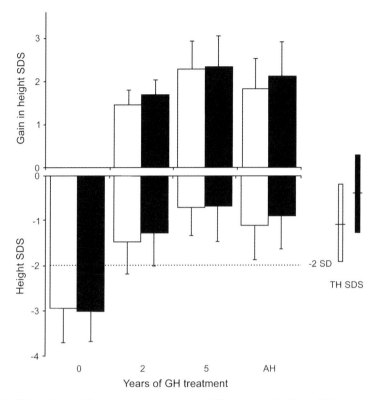

Fig. 5. Effect of growth hormone treatment at two different doses in short children born small for gestational age. (*Upper panel*) Height SDS gains from baseline to 2 years, 5 years, and adult height. (*Lower panel*) Height SDS at baseline, 2 years, 5 years, and adult height. AH, adult height; filled bars, GH dose: 67 µg/kg/d; open bars, GH dose: 33 µg/kg/d; TH, target height. (*From*: van Pareren Y, Mulder P, Houdijk M, et al. Adult height after long-term, continuous growth hormone (GH) treatment in short children born small for gestational age: results of a randomized, double-blind, dose-response GH trial. J Clin Endocrinol Metab 2003;88:3587; with permission.)

GH treatment [230]. Published data from large postmarketing databases for GH-treated children born SGA are limited [231,232]. The most comprehensive report of more than 1900 children born SGA noted increased frequencies of respiratory adverse events in SGA children compared with children with ISS, but no difference in the rates of "abnormal glucose regulation" [232]; no case of diabetes mellitus has been reported during GH treatment in as many as 6000 patient-years of exposure [231–234]. The total number of patient-years of GH treatment is still limited, however, and rates of diabetes mellitus in adulthood remain to be determined. The European product labeling for GH treatment of children born SGA states that "the management of these patients should follow accepted clinical practice and include safety monitoring of fasting insulin and blood glucose before treatment and annually during treatment" [214].

In addition to insulin resistance itself, individuals born SGA have increased risk of metabolic abnormalities in later life, such as dyslipidemia [199], hypertension [233,235] and the metabolic syndrome [236,237]. No information is available on whether GH treatment in childhood increases these risks in adulthood, so long-term follow-up is needed.

Idiopathic short stature

Perhaps the most enigmatic and controversial of all non-GHD growth disorders is so-called "idiopathic short stature," typically referred to as "ISS," although numerous synonyms exist in the literature, including "normal variant short stature," "non-GHD short stature" "constitutional short stature," "small/delay," "GH-dependent growth failure," and "idiopathic growth failure." The complexity of the diagnosis belies the ostensibly innocent etiology inferred by the term "idiopathic," which more accurately denotes a condition that arises "spontaneously" or from an obscure or unknown cause. The diagnosis of ISS is based on a process of exclusion because patients with ISS have no distinguishing clinical or phenotypic features and the probably heterogeneous etiologies are currently not part of standard investigations. By definition, children who have ISS were not born SGA and do not have GH deficiency, TS, CRI, or any other currently diagnosable condition that explains their short stature. Nevertheless, many such children are as short as those who have named growth disorders, so the distinction between idiopathic and defined etiology rests primarily on the clinical acumen of the physician and the depth and sophistication of the diagnostic work-up.

A 1996 consensus meeting among pediatric endocrinologists, auxologists, and statisticians proposed the following criteria for diagnosis of ISS, which remain generally accepted [238]:

> Height more than 2.0 SDS below the mean for age, sex, and population group
> Normal size for gestational age at birth (> -2.0 SD)
> Normal body proportions
> No evidence of chronic organic disease
> No psychiatric disease or severe emotional disturbance
> Normal food intake
> No evidence of endocrine deficiency
> Slow or normal tempo of growth
> Included within this consensus definition are children with so-called constitutional growth delay (also known as constitutional delay of growth and adolescence or constitutional delay of growth and puberty) and those with familial short stature.

GH treatment for ISS has been part of standard practice in Australia since 1993 [14] and was approved by the US FDA in 2003 [239]. Subsequent to the US FDA approval, Canada and a number of other countries have granted approval, but GH treatment for ISS is not yet approved in Europe.

The prevalence of ISS can be estimated on the basis of the following assumptions: (1) about 2% of the population has short stature (height below −2.0 SDS), and (2) approximately 80% of those with short stature have no currently identifiable cause of their growth disorder [240]. With these assumptions, approximately 16 in 1000 individuals may be estimated to have ISS. However, approved height criteria for GH treatment of ISS are more stringent than −2.0 SDS (eg, −2.25 SDS in the United States and Canada), reducing the number who qualify for treatment by 46%, to approximately 9 in 1000. Prescribing data reveal that only about 1% of such children, equivalent to no more than about 1 in 10,000 individuals in the population, is actually prescribed GH treatment for this condition [240].

Untreated growth pattern and adult height

By definition, children who have ISS are of normal size at birth, but they grow slowly during early childhood so average height falls below −2.0 SD by the time they start school. Thereafter, height velocity typically is maintained within the lower half of the normal range, so that children with ISS usually grow below but parallel to the normal percentile channels. At puberty there may be some degree of catch-up growth, but untreated adult height typically remains below the normal range [241–243] and below mid-parental (genetic target) height by about 1 SD [241]. A significantly delayed bone age at presentation should not be taken as reassurance of a good outcome, because height predictions are notoriously inaccurate in this situation [244], and on average patients who remain untreated fail to achieve the adult heights predicted at diagnosis [243,245–248]. The subnormal adult height outcome in untreated ISS is demonstrated most clearly in the placebo-controlled study described in further detail later [249], in which 74% of placebo-treated subjects had adult height below −2.0 SDS and the remaining subjects also were short, with adult heights between the second and fifth percentiles of the general population reference range.

Etiology of the growth disturbance

The term "idiopathic short stature" suggests to some that the diagnosis lacks medical credibility, and the condition is often dismissed as simply a normal variant. In fact, the term has been used for many years as a "black box" for children who have undiagnosed forms of sometimes quite severe short stature. Meanwhile, as molecular methodology has become more sophisticated, and as new growth-related genes have been discovered, many cases of apparent ISS have been found to result from specific defects in one of many genes along the GH-IGF growth plate axis (see Table 2). Counts and Palese [250] reported an association between short stature and a polymorphism in the 5′ untranslated region of the *GH1* gene. Similarly, Horan and colleagues [251] reported polymorphic variations in the promoter region of the *GH1* gene associated with reduced promoter activity in short children. Although mutations of the *GH1* gene are a comparatively rare cause of GH deficiency, more subtle changes may underlie a number of cases of ISS. This possibility is suggested by a study of the *GH1* gene in non-GHD children who had short stature, slow growth, and delayed bone age, in whom a collection of new mutations and polymorphisms was discovered [252]. Indeed, significant evidence for reduced or disorganized GH secretion in children with ISS has existed for more than 20 years [253–257]. In addition, although the hunt for a true bioactive GH was long thought to be futile, the existence of this previously mythical entity was finally verified with the discovery of a mutation that changes a critical cysteine molecule normally involved in the intracellular disulphide bonds to serine, thereby disrupting the tertiary structure of the molecule [258]. The clinical presentation of the affected child was quite compatible with the typical clinical picture of ISS.

Recently, a role for ghrelin and the GH secretagogue receptor (GHSR) in regulation of GH secretion and growth has moved from the speculative to the definitive with the discovery of two unrelated families with a missense mutation in the *GHSR* gene causing an amino acid substitution in the extracellular domain of the GHSR that impairs the constitutive activity of the receptor [259]. Peak stimulated GH in affected short children varied from low (GH peak, 2.5 μg/L) to normal (GH peak, 24 μg/L), indicating that GH synthesis and storage within somatotropes probably is retained, although its spontaneous release may be impaired.

Moving down the GH–IGF axis, there is evidence that the growth disturbance in some children who have normal GH secretion results from reduced responsiveness to endogenous GH on the basis of heterozygous mutations in the *GHR* gene. The finding in cohorts of ISS children of reduced concentrations of GHBP, which is produced by proteolytic cleavage of the extracellular domain of the GHR during cellular processing of the receptor [260–262], suggested the possibility of GHR defects in ISS. Subsequently, this speculation was confirmed in a number of studies [263–266]. In addition to definitive mutations, several *GHR* gene polymorphisms have been reported in children with ISS [263,265,267]. A recent review estimates that as many as 5% of children with the clinical phenotype of ISS may have heterozygous GHR mutations [268].

Downstream of GH secretion and the GHR, data point to heterogenous defects of IGF-I secretion, metabolism, or action. Numerous studies report low IGF-I concentrations in children who have the clinical diagnosis of ISS. In fact overall, about 30% to 50% of children with ISS have subnormal IGF-I concentrations [269,270,249]. In addition, recent reports implicate deficiency of the acid-labile subunit (ALS) of the 150-kD IGF-I/IGFBP-3/ALS ternary complex as a cause of IGF-I deficiency, probably by means of exaggerated metabolism of unbound IGF-I [271–273]. The affected individuals had mildly or moderately subnormal linear growth but profoundly reduced IGF-I, IGFBP-3, and ALS in the face of normal or increased GH responses to stimulation. These findings reflect the physiologic role of ALS in the maintenance of normal IGF-I and IGFBP-3 concentrations.

As indicated by the 2% to 3% prevalence of *SHOX* gene defects in patients who have a clinical phenotype of ISS [103], defects of factors beyond GH, IGF-I, and their receptors also underlie some cases of ISS. This concept is exemplified further by a recent report of a kindred in which heterozygous mutation of the gene encoding natriuretic peptide receptor-B (*NPR2*) was associated with mild short stature, whereas homozygous mutation resulted in a severe form of skeletal dysplasia [274]. This seminal observation demonstrates that haploinsufficiency of a diverse variety of factors previously unsuspected of playing a role in the growth process may explain many cases of currently undiagnosable short stature.

In addition to the variety of genetic defects now demonstrated along the GH–IGF pathway in children who have ISS, there is evidence that acquired disease processes may underlie some cases. Almost one quarter of a group of 60 children with ISS studied by DeBellis and colleagues [275] had antibodies to GH-producing cells. The children with positive antibodies had lower height velocity and lower IGF-I than children who did not have antibodies. This finding suggests that a subset of children diagnosed as having ISS on the basis of normal GH responses to provocative stimuli may in fact represent a developmental stage between normal GH secretion and GH deficiency, similar to children who have subclinical autoimmune thyroid disease or those who have "prediabetes." These findings, in addition to those already described, lend weight to the concept that GH deficiency, GH resistance, and ISS are not separate entities but rather a continuum of variable severity of suboptimal GH secretion or action.

Finally, the important role of nutrition in normal growth cannot be overlooked. Wudy and colleagues [276] reported decreased food responsiveness (using a Child Eating Behavior questionnaire) and a slightly below average body mass index (-0.33 SDS) in 214 children with ISS. Within the group, "poor" eaters had lower body mass indices and more marked food-related behavioral changes than "good" eaters. Because IGF-I secretion is highly nutrition dependent, the low IGF-I concentrations found in many children who have ISS may, at least in part, reflect suboptimal nutritional status (although this may be clinically subtle).

Overall, the weight of the evidence indicates that children who have ISS should not be dismissed as "normal variants" and illustrates the complexity of normal growth and the multiple steps at which the growth process may go awry. Additional insights probably will come from thorough, open-minded investigation of such children, especially those who have very low or high spontaneous GH levels, low GHBP, abnormal concentrations of IGFI, IGFBP-3, or ALS, and additional clinical findings such as mental retardation, disproportion (ie, shorter limbs than trunk, or vice versa), or dysmorphic features.

Efficacy of growth hormone treatment

In the 20 years following the introduction of recombinant GH, more than 40 studies were undertaken in children with ISS. Short-term efficacy was demonstrated by improvements in height velocity, although few of the initial studies followed patients to adult height and none were placebo controlled [277]. In 1987 the Endocrinologic and Metabolic Drugs Advisory Committee of the US FDA provided guidance regarding studies of GH treatment in pediatric patients with non-GHD forms of short stature. The committee indicated that the critical endpoint was final height and that such studies should include a control group. Although there were concerns about the type and feasibility of the control, the committee recommended, "the control group should be a placebo-treated, parallel, randomized group of patients" [278]. Between 1988 and 2001 the National Institutes of Health conducted a study following this recommended design. The results of this study, supported by the data from a European dose-response study, formed the basis for the initial 2003 US FDA approval of GH treatment of ISS in the United States [249,279–282].

The pivotal study randomly assigned 71 patients at an average age of ∼12.4 years, with mean height of −2.8 SDS, to GH (0.22 mg/kg/wk), given in divided doses three times per week [equivalent to 31 μg/kg/d]) or placebo injections [249]. At a mean age of 18.8 years (average treatment duration, 4.4 years) the mean final height of the GH-treated group was within the normal range, at −1.8 SDS, and was significantly greater than that of the placebo-treated group, which remained below the normal range, at −2.3 SDS. By analysis of covariance of final height SDS, with baseline predicted height SDS as the covariate, the mean treatment effect was 0.51 SDS, corresponding to 3.7 cm ($P = 0.017$).

Although this placebo-controlled study provided definitive evidence of GH treatment effect in ISS, the between-group difference of ∼3.7 cm was modest. Certain aspects of the study design probably limited its magnitude, including the relatively low GH dosage (only about 73% of the current standard dosing for GH deficiency in the United States), suboptimal injection frequency (three times per week versus the current standard of daily dosing), and the late age and peripubertal developmental status of the patients at initiation of therapy. Consequently, data from a European dose-response trial were used to complement the data from the placebo-controlled trial, providing a broader perspective of the treatment effect. The European study was a multicenter, two-year, three-arm, open-label, dose-response trial with extension to final height in which 239 children were enrolled (Fig. 6A). Study subjects were assigned randomly, at an average age of 9.8 years, to one of three treatment regimens, using six injections per week at doses equivalent to: 34 μg/kg/d throughout; 34 μg/kg/d for the first year and 53 μg/kg/d thereafter; or 53 μg/kg/d throughout. A dose-response effect was demonstrated for the higher dose compared with the lower dose by (1) a greater increase in height velocity over the first 2 years of treatment ($P = 0.003$) and (2) a 3 cm greater overall

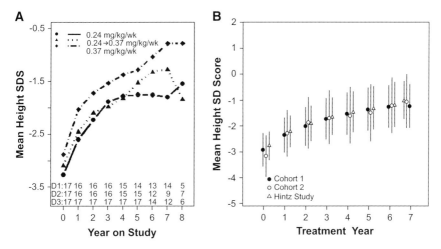

Fig. 6. (*A*) Change in height SDS score of children with idiopathic short stature who received one of three different growth hormone treatment regimens. D1, dosage 1 (0.24 mg/kg/wk [34 μg/kg/d]); D2, dosage 2 (0.24-0.37 mg/kg/wk [34-53 μg/kg/d]); D3, dosage 3 (0.37 mg/kg/wk [53 μg/kg/d]). (*From:* Wit JM, Rekers-Mombarg LT, Cutler GB, et al. Growth hormone (GH) treatment to final height in children with idiopathic short stature: evidence for a dose effect. J Pediatr 2005;146:48; with permission.) (*B*) Changes in height SD scores (mean ± SD) of three cohorts of children with idiopathic short stature treated with growth hormone. Mean ages at treatment initiation: cohort 1, 10.5 ± 2.7; cohort 2, 3.7 ± 0.0; Hintz study, males 10.4 ± 1.8, females 9.7 ± 2.1 [247]. (*From:* Kemp SF, Kuntze J, Attie KM, et al. Efficacy and safety results of long-term growth hormone treatment of idiopathic short stature. J Clin Endocrinol Metab 2005;90:5250; with permission.)

gain from baseline to final height at the higher dosage [280,282]. After an average of ∼6.6 years of treatment, mean gains in final height as compared with baseline predicted height were 5.4 cm and 7.2 cm for the 34 μg/kg/d and 53 μg/kg/d groups, respectively. The greater efficacy demonstrated in the dose-response study compared with the 3.7 cm average height gain of the GH-treated group in the placebo-controlled study probably reflects the combined effects of the higher GH dose, the more physiologic injection frequency (6 times per week), and the younger age at initiation of treatment (average age ∼10 years versus ∼12.5 years). GH treatment did not induce any untoward advancement of skeletal maturation or pubertal development in either of the two ISS registration studies [249,282,283]; similar findings have been reported from a large GH postmarketing database [284,285].

A number of meta-analyses have reviewed of the efficacy of GH treatment in children with ISS [277,286,287]. The meta-analysis performed by Finkelstein and colleagues [277] summarized the data from 38 studies, 4 of which included a concurrent control group to adult height. On the basis of these four studies (weighted mean GH dosage, 0.31 mg/kg/wk in six divided doses (dosage equivalent to 44 μg/kg/d); mean treatment duration, 5.3 years), the GH-induced adult height gain was 5 to 6 cm. Subsequent meta-analyses suggested average height gains between 3 and 7 cm, depending on duration of treatment [286,287]. Therefore, the overall GH-mediated height gain in ISS can be summarized as approximately 1 cm per year of GH treatment across the various registration and nonregistration studies, a magnitude similar to that seen in TS and SGA.

On the basis of the registration studies and the data summarized in the 2002 meta-analysis, the US FDA granted approval for GH treatment of ISS in 2003, as follows: "... long-term treatment of idiopathic short stature, also called non-growth hormone-deficient short-stature, defined by height SDS ≤ -2.25, and associated with growth rates unlikely to permit attainment of adult height in the normal range, in pediatric patients whose epiphyses are not closed and for whom diagnostic evaluation excludes other causes associated with short stature that should be observed or treated by other means" [239]. Two aspects of this language warrant some explanation: the designation of -2.25 SDS as the threshold for approved treatment was based on the height criterion for study entry in the pivotal placebo controlled trial; "unlikely to permit attainment of adult height in the normal range" does not in fact refer to a standard bone age-based Bayley-Pinneau height prediction that falls below the normal height range but rather to the integrated assessment by a pediatric endocrinologist that the child will be unlikely to attain a normal adult height, based on all available clinical and auxologic data.

More recent data from children with ISS enrolled in a large postmarketing study demonstrate the efficacy of GH treatment beyond the confines of small, highly controlled, and somewhat artificial conditions of randomized, clinical trials [288]. The report evaluated data from more than 8000 patients (75% male) treated with an average GH dose of 43 µg/kg/d from a mean age of 10.9 years. A subcohort was selected to be similar in starting age (mean 10.5 years) to patients studied in an earlier clinical trial sponsored by the same GH manufacturer that had demonstrated a 5 to 6 cm increase in adult height compared with baseline predicted adult hight [247]. Height velocity increased markedly in this group of more than 2500 children, from a mean baseline value of 4.0 cm/y to a first-year value of 8.6 cm/y. As expected, height velocity declined somewhat with progressive years of treatment but remained consistently greater than the pretreatment value for as long as 7 years. Accompanying the increases in height velocity, height SDS improved progressively each year from a baseline value of approximately -2.9, to a 7-year value of -1.2. Catch-up occurred rapidly over the initial 4 years of treatment, with a gain of 1.4 SDS over this period (average 0.35 SDS per year); catch-up slowed in the latter years to an average rate of 0.1 SDS per year. When data were examined specifically for the impact of delayed maturation (evidenced by $>$ 2.5 years of bone age delay), no differences were found in short-term or long-term outcome for those who had marked delay compared with those who had milder or no delay (Fig. 6B).

As with all GH treatment indications, responses vary among patients. Accurate prediction of treatment outcome on the basis of pretreatment parameters is not possible currently, although several groups have tried to address this problem. The finding that lower baseline IGF-I values correlate with greater short-term growth response [256,289,290] has led to the concept of titration of GH dose according to IGF-I SDS. In a 2-year, 3-arm study comparing GH administered at a dose of 40 µg/kg/d versus GH doses titrated to maintain IGF-I at either 0 or +2 SDS, the group that received the high GH dose resulting in average IGF-I of +2 SDS had a significantly greater gain in height SDS than the other two groups [26]. However, the change in IGF-I accounted for only about 30 percent of the variability in the response. Although IGF-I serves as a reasonable predictor of short-term response, it seems to play a relatively minor role in the prediction of long-term outcome, as evidenced by regression analyses performed using data collected in the placebo-controlled clinical trial described previously [249]. In these analyses a model for predicting height gain using four variables—the difference between height and mid-parental height, the difference between bone age and chronologic age, pretreatment IGF-I, and pretreatment height velocity—predicted with 100% sensitivity and specificity those patients who gained at least 0.5 SDS in height. Although IGF-I contributed to the model, it did not correlate independently with the height gain. Importantly, while this model was accurate in the context of that specific clinical trial, it cannot be extrapolated for clinical use without independent validation on a large cohort of heterogeneous patients. Other relevant factors in terms of long-term height outcomes include younger age at initiation of treatment and greater dose and duration of GH treatment [248,249,280,291].

Another biologic factor reported by some to influence GH treatment response is the presence of the d3GHR allele. The initial study of the effects of this polymorphism reported significantly greater GH-induced height gains in subjects with ISS who carried at least one copy of this variant receptor, compared with those who were homozygous for the full-length receptor [29]. Subsequent studies in children with a variety of growth disorders (GHD, ISS, SGA) have reported variable correlations between height gain and GH receptor status, some supporting the original findings and others not [292,293].

Safety of growth hormone treatment

Because patients who have ISS are generally healthy, without known increases in any pre-existing comorbidities, they would be expected to be at relatively low risk for side effects of GH treatment. Data from both the registration studies and large, long-term GH-postmarketing studies indeed indicate that rates of adverse events are lower in this patient group than in other GH-treated populations. To evaluate safety of GH treatment in children with ISS, the data from the two registration studies that formed the basis for the US FDA's 2003 approval were compared with those from registration studies in children with GH deficiency and TS [281]. Rates of otitis media, scoliosis, slipped capital femoral epiphysis, hypothyroidism, disturbances of carbohydrate metabolism, and hypertension in patients with ISS were similar to or lower than rates of these conditions in patients with GH deficiency or TS, and no cases of benign intracranial hypertension, edema, pancreatitis, or prepubertal gynecomastia were reported in the ISS studies. Although assessment of carbohydrate metabolism was limited in the ISS registration studies, there was no evidence of a detrimental effect of GH treatment on fasting glucose at doses from 31 to 53 µg/kg/d. Insulin sensitivity, tested only in the placebo-controlled study, showed no GH effect when assessed using the quantitative insulin sensitivity check index [294].

Two cases of neoplasia were reported in the registration studies: after 19 weeks of GH treatment, an 11-year-old boy was diagnosed with Hodgkin disease, which, based on retrospective review of pretreatment data, probably was present subclinically at study entry. The second case was that of a 12-year-old boy diagnosed after 6 years of GH treatment with an abdominal desmoplastic small round cell tumor associated with an 11p13:22q12 chromosome translocation.

Although GH treatment was approved for ISS in the United States only in 2003, there had been prior longstanding use of this treatment in an off-label fashion in children who had unexplained short stature (as is true for other growth disorders). Consequently, between 1985 and 2000 two large postmarketing research programs collected detailed safety data for approximately 9000 patients with ISS, representing approximately 27,000 patient-years of GH exposure. Overall, adverse event rates for these patients were similar to or, in some cases lower than, adverse-event rates for patients with other GH-treated conditions [73,231]. In a report from the Kabi International Growth Study [231], overall rates of adverse events were lower in ISS than in other non-GHD growth disorders. No cases of malignancy were reported. Of the 3493 patients in the database, 2 patients each were reported to have scoliosis and slipped capital femoral epiphysis. Data from the National Cooperative Growth Study provide a similar picture: for example, although the 5671 patients who had ISS comprised 17.1% of the total patients enrolled, they accounted for only 10.1% of the total adverse events, 4.6% of all serious adverse events, and 3.4% of deaths, the fewest of any of the patient populations [73]. There were five deaths (causes not reported); in addition, leukemia, an extracranial malignancy and intracranial hypertension were reported in 1 patient each, diabetes or glucose intolerance in 9 patients, and scoliosis in 12 patients. A recent update from the National Cooperative Growth Study reported 274 adverse events (53 serious) in 8108 patients with ISS [288]. There were two deaths, one caused by Burkitt lymphoma and one suicide. There were five reports of new malignancies, representing 1.4 times the general population background rate. Because the confidence interval (0.35–3.24) includes 1.0, this occurrence rate is not considered significantly elevated. The occurrence rate for diabetes was almost identical with that expected in the general population. As would be expected from its physiologic mechanism of action, administration of GH to non-GH-deficient children who have various growth disorders has been reported to induce increases in fasting insulin; however, insulin concentrations generally have remained within the normal range, and values have returned to baseline after cessation of GH treatment [74,295].

Highlights from recent consensus statements regarding growth hormone treatment in non–growth hormone-deficient growth disorders

A number of societies representing physicians who investigate and treat patients with growth disorders have reviewed the use of GH in childhood and have made specific comments regarding non-GHD growth disorders. A few key points from these publications are summarized here. The reader is referred to the original papers for additional details.

GH Research Society

In 2000 the GH Research Society published a statement regarding safety of GH administration. With respect to non-GHD growth disorders in particular, the statement indicated that "Monitoring of glucose homeostasis in Turner's syndrome and glucose homeostasis and lipid profiles in chronic renal failure on GH treatment should be undertaken at intervals determined by standard clinical practice." No comments were made regarding SGA or ISS because these conditions were not approved for GH treatment at the time [137].

American Association of Clinical Endocrinologists

The American Association of Clinical Endocrinologists published its most recent guidelines regarding GH treatment in early 2003, a few months before FDA approval of GH treatment for ISS [296]. The guidelines discuss specific recommendations for GH dosing and management for TS, CRI, SGA, and PWS, concluding, "GH therapy is best accomplished under the direct supervision of a clinical endocrinologist. Short term treatment is safe in both children and adults. Continued monitoring of side effects and long-term treatment results is needed."

Lawson Wilkins Pediatric Endocrine Society

Shortly after the 2003 US FDA approval of GH for ISS, the Lawson Wilkins Pediatric Endocrine Society published an updated set of guidelines for the use of GH in children [297]. They concluded that GH is "an important pharmacologic agent to stimulate linear growth and improve body composition in children with GHD and to increase linear growth in children with [CRI, TS, PWS and those born SGA]." Regarding the approval by the US

FDA for GH use in ISS, the guidelines state, "the impact of GH treatment on this population remains unclear and this approval should not obviate the need for a thorough investigation of the cause of the short stature."

Summary

Although a large body of data on the efficacy and safety of GH treatment for various non-GHD growth disorders has accumulated from a combination of clinical trial and postmarketing sources in the last 20 years, limitations remain. Clinical trial data have the advantage of directly comparing well-matched, randomized patient groups receiving treatment (or not) under comparable conditions and as such provide the highest-quality evidence of efficacy. Clinical trials, however, typically are too small for any statistically valid assessment for safety, which is more comprehensively addressed using postmarketing data, although underreporting and potential reporting bias remain issues. Consequently, although the efficacy of GH treatment in children who have non-GHD growth disorders has been solidly established and, based the combination of the rigor of the clinical trial data and the numerical power of the postmarketing data, no major concerns exist regarding safety, additional long-term data are required.

One concern raised by a number of commentators relates to the variability of individual patient responses to GH treatment and the challenge of providing the patient and family with any reasonably accurate prediction for long-term outcome at the start of treatment. To address this concern, various models for predicting height gain have been developed for a number of non-GHD conditions. These models typically use baseline and postbaseline variables such as age, bone age, height and mid-parental height, GH dose, IGF-I response, and various combinations of these variables to predict GH-related height gains [26,217,248,249,298–300]. Although these models are appropriate and valid in predicting short- or medium-term height gains in the specific patient populations in whom they were developed, few models have been developed to predict adult height, and because of wide scatter around the regression mean the available models are not generalizable to individual patients seen in normal clinical practice. Furthermore, comparisons between models are problematic, because the outcome measure (ie, the variable designated as the measure of treatment effect) differs among studies; whereas some models use actual height velocity (or height velocity SDS), others use change in height velocity, actual height SDS, or change in height SDS. Nevertheless, based on such studies, the following variables generally can be accepted as important prognosticators of GH treatment outcomes: younger age at initiation of treatment, greater genetic height potential (typically as expressed by mid-parental height or the difference between the patient's baseline height SDS and mid-parental height SDS), lower IGF-I or IGFBP-3 SDS, and greater GH dose and duration. Individual variables, in and of themselves, have limited predictive power outside these statistical models and cannot be used to predict outcome on an individual patient basis.

Finally, it seems appropriate to comment on the somewhat controversial question of monitoring of IGF-I concentrations and maintaining these concentrations within certain limits during GH treatment. Epidemiologic studies have suggested possible associations between IGF-I concentrations in the high-normal range (particularly when accompanied by lower IGFBP-3 concentrations) and cancers of the breast [301–303], prostate [304–306], colon/rectum [307–311], and lung [312,313]. The story becomes more complex given that IGF-I and its major binding protein IGFBP-3, both of which are increased by GH treatment, have essentially opposite effects on cell growth, with IGF-I being anti-apoptotic and IGFBP-3 pro-apoptotic. Indeed, in the study by Ma and colleagues [311] IGF-I and IGFBP-3 seemed to have opposite effects on colon cancer risk. Interpretation of the complex relationships between IGF-I and its binding proteins and neoplasia also requires understanding of the impact of other regulators, such as nutrition, on expression of these proteins, the role of autocrine/paracrine effects of local IGF-I secretion, and the expression of receptors for IGFs on tumor tissues. Discussion of these important considerations is beyond the scope of this article but can be found in a number of comprehensive reviews [314–318].

Despite the generally reassuring data in many thousands of GH-treated patients, the potential relationship among GH, IGF-I, and neoplasia requires careful follow-up of GH-treated patients, particularly at the higher GH doses used in children with non-GHD growth disorders. Consequently, guidance offered by professional societies such as the GH Research Society and the Lawson Wilkins Pediatric Endocrine Society, which have recommended monitoring of IGF-I and IGFBP-3 in patients receiving GH treatment, remains appropriate [137,319].

References

[1] Raben MS. Treatment of a pituitary dwarf with human growth hormone. J Clin Endocrinol Metab 1958;18:901–3.

[2] Underwood LE. Report of the conference on uses and possible abuses of biosynthetic human growth hormone. N Engl J Med 1984;311:606–8.

[3] American Academy of Pediatrics Committee on Drugs and Committee on Bioethics. 1997. Considerations related to the use of recombinant human growth hormone in children. Pediatrics 1997;99:122–9.

[4] American Association of Clinical Endocrinologists. Medical guidelines for clinical practice for growth hormone use in adults and children–2003 Update. Endocr Pract 2003;9:65–76.

[5] Kuczmarski RJ, Ogden CL, Grummer-Strawn LM, et al. CDC growth charts: United States. Adv Data 2000;8:1–27.

[6] Allen DB, Fost NC. Growth hormone therapy for short stature: panacea or Pandora's box? J Pediatr 1990;117:16–21.

[7] Rosenfeld RG, Albertsson-Wikland K, Cassorla F, et al. Diagnostic controversy: the diagnosis of childhood growth hormone deficiency revisited. J Clin Endocrinol Metab 1995;80: 1532–40.

[8] Rosenfeld RG. Is growth hormone deficiency a viable diagnosis? J Clin Endocrinol Metab 1997;82:349–51.
[9] Bright GM, Julius JR, Lima J, et al. Growth hormone stimulation test results as predictors of recombinant human growth hormone treatment outcomes: preliminary analysis of the national cooperative growth study database. Pediatrics 1999;104:1028–31.
[10] Saggese G, Ranke MB, Saenger P, et al. Diagnosis and treatment of growth hormone deficiency in children and adolescents: towards a consensus. Ten Years After the Availability of Recombinant Human Growth Hormone Workshop held in Pisa, Italy, 27–28 March 1998. Horm Res 1998;50:320–40.
[11] Sizonenko PC, Clayton PE, Cohen P, et al. Diagnosis and management of growth hormone deficiency in childhood and adolescence. Part 1: diagnosis of growth hormone deficiency. Growth Horm IGF Res 2001;11:137–65.
[12] Saenger P. Growth hormone in growth hormone deficiency. BMJ 2002;325:58–9.
[13] Badaru A, Wilson DM. Alternatives to growth hormone stimulation testing in children. Trends Endocrinol Metab 2004;15:252–8.
[14] Werther GA, Wang M, Cowell CT. An auxology-based growth hormone program: update on the Australian experience. J Pediatr Endocrinol Metab 2003;16(Suppl 3): 613–8.
[15] Hogler W, Briody J, Moore B, et al. Effect of growth hormone therapy and puberty on bone and body composition in children with idiopathic short stature and growth hormone deficiency. Bone 2005;37:642–50.
[16] Isaksson O, Jansson J, Gause I. Growth hormone stimulates longitudinal bone growth directly. Science 1982;216:1237–9.
[17] Isaksson O, Lindahl A, Nilsson A, et al. Mechanism of the stimulatory effect of growth hormone on longitudinal bone growth. Endocr Rev 1987;8:426–38.
[18] Hokken-Koelega AC, Stijnen T, De Jong MC, et al. Double blind trial comparing the effects of two doses of growth hormone in prepubertal patients with chronic renal insufficiency. J Clin Endocrinol Metab 1994;79(4):1185–90.
[19] Carel JC, Mathivon L, Gendrel C, et al. Near normalization of final height with adapted doses of growth hormone in Turner's syndrome. J Clin Endocrinol Metab 1998;83: 1462–6.
[20] Sas TC, de Muinck Keizer-Schrama SM, Stijnen T, et al. Normalization of height in girls with Turner syndrome after long-term growth hormone treatment: results of a randomized dose-response trial. J Clin Endocrinol Metab 1999;84(12):4607–12.
[21] Sas T, de Waal W, Mulder P, et al. Growth hormone treatment in children with short stature born small for gestational age: 5-year results of a randomized, double-blind, dose-response trial. J Clin Endocrinol Metab 1999;84(9):3064–70.
[22] Lupu F, Terwilliger JD, Lee K, et al. Roles of growth hormone and insulin-like growth factor 1 in mouse postnatal growth. Dev Biol 2001;229:141–62.
[23] van Teunenbroek A, de Muinck Keizer-Schrama SM, Stijnen T, et al. Yearly stepwise increments of the growth hormone dose results in a better growth response after four years in girls with Turner syndrome. Dutch Working Group on Growth Hormone. J Clin Endocrinol Metab 1996;81(11):4013–21.
[24] Cohen P, Bright GM, Rogol AD, et al. Effects of dose and gender on the growth and growth factor response to GH in GH-deficient children: implications for efficacy and safety. J Clin Endocrinol Metab 2002;87:90–8.
[25] Kamp GA, Zwinderman AH, van DJ, et al. Biochemical markers of growth hormone (GH) sensitivity in children with idiopathic short stature: individual capacity of IGF-I generation after high-dose GH treatment determines the growth response to GH. Clin Endocrinol (Ox) 2002;57:315–25.
[26] Cohen P, Weng W, Kappelgaard AM, et al. Multivariate analysis of height outcome from an IGF-1 based GH trial in children with short stature. Abstract OR32-3 presented at the 88th annual meeting of the Endocrine Society. Boston, June 24–27, 2006.

[27] van Pareren YK, de Muinck Keizer-Schrama SM, Stijnen T, et al. Final height in girls with Turner syndrome after long-term growth hormone treatment in three dosages and low dose estrogens. J Clin Endocrinol Metab 2003;88:1119–25.
[28] van Pareren Y, Mulder P, Houdijk M, et al. Adult height after long-term, continuous growth hormone (GH) treatment in short children born small for gestational age: results of a randomized, double-blind, dose-response GH trial. J Clin Endocrinol Metab 2003; 88(8):3584–90.
[29] Dos Santos C, Essioux L, Teinturier C, et al. A common polymorphism of the growth hormone receptor is associated with increased responsiveness to growth hormone. Nat Genet 2004;36(7):720–4.
[30] Warady BA. Growth retardation in children with chronic renal insufficiency. J Am Soc Nephrol 1998;9:S85–9.
[31] Ardissino G, Dacco V, Testa S, et al, for the ItalKid Project. Epidemiology of chronic renal failure in children: data from the ItalKid project. Pediatrics 2003;111(4 Pt 1):e382–7.
[32] Furth SL. Growth and nutrition in children with chronic kidney disease. Adv Chronic Kidney Dis 2005;12(4):366–71.
[33] Furth SL, Stablein D, Fine RN, et al. Adverse clinical outcomes associated with short stature at dialysis initiation: a report of the North American Pediatric Renal Transplant Cooperative Study. Pediatrics 2002;109(5):909–13.
[34] Betts PR, Magrath G, White RH. Role of dietary energy supplementation in growth of children with chronic renal insufficiency. BMJ 1977;1:416–8.
[35] Tonshoff B, Mehls O. Growth retardation in children with chronic renal disease - pathophysiology and treatment. In: Hindmarsh PC, editor. Current indications for growth hormone therapy. 1st edition. Basel: Karger; 1999. p. 128–56.
[36] Schaefer F, Seidel C, Binding A, et al. Pubertal growth in chronic renal failure. Pediatr Res 1990;28(1):5–10.
[37] Scharer K. Growth and development of children with chronic renal failure. Study Group on Pubertal Development in Chronic Renal Failure. Acta Paediatr Scand Suppl 1990;366:90–2.
[38] Vimalachandra D, Hodson EM, Willis NS, et al. Growth hormone for children with chronic kidney disease. Cochrane Database Syst Rev 2001;3:CD003264.
[39] Perrone L, Sinisi AA, Criscuolo T, et al. Plasma and urinary growth hormone and insulin-like growth factor I in children with chronic renal insufficiency. Child Nephrol Urol 1990; 10(2):72–5.
[40] Kohaut EC. Chronic renal disease and growth in childhood. Curr Opin Pediatr 1995;7(2): 171–5.
[41] Fine RN. Growth hormone and the kidney: the use of recombinant human growth hormone (rhGH) in growth-retarded children with chronic renal insufficiency. J Am Soc Nephrol 1991;1(10):1136–45.
[42] Haffner D, Schaefer F, Girard J, et al. Metabolic clearance of recombinant human growth hormone in health and chronic renal failure. J Clin Invest 1994;93(3):1163–71.
[43] Schaefer F, Veldhuis JD, Stanhope R, et al. Alterations in growth hormone secretion and clearance in peripubertal boys with chronic renal failure and after renal transplantation. Cooperative Study Group of Pubertal development in Chronic Renal Failure. J Clin Endocrinol Metab 1994;78(6):1298–306.
[44] Tonshoff B, Veldhuis JD, Heinrich U, et al. Deconvolution analysis of spontaneous nocturnal growth hormone secretion in prepubertal children with preterminal chronic renal failure and with end-stage renal disease. Pediatr Res 1995;37:86–93.
[45] Lanes R. Long-term outcome of growth hormone therapy in children and adolescents. Treat Endocrinol 2004;3:53–66.
[46] Postel-Vinay MC, Tar A, Crosnier H, et al. Plasma growth hormone-binding activity is low in uraemic children. Pediatr Nephrol 1991;5:545–7.
[47] Baumann G. Growth hormone binding protein and free growth hormone in chronic renal failure. Pediatr Nephrol 1996;10(3):328–30.

[48] Tonshoff B, Cronin MJ, Reichert M, et al. Reduced concentration of serum growth hormone (GH)-binding protein in children with chronic renal failure: correlation with GH insensitivity. The European Study Group for Nutritional Treatment of Chronic Renal Failure in Childhood. The German Study Group for Growth Hormone Treatment in Chronic Renal Failure. J Clin Endocrinol Metab 1997;82:1007–13.
[49] Rabkin R, Sun DF, Chen Y, et al. Growth hormone resistance in uremia, a role for impaired JAK/STAT signaling. Pediatr Nephrol 2005;20(3):313–8.
[50] Tonshoff B, Blum WF, Wingen AM, et al. Serum insulin-like growth factors (IGFs) and IGF binding proteins 1, 2, and 3 in children with chronic renal failure: relationship to height and glomerular filtration rate. The European Study Group for Nutritional Treatment of Chronic Renal Failure in Childhood. J Clin Endocrinol Metab 1995;80:2684–91.
[51] Powell DR, Liu F, Baker BK, et al. Insulin-like growth factor-binding protein-6 levels are elevated in serum of children with chronic renal failure: a report of the Southwest Pediatric Nephrology Study Group. J Clin Endocrinol Metab 1997;82:2978–84.
[52] Durham SK, Mohan S, Liu F, et al. Bioactivity of a 29-kilodalton insulin-like growth factor binding protein-3 fragment present in excess in chronic renal failure serum. Pediatr Res 1997;42:335–41.
[53] Koch VH, Lippe BM, Nelson PA, et al. Accelerated growth after recombinant human growth hormone treatment of children with chronic renal failure. J Pediatr 1989;115(3): 365–71.
[54] Rees L, Rigden SP, Ward G, et al. Treatment of short stature in renal disease with recombinant human growth hormone. Arch Dis Child 1990;65:856–60.
[55] Tonshoff B, Mehls O, Heinrich U, et al. Growth-stimulating effects of recombinant human growth hormone in children with end-stage renal disease. J Pediatr 1990;116:561–6.
[56] Fine RN, Pyke-Grimm K, Nelson PA, et al. Recombinant human growth hormone treatment of children with chronic renal failure: long-term (1- to 3-year) outcome. Pediatr Nephrol 1991;5:477–81.
[57] Hokken-Koelega AC, Stijnen T, de Muinck Keizer-Schrama SM, et al. Placebo-controlled, double-blind, cross-over trial of growth hormone treatment in prepubertal children with chronic renal failure. Lancet 1991;338(8767):585–90.
[58] Berard E, Crosnier H, Six-Beneton A, et al. Recombinant human growth hormone treatment of children on hemodialysis. French Society of Pediatric Nephrology. Pediatr Nephrol 1998;12(4):304–10.
[59] Fine RN, Kohaut EC, Brown D, et al. Growth after recombinant human growth hormone treatment in children with chronic renal failure: report of a multicenter randomized double-blind placebo-controlled study. Genentech Cooperative Study Group. J Pediatr 1994;124: 374–82.
[60] Powell DR, Liu F, Baker BK, et al. Modulation of growth factors by growth hormone in children with chronic renal failure. The Southwest Pediatric Nephrology Study Group. Kidney Int 1997;51(6):1970–9.
[61] Nutropin (somatropin [rDNA origin] for injection) [prescribing information]. South San Francisco (CA): Genentech Inc.; 2006.
[62] Fine RN, Sullivan EK, Tejani A. The impact of recombinant human growth hormone treatment on final adult height. Pediatr Nephrol 2000;14:679–81.
[63] Hokken-Koelega A, Mulder P, De Jong R, et al. Long-term effects of growth hormone treatment on growth and puberty in patients with chronic renal insufficiency. Pediatr Nephrol 2000;14(7):701–6.
[64] Fine RN, Brown DF, Kuntze J, et al. Growth after discontinuation of recombinant human growth hormone therapy in children with chronic renal insufficiency. The Genentech Cooperative Study Group. J Pediatr 1996;129(6):883–91.
[65] Haffner D, Schaefer F. Does recombinant growth hormone improve adult height in children with chronic renal failure? Semin Nephrol 2001;21(5):490–7.

[66] Crompton CH. Australian and New Zealand Paediatric Nephrology Association. Long-term recombinant human growth hormone use in Australian children with renal disease. Nephrol 2004;9(5):325–30.
[67] Fine RN, Ho M, Tejani A, et al. Adverse events with rhGH treatment of patients with chronic renal insufficiency and end-stage renal disease. J Pediatr 2003;142(5):539–45.
[68] Lanes R, Gunczler P, Orta N, et al. Changes in bone mineral density, growth velocity and renal function of prepubertal uremic children during growth hormone treatment. Horm Res 1996;46(6):263–8.
[69] Broyer M. Results and side-effects of treating children with growth hormone after kidney transplantation–a preliminary report. Pharmacia & Upjohn Study Group. Acta Paediatr Suppl 1996;417:76–9.
[70] Hokken-Koelega AC. Growth hormone treatment in children before and after renal transplantation. J Pediatr Endocrinol 1996;9(Suppl 3):359–64.
[71] Guest G, Berard E, Crosnier H, et al. Effects of growth hormone in short children after renal transplantation. French Society of Pediatric Nephrology. Pediatr Nephrol 1998;12(6):437–46.
[72] Tonshoff B, Mehls O. Use of rhGH posttransplanation in children. In: Tejani AH, Fine RN, editors. Pediatric renal transplantation. New York: Wiley-Liss; 1994. p. 441–59.
[73] Maneatis T, Baptista J, Connelly K, et al. Growth hormone safety update from the National Cooperative Growth Study. J Pediatr Endocrinol Metab 2000;13:1035–44.
[74] Saenger P, Attie KM, Martino-Nardi J, et al. Carbohydrate metabolism in children receiving growth hormone for 5 years. Chronic renal insufficiency compared with growth hormone deficiency, Turner syndrome, and idiopathic short stature. Genentech Collaborative Group. Pediatr Nephrol 1996;10:261–3.
[75] Turner HH. A syndrome of infantilism, congenital webbed neck, and cubitus valgus. Endocrinology 1938;23:566–74.
[76] Ullrich O. Uber typische Kombinationsbilder multipler Abartung. Z Kinderheilkd 1930;49:271–6.
[77] Nielsen J, Wohlert M. Chromosome abnormalities found among 34,910 newborn children: results from a 13-year incidence study in Arhus, Denmark. Hum Genet 1991;87:81–3.
[78] Ranke MB, Pfluger H, Rosendahl W, et al. Turner syndrome: spontaneous growth in 150 cases and review of the literature. Eur J Pediatr 1983;141(2):81–8.
[79] Karlberg J, Albertsson-Wikland K, Nilsson KO, et al. Growth in infancy and childhood in girls with Turner's syndrome. Acta Paediatr Scand 1991;80:1158–65.
[80] Even L, Cohen A, Marbach N, et al. Longitudinal analysis of growth over the first 3 years of life in Turner's syndrome. J Pediatr 2000;137(4):460–4.
[81] Davenport ML, Punyasavatsut N, Stewart PW, et al. Growth failure in early life: an important manifestation of Turner syndrome. Horm Res 2002;57(5–6):157–64.
[82] Brook CG, Murset G, Zachmann M, et al. Growth in children with 45,XO Turner's syndrome. Arch Dis Child 1974;49:789–95.
[83] Lyon AJ, Preece MA, Grant DB. Growth curve for girls with Turner syndrome. Arch Dis Child 1985;60:932–5.
[84] Rochiccioli P, David M, Malpuech G, et al. Study of final height in Turner's syndrome: ethnic and genetic influences. Acta Paediatr 1994;83:305–8.
[85] Rongen-Westerlaken C, Corel L, van den Broeck J, et al. Reference values for height, height velocity and weight in Turner's syndrome. Swedish Study Group for GH treatment. Acta Paediatr 1997;86(9):937–42.
[86] Holl RW, Kunze D, Etzrodt H, et al. Turner syndrome: final height, glucose tolerance, bone density and psychosocial status in 25 adult patients. Eur J Pediatr 1994;153:11–6.
[87] Massa G, Vanderschueren-Lodeweyckx M, Malvaux P. Linear growth in patients with Turner syndrome: influence of spontaneous puberty and parental height. Eur J Pediatr 1990;149(4):246–50.

[88] Gravholt CH, Weis Naeraa R. Reference values for body proportions and body composition in adult women with Ullrich-Turner syndrome. Am J Med Genet 1997;72: 403–8.
[89] Binder G, Fritsch H, Schweizer R, et al. Radiological signs of Leri-Weill dyschondrosteosis in Turner syndrome. Horm Res 2001;55(2):71–6.
[90] Kim JY, Rosenfeld SR, Keyak JH. Increased prevalence of scoliosis in Turner syndrome. J Pediatr Orthop 2001;21:765–6.
[91] Elder DA, Roper MG, Henderson RC, et al. Kyphosis in a Turner syndrome population. Pediatrics 2002;109:e93.
[92] Ross JL, Long LM, Loriaux DL, et al. Growth hormone secretory dynamics in Turner syndrome. J Pediatr 1985;106:202–6.
[93] Massarano AA, Brook CG, Hindmarsh PC, et al. Growth hormone secretion in Turner's syndrome and influence of oxandrolone and ethinyl oestradiol. Arch Dis Child 1989;64: 587–92.
[94] Massa G, Vanderschueren-Lodeweyckx M, Craen M, et al. Growth hormone treatment of Turner syndrome patients with insufficient growth hormone response to pharmacological stimulation tests. Eur J Pediatr 1991;150:460–3.
[95] Reiter JC, Craen M, Van VG. Decreased growth hormone response to growth hormone-releasing hormone in Turner's syndrome: relation to body weight and adiposity. Acta Endocrinol (Copenh) 1991;125:38–42.
[96] Pirazzoli P, Mazzanti L, Bergamaschi R, et al. Reduced spontaneous growth hormone secretion in patients with Turner's syndrome. Acta Paediatr 1999;88:610–3.
[97] Wit JM, Massarano AA, Kamp GA, et al. Growth hormone secretion in patients with Turner's syndrome as determined by time series analysis. Acta Endocrinol (Copenh) 1992;127:7–12.
[98] Cianfarani S, Vaccaro F, Pasquino AM, et al. Reduced growth hormone secretion in Turner syndrome: is body weight a key factor? Horm Res 1994;41:27–32.
[99] Gravholt CH, Frystyk J, Flyvbjerg A, et al. Reduced free IGF-I and increased IGFBP-3 proteolysis in Turner syndrome: modulation by female sex steroids. Am J Physiol Endocrinol Metab 2001;280:E308–14.
[100] Lebl J, Pruhova S, Zapletalova J, et al. IGF-I resistance and Turner's syndrome. J Pediatr Endocrinol Metab 2001;14:37–41.
[101] Rao E, Weiss B, Fukami M, et al. Pseudoautosomal deletions encompassing a novel homeobox gene cause growth failure in idiopathic short stature and Turner syndrome. Nat Genet 1997;16:54–63.
[102] Kosho T, Muroya K, Nagai T, et al. Skeletal features and growth patterns in 14 patients with haploinsufficiency of SHOX: implications for the development of Turner syndrome. J Clin Endocrinol Metab 1999;84:4613–21.
[103] Rappold G, Werner F, Blum WF, et al. Genotypes and phenotypes in children with short stature: clinical indicators of *SHOX* haploinsufficiency. J Med Genet, in press.
[104] Ogata T, Matsuo N. Turner syndrome and female sex chromosome aberrations: deduction of the principal factors involved in the development of clinical features. Hum Genet 1995; 95:607–29.
[105] Haverkamp F, Wolfle J, Zerres K, et al. Growth retardation in Turner syndrome: aneuploidy, rather than specific gene loss, may explain growth failure. J Clin Endocrinol Metab 1999;84:4578–82.
[106] Buchanan CR, Law CM, Milner RDG. Growth hormone in short, slowly growing children and those with Turner's syndrome. Arch Dis Child 1987;62:912–6.
[107] Rosenfeld RG. Acceleration of growth in Turner syndrome patients treated with growth hormone: summary of three-year results. J Endocrinol Invest 1989;12:49–51.
[108] Vanderschueren-Lodeweyckx M, Massa G, Maes M, et al. Growth-promoting effect of growth hormone and low dose ethinyl estradiol in girls with Turner's syndrome. J Clin Endocrinol Metab 1990;70:122–6.

[109] Werther GA. A multi-centre double-blind study of growth hormone and low-dose estrogen in Turner syndrome: an interim analysis. In: Ranke MB, Rosenfeld RG, editors. Turner syndrome: growth promoting therapies. Philadelphia: Elsevier Science; 1991.
[110] Cave CB, Bryant J, Milne R. Recombinant growth hormone in children and adolescents with Turner syndrome. Cochrane Database Syst Rev 2003:CD003887.
[111] Rosenfeld RG, Attie KM, Frane J, et al. Growth hormone therapy of Turner's syndrome: beneficial effect on adult height. J Pediatr 1998;132:319–24.
[112] Chernausek SD, Attie KM, Cara JF, et al. Growth hormone therapy of Turner syndrome: the impact of age of estrogen replacement on final height. Genentech, Inc., Collaborative Study Group. J Clin Endocrinol Metab 2000;85(7):2439–45.
[113] Quigley CA, Crowe BJ, Anglin DG, et al. Growth hormone and low dose estrogen in Turner syndrome: results of a United States multi-center trial to near-final height. J Clin Endocrinol Metab 2002;87:2033–41.
[114] Stephure DK, on behalf or the Canadian Pediatric Endocrine Group. Impact of growth hormone supplementation on adult height in Turner syndrome: results of the Canadian randomized controlled trial. J Clin Endocrinol Metab 2005;90:3360–6.
[115] Chu CE, Paterson WF, Kelnar CJ, et al. Variable effect of growth hormone on growth and final adult height in Scottish patients with Turner's syndrome. Acta Paediatr 1997;86:160–4.
[116] Dacou-Voutetakis C, Karavanaki-Karanassiou K, Petrou V, et al. The growth pattern and final height of girls with Turner syndrome with and without human growth hormone treatment. Pediatrics 1998;101:663–8.
[117] Taback SP, Van Vliet G. Growth hormone, adult height, and psychosocial outcome in Turner syndrome: the need for controlled studies. In: Albertsson-Wikland K, Ranke MB, editors. Turner syndrome in a life-span perspective. Amsterdam: Elsevier Science B.V.; 1994. p. 183–90.
[118] Taback SP, Collu R, Deal CL, et al. Does growth-hormone supplementation affect adult height in Turner's syndrome? Lancet 1996;348(9019):25–7.
[119] Rosenfield RL, Devine N, Hunold JJ, et al. Salutary effects of combining early very low-dose systemic estradiol with growth hormone therapy in girls with Turner syndrome. J Clin Endocrinol Metab 2005;90:6424–30.
[120] Conway GS. Oestrogen replacement in young women with Turner's syndrome. Clin Endocrinol (Oxf) 2001;54:157–8.
[121] Bondy CA. Care of girls and women with Turner syndrome: a guideline of the Turner Syndrome Study Group. J Clin Endocrinol Metab 2007;92:10.
[122] Carel JC, Elie C, Ecosse E, et al. Self-esteem and social adjustment in young women with Turner syndrome—influence of pubertal management and sexuality: population-based cohort study. J Clin Endocrinol Metab 2006;91:2972–9.
[123] Davenport ML, Crowe BJ, Travers SH, et al. Growth hormone corrects early growth failure in Turner syndrome: results from the randomized, controlled, multi-center, 'Toddler Turner' study. Submitted for publication.
[124] Nielsen J, Johansen K, Yde H. The frequency of diabetes mellitus in patients with Turner's syndrome and pure gonadal dysgenesis. Blood glucose, plasma insulin and growth hormone level during an oral glucose tolerance test. Acta Endocrinol (Copenh) 1969; 62(Suppl 2):251–69.
[125] Caprio S, Boulware D, Tamborlane V. Growth hormone and insulin interactions. Horm Res 1992;38(Suppl 2):47–9.
[126] Bakalov VK, Cooley MM, Quon MJ, et al. Impaired insulin secretion in the Turner metabolic syndrome. J Clin Endocrinol Metab 2004;89(7):3516–20.
[127] Radetti G, Pasquino B, Gottardi E, et al. Insulin sensitivity in Turner's syndrome: influence of GH treatment. Eur J Endocrinol 2004;151(3):351–4.
[128] Wilson DM, Frane JW, Sherman B, et al. Carbohydrate and lipid metabolism in Turner syndrome: effect of therapy with growth hormone, oxandrolone, and a combination of both. J Pediatr 1988;112:210–7.

[129] Weise M, James D, Leitner CH, et al. Glucose metabolism in Ullrich Turner syndrome: long-term effects of therapy with human growth hormone. German Lilly UTS Study Group. Horm Res 1993;39:36–41.
[130] Sas TC, de Muinck Keizer-Schrama SM, Stijnen T, et al. Carbohydrate metabolism during long-term growth hormone (GH) treatment and after discontinuation of GH treatment in girls with Turner syndrome participating in a randomized dose-response study. Dutch Advisory Group on Growth Hormone. J Clin Endocrinol Metab 2000;85:769–75.
[131] Cutfield WS, Wilton P, Bennmarker H, et al. Incidence of diabetes mellitus and impaired glucose tolerance in children and adolescents receiving growth-hormone treatment. Lancet 2000;355(9204):610–3.
[132] Abbott J, Hope K, Nair S, et al. Growth-hormone treatment and risk of diabetes. Lancet 1913;2000:355.
[133] Clayton PE, Cowell CT. Safety issues in children and adolescents during growth hormone therapy—a review. Growth Horm IGF Res 2000;10(6):306–17.
[134] Kohno H, Kuromaru R, Ueyama N, et al. Growth hormone treatment and type 2 diabetes. Growth Horm IGF Res 2001;11:196–7.
[135] Tanaka T, Cohen P, Clayton PE, et al. Diagnosis and management of growth hormone deficiency in childhood and adolescence—part 2: growth hormone treatment in growth hormone-deficient children. Growth Horm IGF Res 2002;12:323–41.
[136] Growth Hormone Research Society. Critical evaluation of the safety of recombinant human growth hormone administration: statement from the Growth Hormone Research Society. J Clin Endocrinol Metab 2001;86:1868–70.
[137] Watkin PM. Otological disease in Turner's syndrome. J Laryngol Otol 1989;103(8):731–8.
[138] Roush J, Quigley CA, Bryant CG, et al. Effect of GH treatment on ear disease in young girls with Turner syndrome (TS) [abstract 902]. Presented at the annual meeting of the Lawson Wilkins Pediatric Endocrine Society and Pediatric Academic Societies. San Francisco, May 1–4, 2004.
[139] Allen DB. Safety of human growth hormone therapy: current topics. J Pediatr 1996;128(5 Pt 2):S8–13.
[140] Blethen SL, Rundle AC. Slipped capital femoral epiphysis in children treated with growth hormone. A summary of the National Cooperative Growth Study experience. Horm Res 1996;46(3):113–6.
[141] Lowenstein EJ, Kim KH, Glick SA. Turner's syndrome in dermatology. J Am Acad Dermatol 2004;50:767–76.
[142] Wyatt D. Melanocytic nevi in children treated with growth hormone. Pediatrics 1999;104(4 Pt 2):1045–50.
[143] Lin AE, Lippe B, Rosenfeld RG. Further delineation of aortic dilation, dissection, and rupture in patients with Turner syndrome. Pediatrics 1998;102(1):e12.
[144] Sybert VP. Cardiovascular malformations and complications in Turner syndrome. Pediatrics 1998;101(1):E11.
[145] van den Berg J, Bannink EM, Wielopolski PA, et al. Aortic distensibility and dimensions and the effects of growth hormone treatment in the turner syndrome. Am J Cardiol 2006;97(11):1644–9.
[146] Belin V, Cusin V, Viot G, et al. SHOX mutations in dyschondrosteosis (Leri-Weill syndrome). Nat Genet 1998;19:67–9.
[147] Shears DJ, Vassal HJ, Goodman FR, et al. Mutation and deletion of the pseudoautosomal gene SHOX cause Leri-Weill dyschondrosteosis. Nat Genet 1998;19:70–3.
[148] Rappold GA, Fukami M, Niesler B, et al. Deletions of the homeobox gene SHOX (short stature homeobox) are an important cause of growth failure in children with short stature. J Clin Endocrinol Metab 2002;87:1402–6.

[149] Binder G, Ranke MB, Martin DD. Auxology is a valuable instrument for the clinical diagnosis of SHOX haploinsufficiency in school-age children with unexplained short stature. J Clin Endocrinol Metab 2003;88:4891–6.
[150] Blum WF, Crowe BJ, Quigley CA, et al. Growth hormone is effective in treatment of short stature associated with short stature homeobox-containing gene deficiency: two year results of a randomized, controlled, multicenter trial. J Clin Endocrinol Metab, 2007;92:219–28.
[151] Falcinelli C, Iughetti L, Percesepe A, et al. SHOX point mutations and deletions in Leri-Weill dyschondrosteosis. J Med Genet 2002;39(6):E33.
[152] Flanagan SF, Munns CF, Hayes M, et al. Prevalence of mutations in the short stature homeobox containing gene (SHOX) in Madelung deformity of childhood. J Med Genet 2002; 39(10):758–63.
[153] Huber C, Rosilio M, Munnich A, et al. High incidence of SHOX anomalies in patients with short stature. J Med Genet 2006;43:735–9.
[154] Léri A, Weill J. Une affection congenital et symetrique du development osseux: la dyschondrosteose. Bull Mém Soc Med Paris 1929;35:1491–4.
[155] Ross JL, Kowal K, Quigley CA, et al. The phenotype of short stature homeobox gene (SHOX) deficiency in childhood: contrasting children with Léri-Weill dyschondrosteosis and Turner syndrome. J Pediatr 2005;147:499–507.
[156] Clement-Jones M, Schiller S, Rao E, et al. The short stature homeobox gene SHOX is involved in skeletal abnormalities in Turner syndrome. Hum Mol Genet 2000;9:695–702.
[157] Ellison JW, Wardak Z, Young MF, et al. PHOG, a candidate gene for involvement in the short stature of Turner syndrome. Hum Mol Genet 1997;6:1341–7.
[158] Rao E, Blaschke RJ, Marchini A, et al. The Léri-Weill and Turner syndrome homeobox gene SHOX encodes a cell-type specific transcriptional activator. Hum Mol Genet 2001; 10:3083–91.
[159] Marchini A, Marttila T, Winter A, et al. The short stature homeodomain protein SHOX induces cellular growth arrest and apoptosis and is expressed in human growth plate chondrocytes. J Biol Chem 2004;279:37103–14.
[160] Munns CJ, Haase HR, Crowther LM, et al. Expression of SHOX in human fetal and childhood growth plate. J Clin Endocrinol Metab 2004;89:4130–5.
[161] Munns CF, Glass IA, LaBrom R, et al. Histopathological analysis of Léri-Weill dyschondrosteosis: disordered growth plate. Hand Surg 2001;6(1):13–23.
[162] Ogata T, Matsuo N, Nishimura G. SHOX haploinsufficiency and overdosage: impact of gonadal function status. J Med Genet 2001;38(1):1–6.
[163] Niesler B, Fischer C, Rappold GA. The human SHOX mutation database. Hum Mutat 2002;20:338–41.
[164] Fischer C, Niesler B. Human short stature gene allelic variant database. Available at: www.shox.uni-hd.de. Accessed January 4, 2007.
[165] Schiller S, Spranger S, Schechinger B, et al. Phenotypic variation and genetic heterogeneity in Leri-Weill syndrome. Eur J Hum Genet 2000;8(1):54–62.
[166] Munns CF, Berry M, Vickers D, et al. Effect of 24 months of recombinant growth hormone on height and body proportions in SHOX haploinsufficiency. J Pediatr Endocrinol Metab 2003;16:997–1004.
[167] Binder G, Renz A, Martinez A, et al. SHOX haploinsufficiency and Léri-Weill dyschondrosteosis: prevalence and growth failure in relation to mutation, sex, and degree of wrist deformity. J Clin Endocrinol Metab 2004;89:4403–8.
[168] Hediger ML, Overpeck MD, Maurer KR, et al. Growth of infants and young children born small or large for gestational age: findings from the Third National Health and Nutrition Examination Survey. Arch Pediatr Adolesc Med 1998;152:1225–31.
[169] Lee PA, Chernausek SD, Hokken-Koelega AC, et al. International Small for Gestational Age Advisory Board consensus development conference statement: management of short children born small for gestational age, April 24-October 1, 2001. Pediatrics 2003;111:1253–61.

[170] Wit JM, Finken MJ, Rijken M, et al. Confusion around the definition of small for gestational age. Pediatr Endocrinol Rev 2005;3:52–3.
[171] Clayton PE, Cianfarani S, Czernichow P, et al. The child born small for gestational age (SGA): a consensus statement on management through to adulthood. J Clin Endocrinol Metab 2007, doi: 10.1210/jc. 2006–2017.
[172] Hokken-Koelega AC, De Ridder MA, Lemmen RJ, et al. Children born small for gestational age: do they catch up? Pediatr Res 1995;38(2):267–71.
[173] Karlberg J, Albertsson-Wikland K. Growth in full-term small-for-gestational-age infants: from birth to final height. Pediatr Res 1995;38(5):733–9.
[174] Karlberg JP, Albertsson-Wikland K, Kwan EY, et al. The timing of early postnatal catch-up growth in normal, full-term infants born short for gestational age. Horm Res 1997;48(Suppl 1):17–24.
[175] Albertsson-WIkland K, Boguszewski M, Karlberg J. Children born small for gestational age: postnatal growth and hormonal status. Horm Res 1998;49(Suppl 2):7–13.
[176] Karlberg J, Albertsson-Wikland K, Kwan CW, et al. Early spontaneous catch-up growth. J Pediatr Endocrinol 2002;15(Suppl 5):1243–55.
[177] Leger J, Levy-Marchal C, Bloch J, et al. Reduced final height and indications for insulin resistance in 20 year olds born small for gestational age: regional cohort study. BMJ 1997;315(7104):341–7.
[178] Chatelain P, Job JC, Blanchard J, et al. Dose-dependent catch-up growth after 2 years of growth hormone treatment in intrauterine growth-retarded children. J Clin Endocrinol Metab 1994;78(6):1454–60.
[179] Albertsson-Wikland K, Karlberg J. Postnatal growth of children born small for gestational age. Acta Paediatr Suppl 1997;423:193–5.
[180] Leger J, Limoni C, Collin D, et al. Prediction factors in the determination of final height in subjects born small for gestational age. Pediatr Res 1998;43:808–12.
[181] de Zegher F, Francois I, Van HM, et al. Clinical review 89: small as fetus and short as child: from endogenous to exogenous growth hormone. J Clin Endocrinol Metab 1997;82:2021–6.
[182] Vicens-Calvet E, Espadero RM, Carrascosa A. Spanish SGA Collaborative Group. Small for gestational age. Longitudinal study of the pubertal growth spurt in children born small for gestational age without postnatal catch-up growth. J Pediatr Endocrinol Metab 2002; 15:381–8.
[183] Hokken-Koelega ACS. Timing of puberty and fetal growth. Best Pract Res Clin Endocrinol Metab 2002;16:65–71.
[184] Job JC, Rolland A. Natural history of intrauterine growth retardation: pubertal growth and adult height. Arch Fr Pediatr 1986;43(5):301–6.
[185] Davies PSW, Valley R, Preece MA. Adolescent growth and pubertal progression in the Silver-Russell syndrome. Arch Dis Child 1988;64:130–5.
[186] Lazar L, Pollak U, Kalter-Leibovici O, et al. Pubertal course of persistently short children born small for gestational age (SGA) compared with idiopathic short children born appropriate for gestational age (AGA). Eur J Endocrinol 2003;149:425–32.
[187] Rogol A. 2006. Growth, puberty and therapeutic interventions [abstract S1–15]. Presented at the 45th European Society for Pediatric Endocrinology annual meeting. Rotterdam, June 30–July 3, 2006.
[188] Wollmann HA. Intrauterine growth restriction: definition and etiology. Horm Res 1998; 49(Suppl 2):1–6.
[189] Lassarre C, Hardouin S, Daffos F, et al. Serum insulin-like growth factors and insulin-like growth factor binding proteins in the human fetus. Relationships with growth in normal subjects and in subjects with intrauterine growth retardation. Pediatr Res 1991;29:219–25.
[190] de Waal WJ, Hokken-Koelega AC, Stijnen T, et al. Endogenous and stimulated GH secretion, urinary GH excretion, and plasma IGF-I and IGF-II levels in prepubertal children

with short stature after intrauterine growth retardation. The Dutch Working Group on Growth Hormone. Clin Endocrinol (Oxf) 1994;41:621–30.
[191] Boguszewski M, Rosberg S, Albertsson-Wikland K. Spontaneous 24-hour growth hormone profiles in prepubertal small for gestational age children. J Clin Endocrinol Metab 1995;80:2599–606.
[192] Ogilvy-Stuart AL, Hands SJ, Adcock CJ, et al. Insulin, insulin-like growth factor I (IGF-I), IGF-binding protein-1, growth hormone, and feeding in the newborn. J Clin Endocrinol Metab 1998;83:3550–7.
[193] Audi L, Esteban C, Espadero R, et al. D-3 growth hormone receptor polymorphism (d3-GHR) genotype frequencies differ between short small for gestational age children (SGA) (n = 247) and a control population with normal adult height (n = 289) (CPNAH) [abstract S1–15]. Presented at the 45th European Society for Pediatric Endocrinology annual meeting. Rotterdam, June 30–July 3, 2006.
[194] Jensen RB, Viewerth S, Larsen T, et al. High prevalence of d3-growth hormone gene polymorphism in adolescents born SGA/IUGR: association with intrauterine growth velocity, birth weight, postnatal catch-up growth and near-final height [abstract FC1–42]. Presented at the 45th European Society for Pediatric Endocrinology annual meeting. Rotterdam, June 30–July 3, 2006.
[195] Woods KA, Camacho-Hubner C, Savage MO, et al. Intrauterine growth retardation and postnatal growth failure associated with deletion of the insulin-like growth factor I gene. N Engl J Med 1996;335:1363–7.
[196] Bonapace G, Concolino D, Formicola S, et al. A novel mutation in a patient with insulin-like growth factor 1 (IGF1) deficiency. J Med Genet 2003;40:913–7.
[197] Walenkamp MJ, Karperien M, Pereira AM, et al. Homozygous and heterozygous expression of a novel insulin-like growth factor-I mutation. J Clin Endocrinol Metab 2005;90: 2855–64.
[198] Cianfarani S, Maiorana A, Geremia C, et al. Blood glucose concentrations are reduced in children born small for gestational age (SGA), and thyroid-stimulating hormone levels are increased in SGA with blunted postnatal catch-up growth. J Clin Endocrinol Metab 2003; 88:2699–705.
[199] Tenhola S, Halonen P, Jaaskelainen J, et al. Serum markers of GH and insulin action in 12-year-old children born small for gestational age. Eur J Endocrinol 2005;152: 335–40.
[200] Arends N, Johnston L, Hokken-Kolega A. Polymorphism in the IGF-I gene: clinical relevance for short children born small for gestational age (SGA). J Clin Endocrinol Metab 2002;87:2720.
[201] Vaessen N, Janssen JA, Heutink P, et al. Association between genetic variation in the gene for insulin-like growth factor-I and low birthweight. Lancet 2002;359:1036–7.
[202] Johnston LB, Dahlgren J, Leger J, et al. Association between insulin-like growth factor I (IGF-I) polymorphisms, circulating IGF-I, and pre- and postnatal growth in two European small for gestational age populations. J Clin Endocrinol Metab 2003;88:4805–10.
[203] Netchine I, Rossignol S, Dufourg MNC, et al. Epimutation of the telomeric domain on chromosome 11p15 is a major and specific cause of Silver-Russell syndrome [abstract HA-8]. Presented at the 45th European Society for Pediatric Endocrinology annual meeting. Rotterdam, June 30–July 3, 2006.
[204] Abuzzahab MJ, Schneider A, Goddard A, et al. IGF-I receptor mutations resulting in intrauterine and postnatal growth retardation. N Engl J Med 2003;349:2211–22.
[205] Tanner JM, Ham TJ. Low birthweight dwarfism with asymmetry (Silver's syndrome): treatment with human growth hormone. Arch Dis Child 1969;44:231–43.
[206] Lee PA, Blizzard RM, Cheek DB, et al. Growth and body composition in intrauterine growth retardation (IUGR) before and during human growth hormone administration. Metabolism 1974;23:913–9.

[207] de Zegher F, Du Caju MV, Heinrichs C, et al. Early, discontinuous, high dose growth hormone treatment to normalize height and weight of short children born small for gestational age: results over 6 years. J Clin Endocrinol Metab 1999;84:1558–61.
[208] Hokken-Koelega A, van Pareren Y, Arends N, et al. Efficacy and safety of long-term continuous growth hormone treatment of children born small for gestational age. Horm Res 2004;62(Suppl 3):149–54.
[209] de Zegher F, Hokken-Koelega A. Growth hormone therapy for children born small for gestational age: height gain is less dose dependent over the long term than over the short term. Pediatrics 2005;115:e458–62.
[210] Rosilio M, Carel JC, Ecosse E, et al. Adult height of prepubertal short children born small for gestational age treated with GH. Eur J Endocrinol 2005;152(6):835–43.
[211] de Zegher F, Ong KK, Ibanez L, et al. Growth hormone therapy in short children born small for gestational age. Horm Res 2006;65(Suppl 3):145–52.
[212] Genotropin (somatropin [rDNA origin] for injection) [prescribing information]. New York: Pfizer, Inc.; 2006.
[213] Job JC, Chaussain JL, Job B, et al. Follow-up of three years of treatment with growth hormone and of one post-treatment year, in children with severe growth retardation of intrauterine onset. Pediatr Res 1996;39:354–9.
[214] Ong K, Beardsall K, de Zegher F. Growth hormone therapy in short children born small for gestational age. Early Hum Devel 2005;81:973–80.
[215] Norditropin (somatropin [rDNA origin] for injection) [statement of product characteristics]. EMEA/CPMP/1404/03. Available at: http://www.emea.eu.int/pdfs/human/referral/norditropin/347803en.pdf. Accessed January 4, 2007.
[216] Carel JC, Chatelain P, Rochiccioli P, et al. Improvement in adult height after growth hormone treatment in adolescents with short stature born small for gestational age: results of a randomized controlled study. J Clin Endocrinol Metab 2003;88:1587–93.
[217] Ranke MB, Lindberg A, Cowell CT, et al. Prediction of response to growth hormone treatment in short children born small for gestational age: analysis of data from KIGS (Pharmacia International Growth Database). J Clin Endocrinol Metab 2003;88:125–31.
[218] Binder G, Baur F, Schweizer R, et al. The d3-growth hormone (GH) receptor polymorphism is associated with increased responsiveness to GH in Turner syndrome and short small-for-gestational-age children. J Clin Endocrinol Metab 2006;91:659–64.
[219] de Zegher F, Albertsson-Wikland K, Wollmann HA, et al. Growth hormone treatment of short children born small for gestational age: growth responses with continuous and discontinuous regimens over 6 years. J Clin Endocrinol Metab 2000;85:2816–21.
[220] Fjellestad-Paulsen A, Simon D, Czernichow P. Short children born small for gestational age and treated with growth hormone for three years have an important catch-down five years after discontinuation of treatment. J Clin Endocrinol Metab 2004;89:1234–9.
[221] Barker DJ, Hales CN, Fall CH, et al. Type 2 (non-insulin-dependent) diabetes mellitus, hypertension and hyperlipidaemia (syndrome X): relation to reduced fetal growth. Diabetologia 1993;36:62–7.
[222] Hofman PL, Cutfield WS, Robinson EM, et al. Insulin resistance in short children with intrauterine growth retardation. J Clin Endocrinol Metab 1997;82:402–6.
[223] Crowther NJ, Cameron N, Trusler J, et al. Association between poor glucose tolerance and rapid post natal weight gain in seven-year old children. Diabetologia 1998;41:1163–7.
[224] Jaquet D, Gaboriau A, Czernichow P, et al. Insulin resistance early in adulthood in subjects born with intrauterine growth retardation. J Clin Endocrinol Metab 2000;85:1401–6.
[225] Veening MA, Van Weissenbruch MM, Delemarre-Van De Waal HA. Glucose tolerance, insulin sensitivity, and insulin secretion in children born small for gestational age. J Clin Endocrinol Metab 2002;87:4657–61.
[226] Soto N, Bazaes RA, Pena V, et al. Insulin sensitivity and secretion are related to catch-up growth in small-for-gestational-age infants at age 1 year: results from a prospective cohort. J Clin Endocrinol Metab 2003;88:3645–50.

[227] de Zegher F, Maes M, Gargosky SE, et al. High-dose growth hormone treatment of short children born small for gestational age. J Clin Endocrinol Metab 1996;81(5):1887–92.

[228] Sas T, Mulder P, Aanstoot HJ, et al. Carbohydrate metabolism during long-term growth hormone treatment in children with short stature born small for gestational age. Clin Endocrinol (Oxf) 2001;54(2):243–51.

[229] de Zegher F, Ong K, van Helvoirt M, et al. High-dose growth hormone (GH) treatment in non-GH-deficient children born small for gestational age induces growth responses related to pretreatment GH secretion and associated with a reversible decrease in insulin sensitivity. J Clin Endocrinol Metab 2002;87(1):148–51.

[230] van Pareren Y, Mulder P, Houdijk M, et al. Effect of discontinuation of growth hormone treatment on risk factors for cardiovascular disease in adolescents born small for gestational age. J Clin Endocrinol Metab 2003;88(1):347–53.

[231] Wilton P. Adverse events during GH treatment: 10 years' experience in KIGS, a pharmacoepidemiological survey. In: Ranke MB, Wilton P, editors. Growth hormone therapy in KIGS. 10 years' experience. Heidelberg (Germany): Johann Ambrosius Barth Verlag; 1999. p. 349–64.

[232] Cutfield WS, Lindberg A, Rapaport R, et al. Safety of growth hormone treatment in children born small for gestational age: the US trial and KIGS analysis. Horm Res 2006; 65(Suppl 3):153–9.

[233] Sas T, Mulder P, Hokken-Koelega A. Body composition, blood pressure, and lipid metabolism before and during long-term growth hormone (GH) treatment in children with short stature born small for gestational age either with or without GH deficiency. J Clin Endocrinol Metab 2000;85(10):3786–92.

[234] Czernichow P. Treatment with growth hormone in short children born with intrauterine growth retardation. Endocrine 2001;15(1):39–42.

[235] Simeoni U, Zetterstrom R. Long-term circulatory and renal consequences of intrauterine growth restriction. Acta Paediatr 2005;94(7):819–24.

[236] Chatelain P. Children born with intra-uterine growth retardation (IUGR) or small for gestational age (SGA): long term growth and metabolic consequences. Endocr Regul 2000; 34(1):33–6.

[237] Levy-Marchal C, Czernichow P. Small for gestational age and the metabolic syndrome: which mechanism is suggested by epidemiological and clinical studies? Horm Res 2006; 65(Suppl 3):123–30.

[238] Ranke MB. Towards a consensus on the definition of idiopathic short stature. Horm Res 1996;45(Suppl 2):64–6.

[239] Humatrope (somatropin [rDNA origin] for injection) {prescribing information}. Indianapolis (IN): Eli Lilly and Company; 2006.

[240] Finkelstein BS, Silvers JB, Marrero U, et al. Insurance coverage, physician recommendations, and access to emerging treatments: growth hormone therapy for childhood short stature. JAMA 1998;279(9):663–8.

[241] Price DA. Spontaneous adult height in patients with idiopathic short stature. Horm Res 1996;45(Suppl 2):59–63.

[242] Wit JM, Kamp GA, Rikken B. Spontaneous growth and response to growth hormone treatment in children with growth hormone deficiency and idiopathic short stature. Pediatr Res 1996;39:295–302.

[243] Buchlis JG, Irizarry L, Crotzer BC, et al. Comparison of final heights of growth hormone-treated vs. untreated children with idiopathic growth failure. J Clin Endocrinol Metab 1998;83:1075–9.

[244] Crowne EC, Shalet SM, Wallace WH, et al. Final height in boys with untreated constitutional delay in growth and puberty. Arch Dis Child 1990;65:1109–12.

[245] Bramswig JH, Fasse M, Holthoff ML, et al. Adult height in boys and girls with untreated short stature and constitutional delay of growth and puberty: accuracy of five different methods of height prediction. J Pediatr 1990;117:886–91.

[246] Ranke MB, Grauer ML, Kistner K, et al. Spontaneous adult height in idiopathic short stature. Horm Res 1995;44(4):152–7.
[247] Hintz RL, Attie KM, Baptista J, et al. Effect of growth hormone treatment on adult height of children with idiopathic short stature. N Engl J Med 1999;340:502–7.
[248] Rekers-Mombarg LT, Kamp GA, Massa GG, et al. Influence of growth hormone treatment on pubertal timing and pubertal growth in children with idiopathic short stature. Dutch Growth Hormone Working Group. J Pediatr Endocrinol Metab 1999;12:611–22.
[249] Leschek EW, Rose SR, Yanovski JA, et al. Effect of growth hormone treatment on adult height in peripubertal children with idiopathic short stature: a randomized, double-blind, placebo-controlled trial. J Clin Endocrinol Metab 2004;89:3140–8.
[250] Counts D, Palese T. A novel polymorphism in exon 4 of the growth hormone gene and association of short stature with a known 5′ UTR polymorphism. J Endocr Genet 2001;2:55–60.
[251] Horan M, Millar DS, Hedderich J, et al. Human growth hormone 1 (GH1) gene expression: complex haplotype-dependent influence of polymorphic variation in the proximal promoter and locus control region. Hum Mutat 2003;21:408–23.
[252] Millar DS, Lewis MD, Horan M, et al. Novel mutations of the growth hormone 1 (GH1) gene disclosed by modulation of the clinical selection criteria for individuals with short stature. Hum Mutat 2003;4:424–40.
[253] Spiliotis BE, August GP, Hung W, et al. Growth hormone neurosecretory dysfunction. A treatable cause of short stature. JAMA 1984;251:2223–30.
[254] Zadik Z, Chalew SA, Raiti S, et al. Do short children secrete insufficient growth hormone? Pediatrics 1985;76:355–60.
[255] Spadoni GL, Cianfarani S, Bernardini S, et al. Twelve-hour spontaneous nocturnal growth hormone secretion in growth retarded patients. Clin Pediatr (Phila) 1988;27:473–8.
[256] Rogol AD, Blethen SL, Sy JP, et al. Do growth hormone (GH) serial sampling, insulin-like growth factor-I (IGF-I) or auxological measurements have an advantage over GH stimulation testing in predicting the linear growth response to GH therapy? Clin Endocrinol (Oxf) 2003;58:229–37.
[257] Blair JC, Camacho-Hubner C, Miraki MF, et al. Standard and low-dose IGF-I generation tests and spontaneous growth hormone secretion in children with idiopathic short stature. Clin Endocrinol (Oxf) 2004;60:163–8.
[258] Besson A, Salemi S, Deladoey J, et al. Short stature caused by a biologically inactive mutant growth hormone (GH-C53S). J Clin Endocrinol Metab 2005;90:2493–9.
[259] Pantel J, Legendre M, Cabrol S, et al. Loss of constitutive activity of the growth hormone secretagogue receptor in familial short stature. J Clin Invest 2006;116(3):760–8.
[260] Carlsson LM, Attie KM, Compton PG, et al. Reduced concentration of serum growth hormone-binding protein in children with idiopathic short stature. National Cooperative Growth Study. J Clin Endocrinol Metab 1994;78:1325–30.
[261] Mauras N, Carlsson LM, Murphy S, et al. Growth hormone-binding protein levels: studies of children with short stature. Metabolism 1994;43:357–9.
[262] Goddard AD, Covello R, Luoh SM, et al. Mutations of the growth hormone receptor in children with idiopathic short stature. The Growth Hormone Insensitivity Study Group. N Engl J Med 1995;333:1093–8.
[263] Goddard AD, Dowd P, Chernausek S, et al. Partial growth-hormone insensitivity: the role of growth-hormone receptor mutations in idiopathic short stature. J Pediatr 1997;131:S51–5.
[264] Rosenbloom AL. Physiology and disorders of the growth hormone receptor (GHR) and GH-GHR signal transduction. Endocrine 2000;12:107–19.
[265] Salerno M, Balestrieri B, Matrecano E, et al. Abnormal GH receptor signaling in children with idiopathic short stature. J Clin Endocrinol Metab 2001;86:3882–8.
[266] Bonioli E, Taro M, Rosa CL, et al. Heterozygous mutations of growth hormone receptor gene in children with idiopathic short stature. Growth Horm IGF Res 2005;15:405–10.

[267] Sjoberg M, Salazar T, Espinosa C, et al. Study of GH sensitivity in Chilean patients with idiopathic short stature. J Clin Endocrinol Metab 2001;86:4375–81.
[268] Savage MO, Attie KM, David A, et al. Endocrine assessment, molecular characterization and treatment of growth hormone insensitivity disorders. Nat Clin Pract Endocrinol Metab 2006;2:395–407.
[269] Bussieres L, Souberbielle JC, Pinto G, et al. The use of insulin-like growth factor 1 reference values for the diagnosis of growth hormone deficiency in prepubertal children. Clin Endocrinol (Oxf) 2000;52:735–9.
[270] Mauras N, Walton P, Nicar M, et al. Growth hormone stimulation testing in both short and normal statured children: use of an immunofunctional assay. Pediatr Res 2000;48:614–8.
[271] Domene HM, Bengolea SV, Martinez AS, et al. Deficiency of the circulating insulin-like growth factor system associated with inactivation of the acid-labile subunit gene. N Engl J Med 2004;350:570–7.
[272] Domene HM, Bengolea SV, Jasper HG, et al. Acid-labile subunit deficiency: phenotypic similarities and differences between human and mouse. J Endocrinol Invest 2005;28:43–6.
[273] Hwa V, Haeusler G, Pratt KL, et al. Total absence of functional acid labile subunit, resulting in severe insulin-like growth factor deficiency and moderate growth failure. J Clin Endocrinol Metab 2006;91:1826–31.
[274] Olney RC, Bukulmez H, Bartels CF, et al. Heterozygous mutations in natriuretic peptide receptor-B (NPR2) are associated with short stature. J Clin Endocrinol Metab 2006;91(4):1229–32.
[275] De Bellis A, Salerno M, Conte M, et al. Antipituitary antibodies recognizing growth hormone (GH)-producing cells in children with idiopathic GH deficiency and in children with idiopathic short stature. J Clin Endocrinol Metab 2006;91(7):2484–9.
[276] Wudy SA, Hagemann S, Dempfle A, et al. Children with idiopathic short stature are poor eaters and have decreased body mass index. Pediatrics 2005;116:e52–7.
[277] Finkelstein BS, Imperiale TF, Speroff T, et al. Effect of growth hormone therapy on height in children with idiopathic short stature: a meta-analysis. Arch Pediatr Adolesc Med 2002;156(3):230–40.
[278] Food and Drug Administration. Transcript of the meeting of the Endocrinologic and Metabolic Drugs Advisory Committee, September 28–29, 1987. Rockville (MD): Food and Drug Administration; 1987. p. 169.
[279] Ross JL, Sandberg DE, Rose SR, et al. Psychological adaptation in children with idiopathic short stature treated with growth hormone or placebo. J Clin Endocrinol Metab 2004;89(10):4873–8.
[280] Wit JM, Rekers-Mombarg LT, Cutler GB, et al. Growth hormone (GH) treatment to final height in children with idiopathic short stature: evidence for a dose effect. J Pediatr 2005;146:45–53.
[281] Quigley CA, Gill AM, Crowe BJ, et al. Safety of GH treatment in patients with idiopathic short stature. J Clin Endocrinol Metab 2005;90:5188–96.
[282] Crowe BJ, Rekers-Mombarg LT, et al. European Idiopathic Short Stature Group. Effect of growth hormone dose on bone maturation and puberty in children with idiopathic short stature. J Clin Endocrinol Metab 2006;91(1):169–75.
[283] Leschek EW, Troendle JF, Yanovski JA, et al. Effect of growth hormone treatment on testicular function, puberty, and adrenarche in boys with non-growth hormone-deficient short stature: a randomized, double-blind, placebo-controlled trial. J Pediatr 2001;138(3):406–10.
[284] Lindgren AC, Chatelain P, Lindberg A, et al. Normal progression of testicular size in boys with idiopathic short stature and isolated growth hormone deficiency treated with growth hormone: experience from the KIGS. Horm Res 2002;58:83–7.

[285] Darendeliler F, Ranke MB, Bakker B, et al. Bone age progression during the first year of growth hormone therapy in pre-pubertal children with idiopathic growth hormone deficiency, Turner syndrome or idiopathic short stature, and in short children born small for gestational age: analysis of data from KIGS (Pfizer International Growth Database). Horm Res 2005;63:40–7.

[286] Bryant J, Cave C, Milne R. Recombinant growth hormone for idiopathic short stature in children and adolescents. Cochrane Database Syst Rev 2003;27:CD004440.

[287] Weise KL, Nahata MC. Growth hormone use in children with idiopathic short stature. Ann Pharmacother 2004;38:1460–8.

[288] Kemp SF, Kuntze J, Attie KM, et al. Efficacy and safety results of long-term growth hormone treatment of idiopathic short stature. J Clin Endocrinol Metab 2005;90: 5247–53.

[289] Ranke MB, Schweizer R, Elmlinger MW, et al. Relevance of IGF-I, IGFBP-3, and IGFBP-2 measurements during GH treatment of GH-deficient and non-GH-deficient children and adolescents. Horm Res 2001;55(3):115–24.

[290] Park P, Cohen P. Insulin-like growth factor I (IGF-I) measurements in growth hormone (GH) therapy of idiopathic short stature (ISS). Growth Horm IGF Res 2005;15(Suppl A): S13–20.

[291] Wit JM, Rekers-Mombarg LT. Dutch Growth Hormone Advisory Group. Final height gain by GH therapy in children with idiopathic short stature is dose dependent. J Clin Endocrinol Metab 2002;87(2):604–11.

[292] Blum WF, Machinis K, Shavrikova EP, et al. The growth response to growth hormone (GH) treatment in children with isolated GH deficiency is independent of the presence of the exon 3-minus isoform of the GH receptor. J Clin Endocrinol Metab 2006;91:4171–4.

[293] Carrascosa A, Esteban C, Espadero R, et al, for the Spanish SGA Study Group. The d3/fl-growth hormone (GH) receptor polymorphism does not influence the effect of GH treatment (66 microg/kg per day) or the spontaneous growth in short non-GH-deficient small-for-gestational-age children: results from a two-year controlled prospective study in 170 Spanish patients. J Clin Endocrinol Metab 2006;91(9):3281–6.

[294] Katz A, Nambi SS, Mather K, et al. Quantitative insulin sensitivity check index: a simple, accurate method for assessing insulin sensitivity in humans. J Clin Endocrinol Metab 2000; 85:2402–10.

[295] Saenger P, Attie KM, Martino-Nardi J, et al. Metabolic consequences of 5-year growth hormone (GH) therapy in children treated with GH for idiopathic short stature. Genentech Collaborative Study Group. J Clin Endocrinol Metab 1998;83:3115–20.

[296] American Association of Clinical Endocrinologists. Medical guidelines for clinical practice for growth hormone use in adults and children—2003 Update. Endocr Pract 2003;9(1): 65–76.

[297] Wilson TA, Rose SR, Cohen P, et al, for the The Lawson Wilkins Pediatric Endocrinology Society Drug and Therapeutics Committee. Update of guidelines for the use of growth hormone in children: the Lawson Wilkins Pediatric Endocrinology Society Drug and Therapeutics Committee. J Pediatr 2003;143(4):415–21.

[298] Hindmarsh PC, Cole TJ. Evidence-based growth hormone therapy prediction models. J Pediatr Endocrinol Metab 2000;13(Suppl 6):1359–64.

[299] Ranke MB. New paradigms for growth hormone treatment in the 21st century: prediction models. J Pediatr Endocrinol 2000;13(Suppl 6):1365–9.

[300] Blum WF, Shavrikova EP, Keller A, et al. New methods for adult height prediction in children with idiopathic short stature [abstract]. Presented at the Endocrine Society's 87th Annual Meeting. San Diego, CA, June 4–7, 2005.

[301] Peyrat JP, Bonneterre J, Hecquet B, et al. Plasma insulin-like growth factor-1 (IGF-1) concentrations in human breast cancer. Eur J Cancer 1993;29A:492–7.

[302] Bruning PF, van DJ, Bonfrer JM, et al. Insulin-like growth-factor-binding protein 3 is decreased in early-stage operable pre-menopausal breast cancer. Int J Cancer 1995;62:266–70.

[303] Hankinson SE, Willett WC, Colditz GA, et al. Circulating concentrations of insulin-like growth factor-I and risk of breast cancer. Lancet 1998;351:1393–6.
[304] Mantzoros CS, Tzonou A, Signorello LB, et al. Insulin-like growth factor 1 in relation to prostate cancer and benign prostatic hyperplasia. Br J Cancer 1997;76:1115–8.
[305] Chan JM, Stampfer MJ, Giovannucci E, et al. Plasma insulin-like growth factor-I and prostate cancer risk: a prospective study. Science 1998;279:563–6.
[306] Wolk A, Mantzoros CS, Andersson SO, et al. Insulin-like growth factor 1 and prostate cancer risk: a population-based, case-control study. J Natl Cancer Inst 1998;90:911–5.
[307] Brunner JE, Johnson CC, Zafar S, et al. Colon cancer and polyps in acromegaly: increased risk associated with family history of colon cancer. Clin Endocrinol (Oxf) 1990;32:65–71.
[308] Ron E, Gridley G, Hrubec Z, et al. Acromegaly and gastrointestinal cancer. Cancer 1991; 68:1673–7.
[309] Jenkins PJ, Fairclough PD, Richards T, et al. Acromegaly, colonic polyps and carcinoma. Clin Endocrinol (Oxf) 1997;47:17–22.
[310] Kaaks R, Toniolo P, Akhmedkhanov A, et al. Serum C-peptide, insulin-like growth factor (IGF)-I, IGF-binding proteins, and colorectal cancer risk in women. J Natl Cancer Inst 2000;92:1592–600.
[311] Ma J, Pollak M, Giovannucci E, et al. A prospective study of plasma levels of insulin-like growth factor I (IGF-I) and IGF-binding protein-3, and colorectal cancer risk among men. Growth Horm IGF Res 2000;10(Suppl A):S28–9.
[312] Yu H, Spitz MR, Mistry J, et al. Plasma levels of insulin-like growth factor-I and lung cancer risk: a case-control analysis. J Natl Cancer Inst 1999;91:151–6.
[313] Lukanova A, Toniolo P, Akhmedkhanov A, et al. A prospective study of insulin-like growth factor-I, IGF-binding proteins-1, -2 and -3 and lung cancer risk in women. Int J Cancer 2001;92:888–92.
[314] Khandwala HM, McCutcheon IE, Flyvbjerg A, et al. The effects of insulin-like growth factors on tumorigenesis and neoplastic growth. Endocr Rev 2000;21:215–44.
[315] Furstenberger G, Senn HJ. Insulin-like growth factors and cancer. Lancet Oncol 2002;3:298–302.
[316] Ali O, Cohen P, Lee KW. Epidemiology and biology of insulin-like growth factor binding protein-3 (IGFBP-3) as an anti-cancer molecule. Horm Metab Res 2003;35:726–33.
[317] Ogilvy-Stuart AL, Gleeson H. Cancer risk following growth hormone use in childhood: implications for current practice. Drug Saf 2004;27:369–82.
[318] Renehan AG, Zwahlen M, Minder C, et al. Insulin-like growth factor (IGF)-I, IGF binding protein-3, and cancer risk: systematic review and meta-regression analysis. Lancet 2004; 363:1346–53.
[319] Sperling MA, Saenger PH, Hintz R, et al. Growth hormone treatment and neoplasia—coincidence or consequence? J Clin Endocrinol Metab 2002;87:5351–2.
[320] Wang HJ, Geller F, Dempfle A, et al. Ghrelin receptor gene: indentification of several sequence variants in extremely obese children and adolescents, healthy normal-weight and underweight students, and children with normal stature. J Clin Endocrinol Metab 2004; 89(1):157–62.
[321] Lewis MD, Horan M, Millar DS, et al. A novel dysfunction growth hormone variant (Ile179Met) exhibits a decreased ability to activate the extracellular signal-regulated kinase pathway. J Clin Endocrinol Metab 2004;89(3):1068–75.
[322] Takahashi Y, Shirono H, Arisaka O, et al. Biologically inactive growth hormone caused by an amino acid substitution. J Clin Invest 1997;100(5):1156–65.
[323] Sjoberg M, Salazar T, Espinosa C, et al. Study of GH sensitivity in Chilean patients with idiopathic stature. J Clin Endocrinol Metab 2001;86(9):4375–81.
[324] Sanchez JE, Perera E, Baumbach L, et al. Growth hormone receptor mutations in children with idiopathic short stature. J Clin Endocrinol Metab 1998;83(11):4079–83.
[325] Ayling RM, Ross RJ, Towner P, et al. A dominant-negative mutation of the growth hormone receptor causes familial short stature. Nature Genet 1997;16(1):13–4.

[326] Ayling RM, Ross RJ, Towner P, et al. New growth hormone receptor exon 9 mutation causes genetic short stature. Acta Paediatrica Suppl 1999;88(428):168–72.
[327] Kofoed EM, Hwa V, Little B, et al. Growth hormone insensitivity associated with a Stat5b mutation. N Engl J Med 2003;349(12):1139–47.
[328] Kawashima Y, Kanzaki S, Yang F, et al. Mutation at cleavage site of insulin-like growth factor receptor in a short-stature child born with intrauterine growth retardation. J Clin Endocrinol Metab 2005;90(8):4679–87.

Growth Hormone and the Transition from Puberty into Adulthood

Andrea F. Attanasio, MD[a],*, Stephen M. Shalet, MD[b]

[a]Cascina del Rosone, Via Monsarinero 45, 14041 Agliano Terme, Italy
[b]Department of Endocrinology, Christie Hospital, NHS Trust, Wilmslow Road, Manchester M20 4BX, United Kingdom

With modern growth hormone (GH) replacement algorithms, children with a diagnosis of GH deficiency (GHD) achieve at the end of pediatric GH treatment an adult height that is on the average in the normal range. Recent experience with GH replacement in young adults with childhood onset (CO) GHD has shown, however, that these patients, specifically those with severe GHD, present with variable degrees of somatic immaturity [1,2]. As childhood GH treatment is discontinued when final height is attained, attention moved on to the phase of somatic development, which follows the end of longitudinal growth. This phase, somewhat arbitrarily called "transition," has been excluded previously from consideration for either pediatric or adult GH replacement. In a recently published Consensus Statement on the Management of the Growth Hormone Treated Adolescent in the Transition to Adult Care, the transition has been defined as a period starting in late puberty and ending with full adult maturation (ie, from mid to late teenage years until 6–7 years after achievement of final height) [3].

This article reviews the changes that take place during this phase of development and their relevance for the attainment of adult body maturation. The critical role of GH in this process is described.

The physiology of transition

As a rule, longitudinal growth is considered completed when the height velocity of a subject has fallen below a certain value (usually 2 or 1.5 cm/y) and bone maturation is almost (95%–98%) completed. At this point only

---

* Corresponding author.
E-mail address: andreaattanasio1@alice.it (A.F. Attanasio).

a small residual growth potential remains; however, maturation of body mass components (lean [LBM]; fat [FM]; and bone [BMC]) is not yet complete. Peak bone mass is not achieved before age 20 to 25 years [4–6] and end-stage maturation of the other body components, LBM and FM, is also achieved after the end of the pubertal growth spurt [7]. Boys continue gaining LBM well beyond age 20, and girls until age 14, without further significant increases thereafter [7,8]. For FM, the gain in boys is negligible after the end of puberty, whereas girls continue gaining FM until age 20. After the end of pubertal growth boys continue gaining preferentially muscle mass and girls continue gaining FM [7–10].

The timing of the overall changes in bone, muscle, and fat mass may differ individually, but it is important to realize that both females and males continue tracking their own adult target body composition well into the third decade of life.

These gender dimorphic changes take place in the presence of both declining GH secretion and serum insulin-like growth factor (IGF-I) levels. Endogenous GH secretion and serum IGF-I levels increase until peak height velocity and decrease thereafter [11,12]. Less GH action is required for end-stage somatic maturation during transition, but how in the individual subject the timing of the declining pattern of GH secretion relates to the described changes in body composition is unclear.

Effect of discontinuing pediatric growth hormone treatment at final height

In early studies on adult GH replacement, CO patients were found to be smaller, to have less muscle and bone mass, and lower IGF-I and IGFBP-3 serum concentrations than patients with adult onset (AO) GHD [1]. The authors confirmed these findings in a group of young adults with severe CO-GHD who had received state-of-the-art GH replacement during childhood [13]. Compared with GH-untreated AO-GHD patients of the same age range, they were on the average about 1 height SD shorter, and as shown in Fig. 1, body mass index values were significantly lower, indicating that on the average CO patients had less body mass per unit height than AO patients. A height-normalized comparison of the individual components of body mass assessed by dual energy X-ray absorptiometry (DXA) showed that they had about 80% LBM, FM, and BMC compared with the age-matched AO patients and had not reached mature adult body mass. If pediatric GH replacement had been optimal, then in the GH-unreplaced state their body composition should have been at least comparable with that of AO-GHD subjects, who had previously gone through a normal prepubertal and pubertal development.

Another important confirmatory finding from this study was the very significant difference between AO and CO subjects in their IGF-I and IGFBP-3 levels. A difference in IGF-I levels between AO and CO had been noted in

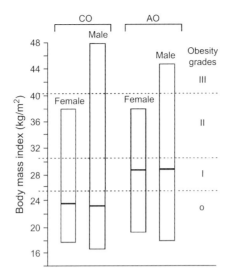

Fig. 1. Mean and range of body mass index values in CO and AO GHD patients, with obesity grades shown by dashed lines. (*From* Attanasio AF, Howell S, Bates PC, et al. Body composition, IGF-I and IGFBP-3 concentrations as outcome measures in severely GH-deficient (GHD) patients after childhood GH treatment: a comparison with adult onset GHD patients. J Clin Endocrinol Metab 2002;87:3368–72; with permission.)

other studies [1,14]; in this study differences up to 3 to 4 standard deviation (SD) were found between CO and AO patients who were comparable for age and severity of GHD. IGF-I and IGFBP-3 levels significantly correlated with height and LBM in CO and AO, and with BMC only in CO, indicating that the developmental delay affected the IGF-I system to the same extent as the components of body mass.

Young adults with CO GHD have not reached adult targets, and the question arises if the contribution of GH action during the transition phase may account for these differences. In early studies a progressive reduction in muscle strength and size with an increase in FM was described after termination of pediatric GH replacement [15,16]. More recent studies performed in patients with severe CO GHD followed up for periods up to 2 years after withdrawal of pediatric GH confirmed a progressive increase in overall and abdominal FM (Fig. 2), a halt or decline in accrual of LBM, and BMC and a deterioration of the lipid profile [17]. In a cohort of 149 patients with severe GHD the authors found a statistically significant inverse correlation between the time since the last pediatric GH injection and the baseline serum concentrations of bone alkaline phosphatase and the type I collagen C-terminal telopeptide/creatinine ratio (Fig. 3) [18]. In addition, in the 32 patients who had been randomized to no GH treatment, a further progressive decrease in bone turnover markers was seen during the 2 years of longitudinal follow-up. In the same group of patients, rates of accrual per year of

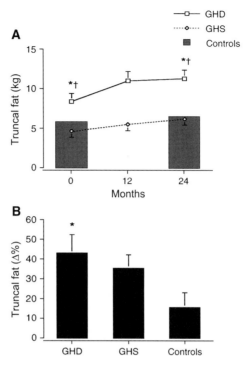

Fig. 2. (*A*) Truncal fat mass before and 12 and 24 months after the discontinuation of GH treatment in adolescents. One group comprised 21 adolescents with severe GHD continuing into adulthood (*solid line*), the second group comprised 19 adolescents with their own GH secretion (growth hormone sufficient [GHS] [*broken line*]), and the third group consisted of 16 healthy controls studied on two occasions with 2 years between studies. (*B*) Percent change in truncal fat mass between baseline and 2 years after discontinuing GH treatment in the 2 patient groups and the controls. *, $P < 0.05$ compared with controls; †, $P < 0.05$ compared with the GHS group. (*From* Johannsson G, Albertsson-Wikland K, Bengtsson B-Å. Discontinuation of growth hormone (GH) treatment: metabolic effects in GH-deficient and GH-sufficient adolescent patients compared with control subjects. J Clin Endocrinol Metab 1999;84:4516–24; with permission.)

BMC and LBM were significantly reduced compared with normal-developing transition subjects. Comparable changes with GH withdrawal have also been described in patients with less severe GHD, and Tauber and coworkers [19] found an increased total body fat and decreased LBM in 91 adolescents with partial GHD 1 year after completion of childhood GH treatment.

Taken together, these studies show that with the abrupt termination of GH-dependent somatic maturation GHD patients in transition start developing the typical features of the adult GHD syndrome, characterized by obesity, reduced muscle mass, and dyslipidemia. These findings predispose them prematurely to an increased risk of cardiovascular mortality and an adverse cardiovascular profile, both of which occur in adult GHD patients [20,21].

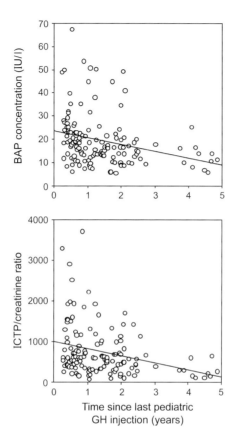

Fig. 3. Correlation of serum Bone Alkaline Phosphatase (BAP) (*upper panel*) and urinary type I collagen α-cross-linked C-terminal telopeptide (ICTP)/creatinine ratio (*lower panel*) with time since stopping pediatric GH treatment in patients with childhood GHD. (*From* Shalet SM, Shavrikova E, Cromer M, et al. Effect of growth hormone (GH) treatment on bone in postpubertal GH-deficient patients: a 2-year randomized, controlled, dose-ranging study. J Clin Endocrinol Metab 2003;88:4124–9; with permission.)

Growth hormone replacement during the transition phase

Re-establishing the diagnosis of growth hormone deficiency

Over the last 20 years there have been a number of studies in which the GH status of children and young adults who had received GH replacement during childhood was reassessed after completion of growth and puberty [22–29]. In these studies, GH status was found to be normal in 20% to 87% [23–29], and the etiologic classification of the childhood diagnosis in most was isolated idiopathic GHD. Whereas childhood GH replacement is offered to children with all grades of GHD ranging from complete absence of GH secretion to GH insufficiency, adult GH replacement therapy has been recommended only for severe GHD.

In the past it has been usual to use more than one test to establish GH status at reassessment, either two GH provocative tests [27] or a single GH provocative test plus an IGF-I estimation [30]. With either approach discordance between the two test results may be seen especially if the only categories being defined are normal and severely GHD. The guidelines issued by ESPE Consensus Workshop on the management of GHD during the transition [3] propose, for the first time ever, two different strategies of investigative approach dependent on the likelihood of the patient being severely GHD. After an interval of at least 1 month since discontinuation of GH, GH reserve is assessed by measurement of IGF-I or a dynamic test of GH status. Tests currently recommended are the insulin tolerance test, with the arginine or glucagon tests as alternatives. The GH-releasing hormone–arginine test may be unreliable in patients with suspected hypothalamic disease; GH-releasing hormone alone or clonidine is not recommended.

Patients with a high likelihood of GHD are defined as those with severe GHD in childhood with or without two or three additional hormone deficiencies, which may be caused by a defined genetic cause; those with severe GHD caused by structural hypothalamic-pituitary abnormalities; central nervous system tumors; or patients having received high-dose cranial irradiation. In this group an IGF-I value ≤ -2 SD following discontinuation of GH is considered sufficient evidence of severe GHD. If the IGF-I level is > -2 SD a GH provocation test should be performed. If this shows a "pathologically" low GH response the diagnosis of severe GHD is confirmed; if the peak GH response is above the appropriate cutoff value, the diagnosis should be reconsidered.

The second group with a low likelihood of severe GHD represents the remaining patients, a group dominated by the diagnostic category of idiopathic GHD, either isolated or with one additional pituitary hormonal deficit. This patient group requires an IGF-I measurement and one GH provocation test. If both are low, the diagnosis of severe GHD is confirmed; if both are normal, the patient can be discharged unless they are at risk of evolving endocrinopathy, such as those who received prior central nervous system irradiation. If the tests are discordant, the patient should receive follow-up.

The second major diagnostic recommendation in the ESPE guidelines relates to the biochemical threshold for the GH provocation test used to define severe GHD at reassessment. Up until this point, the diagnostic threshold for a 16 year old at reassessment was exactly the same as that for defining severe GHD in a 50 year old with pituitary disease (ie, a peak GH response to an insulin tolerance test of less than 3 µg/L). Now, however, it was believed that much greater weight should be placed on the long accepted observation that under normal circumstances a 16 year old produces far more GH than a 50 year old and the threshold for defining severe GHD in a 16 year old should be significantly greater than that for a 50-year-old patient. At reassessment the ESPE guidelines recommend a diagnostic threshold of 5 µg/L for the peak GH response to any provocative test.

Growth hormone dosing

Pediatric GH dosing is a body weight–based algorithm, and the average pediatric GH dose for GH-deficient subjects is 25 µg/kg/d, although in clinical practice higher doses are not uncommon. Adult GH replacement uses a different algorithm, because the final replacement dose is up-titrated according to clinical response, IGF-I levels, and possible side effects, resulting in significantly lower dosage, very seldom exceeding 1 mg/day [31]. Published studies with the primary objective of evaluating the effects of GH replacement during transition, however, have used a fixed weight-based dose regimen [18,32–34]. Doses in the range from 12.5 to 25 µg/kg/d were given for periods of 1 to 2 years. Two studies, by Shalet and coworkers [18] and Underwood and coworkers [33], were dose ranging and randomized patients to placebo or control and to two identical GH treatment arms, one with a dose of 12.5 and one with a dose of 25 µg/kg/d for 2 years. Although in the Underwood and coworkers [33] study a clear dose effect on the IGF-I levels and body composition measures was seen, the dose effect on the same parameters was much less pronounced in the Shalet and coworkers [18] study.

Nørrelund and coworkers [32] have shown that a fixed GH dose of 16 µg/kg/d causes pronounced insulin resistance even if patients lose body fat and increase fat-free mass, suggesting that beneficial effects of GH treatment on body composition should be carefully evaluated against the direct insulin antagonistic effects. When starting GH replacement in a transition patient, an adult GH dosing algorithm should be used and the ESPE consensus guidelines [3] recommend a restart with a total daily dose of 0.2 to 0.5 mg, with subsequent dose-adjustment on the basis of IGF-I measurements.

With GH reinstitution IGF-I levels may not normalize despite good clinical response. Such factors as gonadal hormone replacement may affect baseline and GH-stimulated IGF-I levels, because it is known that oral but not transdermal estrogen inhibits IGF-I production in the liver. No studies addressing this aspect during transition, however, have been performed.

Effect of growth hormone replacement on lean body mass and fat mass

Four different randomized, controlled studies have demonstrated a significant effect of GH replacement on body composition in transition patients. Vahl and coworkers [35] found a significant increase in thigh muscle mass and a significant decrease in abdominal FM after 12 months of GH treatment at a dose of 16 µg/kg/d. In two studies, the authors' [36] and that of Underwood and coworkers [33], patients were randomized to the same dose regimens, either 12.5 or 25 µg/kg/d, or to control. In the Underwood and coworkers [33] study, in which 65 patients with a maximum age of 35 years were treated, significant dose-dependent increases in LBM (DXA) and decreases in FM (DXA) were seen after 2 years. In the authors' multinational study [36], which had enrolled 149 subjects, changes in body composition were significant but without a demonstrable GH dose-effect. The fourth study, a randomized, placebo-controlled study by Mauras and coworkers [34] in 40 subjects, used a GH dose of 20 μg/kg/d and failed to show any significant effect of GH versus placebo on LBM and FM (DXA) after 2 years of treatment.

The effect of GH on body composition measures could be documented in three out of four studies, and a dose effect could be demonstrated in one study but not in the other. The large variability that exists in developmental status between end of pubertal development, transition and young adulthood, age and degree of severity of GHD, and the sample size of the study may well explain the discrepancies in the results described. It is not surprising that the most clear-cut results were seen in the Underwood and coworkers [33] study, in which most patients were beyond the age of transition and had a very homogenous baseline GH status (severe GHD).

In contrast to the other studies reported, the authors also analyzed the changes in body composition by gender. This analysis shows that while females gain a small amount of LBM without significant changes in FM; males gain a very significant amount of LBM and lose a significant amount of FM. For LBM, this gender dimorphic pattern is best shown by the relationship between LBM and height change. Fig. 4 shows that the changes in the LBM/height ratio under GH treatment with both dosages used are significantly greater in males than in females. The different changes seen in the LBM/height ratio and the contrasting changes in FM seen in males and females suggest that the pattern of response to GH replacement in patients reflects primary developmental gender dimorphism.

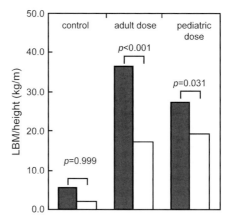

Fig. 4. Changes from baseline in LBM/height after 2 years of GH treatment in male (■) and female (□) patients with CO GHD, with P values for differences between genders. (*From* Attanasio AF, Shavrikova E, Blum WF, et al. Continued growth hormone (GH) treatment after final height is necessary to complete somatic development in childhood-onset GH-deficient patients. J Clin Endocrinol Metab 2004;89:4857–62; with permission.)

Effect of growth hormone replacement on bone

AO-GHD adults have normal bone density and morphometry, and lesser degrees of GHD, such as GH insufficiency, whether present during childhood or acquired during adult life, have a negligible impact on the skeleton. CO-GHD adults have normal trabecular density, marginally reduced cortical density, but significant reductions in cortical thickness, cortical cross-sectional area, and overall cortical content, which along with the smaller bone size account for the reduced bone mineral density (BMD) observed in DEXA studies and place these patients at increased risk of fracture [37,38]. Reduced values for the biomechanical measures of bone strength also suggest an increased risk of fracture in this subset of GHD adults [14], a prediction that has now been borne out in clinical practice [39].

Over the last few years a small number of studies have investigated the impact of GH replacement on bone in the transition period. A United Kingdom 1-year study of bone mineral accretion in 24 adolescents aged between 14 and 20 years suggested that in adolescent patients with severe GHD, discontinuation of GH at completion of growth may limit the attainment of peak bone mass, predisposing the individual to clinically significant osteopenia in later adult life [40]. In the 12 patients who continued GH treatment, on a dose of 17 µg/kg/d, median BMC increased by 3.8% at 6 months and by 6% at 12 months ($P < .001$) compared with baseline. In those 12 patients who discontinued GH treatment, median BMC was unchanged at 6 and 12 months [40].

In the Underwood and coworkers [33] study with two treatment arms using 12.5 and 25 µg/kg/d of GH versus a placebo arm, spinal BMD did not change significantly in the placebo group but did so in both treatment arms; furthermore, the increase in spinal BMD in the treatment arms was dose-related (3.3% [12.5 µg] versus 5.2% [25 µg]).

The atypical aspect of this study in relation to transition is that the mean age of the patients was 23.8 years and they had been off pediatric GH for a mean time of 5.6 years.

In the authors' study [18], performed in patients with severe GHD (baseline IGF-I SD below first centile) who had terminated pediatric GH at final height, significant increases in BMC and BMD (DEXA) measured centrally were seen with the dose of 12.5 µg/kg/d (N = 59) and with the dose of 25 µg/kg/d (N = 58) compared with no GH treatment (N = 32). Total BMC increased by 9.5% in the adult-dose group, 8.1% in the pediatric-dose group, and 5.6% in controls; there was a significant treatment effect but no dose effect. There were no gender-related differences in BMC changes with either dose [18].

In the United States randomized multicenter study of Mauras and coworkers [34] the study BMD z-scores were assessed at baseline and after 2 years in a GH treatment group (20 µg/kg/d, N = 25); a placebo group (N = 15); and an additional control group of 18 subjects who proved to be of normal GH status at retest. No significant differences in BMD z-scores at baseline across the three groups and no differences in the rate of increase in BMD over 2 years follow-up in any of the three treatment groups were found. The main diagnosis responsible for the GHD in this population of patients seemed to be idiopathic isolated GHD, however, and closer inspection of the GH–IGF-I parameters suggests that this is a cohort of patients with a much less severe form of GHD. The results are in line with the observations of Murray and coworkers [38] and consistent with the view that a significant impact of GH status on the skeleton at this age is only seen in transitional patients with severe GHD.

Effects of growth hormone replacement on lipid status and glucose homeostasis

Compared with normal controls or with subjects who had a normal GH response at retest after pediatric GH treatment, variable lipid abnormalities had been described in GHD transition patients. Johannsson and coworkers [17] performed a parallel, 2-year follow-up of 40 GHD subjects who had terminated GH treatment and of 16 postadolescent normals matched for age, gender, height, and body weight. At baseline, before pediatric GH treatment was stopped, serum concentrations of total cholesterol, low-density lipoprotein cholesterol, and apolipoprotein B were higher in the GHD patients than in the controls, and progressively increased in the former but not in the latter after discontinuation of GH treatment. Serum concentrations of high-density lipoprotein cholesterol increased in the healthy controls, and a slight reduction was seen in the GHD patients, the overall change over 2 years being statistically significant between the 2 groups.

Other studies have reported variable effects of GH replacement on lipid status. In the previously mentioned randomized controlled studies, no changes during 2 years of GH treatment in total low-density lipoprotein and high-density lipoprotein cholesterol, and triglycerides were seen by Mauras and coworkers [34], and an increase in high-density lipoprotein cholesterol only was seen by Vahl and coworkers [35]. In the Underwood study [33] a significant decrease in low-density lipoprotein cholesterol and the low-density lipoprotein/high-density lipoprotein ratio was seen only with the higher (25 µg/kg/d) GH dose. In the authors' randomized, controlled study, a statistically significant increase in total cholesterol levels from baseline to the 2-year end point in the GH-untreated control group was found, and no change in the GH-treated groups was found [36].

The discrepancies in response to GH intervention in the mentioned studies can be ascribed to differences in the population studied and in baseline lipid status. Lipid status is not only affected by GHD but also by other factors, such as obesity and diet. Overall, however, there is consistent evidence that the abnormalities in lipid status typically described for the adult GHD syndrome start developing in GHD subjects already at a relatively young age, before they become adult.

Not many studies have examined glucose and insulin homeostasis. Fasting insulin levels significantly decrease after withdrawal of GH at final height [17] and significantly increase with GH reinstitution [32]. Insulin resistance while on GH replacement has been quantitatively assessed in two controlled studies. Using measures of insulin resistance (HOMA) and sensitivity (QUICKI) Mauras and coworkers [34] found no change after 2 years of GH replacement at a dose of 20 μg/kg/d. Nørrelund and coworkers [32] evaluated insulin sensitivity by the euglycemic glucose clamp method in a group of transition GHD subjects taking GH at a dose of 16 μg/kg/d and found a significant decrease in insulin sensitivity despite a significant loss of FM and gain in LBM. The data described must be interpreted against the background of the physiologic decrease in insulin sensitivity that occurs during normal puberty [41]. Because insulin resistance during normal puberty has been causally linked to the concomitant increase in GH secretion [42], the GH-insulin relationship is individually difficult to evaluate in a subject during transition when GH levels start to decline, even more if this subject is GHD. It follows that the relationship between GH dose and glucose homeostasis should be carefully monitored during transition.

Quality of life aspects

A number of studies have tried to determine which patients are most likely to respond to GH replacement with a significant improvement in quality of life. No effect of gender, age, duration of GHD, severity of GHD, number of additional anterior pituitary hormone deficits, concurrent changes in body composition, or IGF-I has been found to predict quality of life in most studies. To date, the only factor consistently able to predict the degree of improvement in quality of life with GH replacement has been baseline quality of life. Females have been reported to show a greater improvement in quality of life with GH replacement; however, this has not been reported consistently.

Following the study by Attanasio and coworkers [1] showing improvements in the Nottingham Health Profile with GH therapy in AO patients but not CO patients, there has been considerable debate as to the effect of GH on quality of life of CO patients. When CO patients with subjectively impaired quality of life are specifically selected, these patients show at least equal improvements in quality of life to AO patients also selected for poor quality of life [43].

Up until recently, quality of life had not been specifically studied in GHD transition patients. In the authors' previously mentioned transition study, they assessed quality of life at baseline, and after 1 and 2 years of GH treatment in a subgroup of 66 patients using an adult GHD-specific questionnaire, the QLS-H [44]. The mean age of the females was 17.5 years and the males 18.9 years; the time interval since the last pediatric GH treatment was 1.7 and 1.4 years for females and males, respectively.

SD scores for QLS-H were calculated from normative data, specific to country of origin, gender, and age range of the patients. Baseline QLS-H SD scores were -0.35 ± 1.17 in females and -0.70 ± 1.05 in males ($P = 0.280$), without differences between isolated GHD and multiple pituitary hormone deficiency. SD scores for individual dimensions of ability to become sexually aroused, ability to tolerate stress, body shape, concentration, initiative and drive, physical stamina, and self-confidence were significantly lower than the normal average. Particularly affected were body shape (SD score -0.80) and sexual arousal (SD score -0.41). Total QLS-H SD score increased slightly but not significantly for combined GH-treatment groups compared with control at year 1 but not after year 2; no dose effect of GH was observed. GH treatment significantly increased SD score from baseline to year 2 for sexual arousal and body shape (0.23 and 0.46, respectively), without differences between isolated GHD and multiple pituitary hormone deficiency. Although overall baseline quality of life was not compromised in severely GHD patients during the transition period, dimensions related to age-specific psychologic problems were significantly worse than in healthy subjects and seemed to respond positively to GH treatment [44].

Summary

Somatic development is not terminated at final height and available studies with GH replacement in severely GHD individuals provide evidence that GH action is necessary in the postpubertal transition phase to achieve normal adult status. To this purpose, the ESPE consensus guidelines have created a suitable framework to be used in the clinical care of these patients. Clinical experience in treating these patients is still limited and many aspects require further research. During transition there is a shift in GH action from promoting body growth to maintaining body proportions and metabolic balance. The timely quantitative and qualitative pattern of this shift is not sufficiently defined to provide a background to individually optimize GH dose and efficacy for the transition period. Also, very significant psychosocial adjustment takes place during transition. CO patients are not, and should not be, declared adults with the last pediatric GH injection at final height, but should receive specific care and continue receiving GH treatment to complete adult body shape and avoid developing the adult GHD syndrome.

References

[1] Attanasio AF, Lamberts SWJ, Matranga AMC, et al. Adult growth hormone (GH)-deficient patients demonstrate heterogeneity between childhood onset and adult onset before and during human GH treatment. J Clin Endocrinol Metab 1997;82:82–8.

[2] De Boer H, van der Veen EA. Why retest young adults with childhood-onset growth deficiency? J Clin Endocrinol Metab 1997;82:2032–6.
[3] Clayton PE, Cuneo RC, Juul A, et al. Consensus statement on the management of the GH-treated adolescent in the transition to adult care. Eur J Endocrinol 2005;152:165–70.
[4] Theintz G, Buchs B, Rizzoli R, et al. Longitudinal monitoring of bone mass accumulation in healthy adolescents: evidence for a marked reduction after 16 years of age at the levels of lumbar spine and femoral neck in female subjects. J Clin Endocrinol Metab 1992;75:1060–5.
[5] Rizzoli R, Bonjour JP. Determinants of peak bone mass and mechanisms of bone loss. Osteoporos Int 1999;9(Suppl 2):17–23.
[6] Ngyuen TV, Maynard LM, Towne B, et al. Sex differences in bone mass acquisition during growth: the Fels Longitudinal Study. J Clin Densitom 2001;4:147–57.
[7] Guo SS, Chumlea WC, Roche AF, et al. Age- and maturity-related changes in body composition during adolescence into adulthood: the Fels Longitudinal Study. Int J Obes Relat Metab Disord 1997;21:1167–75.
[8] Forbes GB. Relation of lean body mass to height in children and adolescents. Pediatr Res 1972;6:32–7.
[9] Barlett HL, Puhl SM, Hodgson JL, et al. Fat-free mass in relation to stature: ratios of fat-free mass to height in children, adults, and elderly subjects. Am J Clin Nutr 1991;53:1112–6.
[10] Forbes GB. Body composition in adolescence. Prog Clin Biol Res 1981;61:55–72.
[11] Zadik Z, Chalew SA, McCarter RJ Jr, et al. The influence of age on the 24-hour integrated concentration of growth hormone in normal individuals. J Clin Endocrinol Metab 1985;605:513–6.
[12] Blum WF, Breier BH. Radioimmunoassays for IGFs and IGFBPs. Growth Regul 1994;4:11–9.
[13] Attanasio AF, Howell S, Bates PC, et al. Body composition, IGF-I and IGFBP-3 concentrations as outcome measures in severely GH-deficient (GHD) patients after childhood GH treatment: a comparison with adult onset GHD patients. J Clin Endocrinol Metab 2002;87:3368–72.
[14] Lissett CA, Jonsson P, Monson JP, et al. Determinants of IGF-I status in a large cohort of growth hormone-deficient (GHD) subjects: the role of timing of onset of GHD. Clin Endocrinol (Oxf) 2003;59:773–8.
[15] Rutherford OM, Jones DA, Round JM, et al. Changes in skeletal muscle after discontinuation of growth hormone treatment in growth hormone deficient young adults. Clin Endocrinol (Oxf) 1991;34:469–75.
[16] Ogle GD, Moore B, Lu PW, et al. 1994 Changes in body composition and bone density after discontinuation of growth hormone therapy in adolescence: an interim report. Acta Paediatr Suppl 1994;399:3–7.
[17] Johannsson G, Albertsson-Wikland K, Bengtsson B-Å. Discontinuation of growth hormone (GH) treatment: metabolic effects in GH-deficient and GH-sufficient adolescent patients compared with control subjects. J Clin Endocrinol Metab 1999;84:4516–24.
[18] Shalet SM, Shavrikova E, Cromer M, et al. Effect of growth hormone (GH) treatment on bone in postpubertal GH-deficient patients: a 2-year randomized, controlled, dose-ranging study. J Clin Endocrinol Metab 2003;88:4124–9.
[19] Tauber M, Jouret B, Cartault A, et al. Adolescents with partial growth hormone (GH) deficiency develop alterations of body composition after GH discontinuation and require follow-up. J Clin Endocrinol Metab 2003;88:5101–6.
[20] Rosén T, Bengtsson B-A. Premature mortality due to cardiovascular diseases in hypopituitarism. Lancet 1990;336:285–8.
[21] Bülow B, Hagmar L, Mikoczy Z, et al. Increased cerebrovascular mortality in patients with hypopituitarism. Clin Endocrinol (Oxf) 1997;46:75–81.
[22] DeBoer H, Blok CJ, Van der Veen EA. Clinical aspects of growth hormone deficiency in adults. Endocr Rev 1995;16:63–86.

[23] Clayton PE, Price DA, Shalet SM. Growth hormone state after completion of treatment with growth hormone. Arch Dis Child 1987;62:222–6.
[24] Cacciari E, Tassoni P, Parisi G, et al. Pitfalls in diagnosing impaired growth hormone (GH) secretion: retesting after replacement therapy of 63 patients defined as GH deficient. J Clin Endocrinol Metab 1992;74:1284–9.
[25] Wacharasindhu S, Cotterill AM, Camacho-Hubner E, et al. Normal growth hormone secretion in growth hormone insufficient children retested after completion of linear growth. Clin Endocrinol (Oxf) 1996;45:553–6.
[26] Longobardi S, Merola B, Pivonello R, et al. Re-evaluation of growth hormone (GH) secretion in 69 adults diagnosed as GH deficient patients during childhood. J Clin Endocrinol Metab 1996;81:1244–7.
[27] Nicolson A, Toogood AA, Rahim A, et al. The prevalence of severe growth hormone deficiency in adults who received growth hormone replacement in childhood. Clin Endocrinol (Oxf) 1996;44:311–6.
[28] Tauber M, Moulin P, Pienkowski C, et al. Growth hormone (GH) retesting and auxological data in 131 GH-deficient patients after completion of treatment. J Clin Endocrinol Metab 1997;82:352–6.
[29] Juul A, Kastrup KW, Pedersen SA, et al. Growth hormone (GH) provocative retesting of 108 young adults with childhood-onset GH deficiency and the diagnostic value of insulin-like growth factor 1 (IGF-I) and IGF-binding protein-3. J Clin Endocrinol Metab 1997;82:1195–201.
[30] Attanasio AF, Howell S, Bates PC, et al. Confirmation of severe GH deficiency after final height in patients diagnosed as GH deficient during childhood. Clin Endocrinol (Oxf) 2002;56:503–7.
[31] Growth Hormone Research Society. Consensus guidelines for the diagnosis and treatment of adults with growth hormone deficiency: summary statement of the Growth Hormone Research Society workshop on adult growth hormone deficiency. J Clin Endocrinol Metab 1998;83:379–81.
[32] Nørrelund H, Vahl N, Juul A, et al. Continuation of growth hormone (GH) therapy in GH-deficient patients during transition from childhood to adulthood: impact on insulin sensitivity and substrate metabolism. J Clin Endocrinol Metab 2000;85:1912–7.
[33] Underwood LE, Attie KM, Baptista J, et al. Growth hormone (GH) dose-response in young adults with childhood-onset GH deficiency: a two-year, multicenter, multiple-dose, placebo-controlled study. J Clin Endocrinol Metab 2003;88:5273–80.
[34] Mauras N, Pescovitz OH, Allada V, et al. Limited efficacy of growth hormone (GH) during transition of GH-deficient patients from adolescence to adulthood: a phase III multicenter, double-blind, randomized two-year trial. J Clin Endocrinol Metab 2005;90:3946–55.
[35] Vahl N, Juul A, Jørgensen JOL, et al. Continuation of growth hormone (GH) replacement in GH-deficient patients during transition from childhood to adulthood: a two-year placebo-controlled study. J Clin Endocrinol Metab 2000;85:1874–81.
[36] Attanasio AF, Shavrikova E, Blum WF, et al. Continued growth hormone (GH) treatment after final height is necessary to complete somatic development in childhood-onset GH-deficient patients. J Clin Endocrinol Metab 2004;89:4857–62.
[37] Murray RD, Columb B, Adams JE, et al. Low bone mass is an infrequent feature of the adult growth hormone deficiency syndrome in middle-age adults and the elderly. J Clin Endocrinol Metab 2004;89:1124–30.
[38] Murray RD, Adams JE, Shalet SM. A densitometric and morphometric analysis of the skeleton in adults with varying degrees of growth hormone deficiency. J Clin Endocrinol Metab 2005;91:432–8.
[39] Bouillon R, Koledova E, Bezlepkina O, et al. Bone status and fracture in Russian adults with childhood-onset growth hormone deficiency. J Clin Endocrinol Metab 2004;89:4993–8.

[40] Drake WM, Carroll PV, Maher KT, et al. The effect of cessation of growth hormone (GH) therapy on bone mineral accretion in GH-deficient adolescents at the completion of linear growth. J Clin Endocrinol Metab 2005;88:1658–63.
[41] Caprio S, Plewe G, Diamond MP, et al. Increased insulin secretion in puberty: a compensatory response to reductions in insulin sensitivity. J Pediatr 1989;114:963–7.
[42] Caprio S, Boulware D, Tamborlane W. Growth hormone and insulin interactions. Horm Res 1992;38(Suppl 2):47–9.
[43] Murray RD, Skillicorn CJ, Howell SJ, et al. Influences on quality of life in GH deficient adults and their effect on response to treatment. Clin Endocrinol 1999;51:565–73.
[44] Attanasio AF, Shavrikova EP, Blum WF, et al. Quality of life in childhood onset growth hormone deficient patients in the transition phase from childhood to adulthood. J Clin Endocrinol Metab 2005;90:4525–9.

Management of Adult Growth Hormone Deficiency

Gudmundur Johannsson, MD, PhD

Department of Endocrinology, Sahlgrenska University Hospital and the Sahlgrenska Academy, Gothenburg University, SE-413 45 Gothenburg, Sweden

Growth hormone (GH) deficiency (GHD) in adults is characterized by perturbations in body composition, carbohydrate and lipid metabolism, bone mineral density, cardiovascular risk profile, and quality of life (QOL). Furthermore, it is likely, although unproven, that GHD contributes to the increase in cardiovascular morbidity and mortality that has repeatedly been observed in hypopituitary adults [1,2].

The availability of recombinant human GH from 1985 onward stimulated the investigation of the role of GH in adult life and in particular the effects of GH replacement in adults who have GHD. In the absence of any pilot data to indicate the most appropriate dose of GH, initial studies were performed using a dose based on weight or body surface area, derived from experience in the pediatric setting [3–5]. Although these studies provided evidence of the beneficial effects of GH replacement, they were also associated with a high incidence of side effects related to fluid retention, most commonly arthralgia and peripheral edema, which were demonstrated to be dose-related side effects of treatment.

This article summarizes current knowledge of managing GH replacement therapy in adults and mentions the experience of long-term treatment and safety.

Diagnosis of growth hormone deficiency in adults

The probability of the diagnosis of severe GHD in adults is high in patients who have gross hypothalamic-pituitary diseases and two or more other pituitary hormone deficiencies [6]. The tools for defining severe GHD in adults are well validated in patients who have well-defined hypothalamic-pituitary

E-mail address: gudmundur.johannsson@medic.gu.se

disease with and without other pituitary hormone deficiency [7]. The performance of the tests used to define GHD in adults have been less validated in patients who do not have major well-defined pituitary pathology, which is often the case in patients who have history of major head trauma without the presence of other pituitary hormone deficiency. A similar situation is the adolescent patient who has idiopathic isolated GHD diagnosed in childhood at the time of re-evaluation for continuing care and treatment into adulthood or the middle-aged or elderly patient who has symptoms and signs of severe GHD without any known hypothalamic-pituitary disease. Moreover, the performance of tests used to define GHD in adults has not been properly validated in obesity. The performance of the tests and the assay used to defined GHD is therefore of utmost importance [8].

GHD should be suspected in patients who have any form of hypothalamic-pituitary disease, patients who have any other pituitary hormone deficiency, patients who have received cranial irradiation or experienced severe head trauma, and young adults who have been diagnosed with GHD in childhood. Serum insulinlike growth factor (IGF)–I is GH dependent and useful in the screening and diagnostic procedure for GHD. IGF-I serum level is also affected by other factors, such as nutritional status and size of lean body mass [9], and with increasing age the overlap between those who have severe GHD and normative data becomes greater. Serum IGF-I concentration is a useful screening test, but a normal serum concentration does not rule out the presence of severe GHD at any age.

Various tests to stimulate GH secretion to determine a peak response have been used for diagnostic purposes in hypopituitarism and GHD. Among the validated tests, the insulin tolerance test (ITT), arginine-GH releasing hormone (Arg-GHRH), GHRH-GH releasing peptide (GHRP)–6, and GHRH-pyridostigmine are excellent, whereas clonidine and L-Dopa are inadequate [10]. The ITT has been used in the diagnosis of GHD since 1963 [11] and has in recent years been re-evaluated and found to have high sensitivity and specificity in all age groups [12]. The peak is reduced, however, in obesity [13] and may also be lower in women than in men [14], but this has not been considered in the cutoff levels. The ITT has been recommended as the test of choice in adults who have suspected GHD. The test is potentially hazardous, however, especially in patients who have heart disease and seizure disorders, but it has been found to be safe if performed in specialized units under adequate supervision [15].

The Arg-GHRH stimulation test benefits from the potentiating of both substances and a reproducible stimulation of the GH secretion is achieved [16]. The peak GH response to Arg-GHRH seems to be independent of age and sex but is reduced by obesity; therefore, BMI-related reference ranges have been established [17]. This test is safe and is associated with less discomfort than the ITT; it is therefore a good alternative for the ITT in most situations, except in patients who have a primary hypothalamic lesion. This phenomenon was clearly demonstrated in patients who had received irradiation to the hypothalamic-pituitary area for a nonpituitary disorder who were defined to have severe GHD using the ITT and not by Arg-GHRH within 5 years after radiotherapy, whereas a consistency was obtained between the two tests thereafter [18]. The GHRH-GHRP-6 stimulation test has a peak GH response that is not influenced by age or body composition and as such is a promising test [19].

The diagnosis of severe GHD in adults has been considered to be a peak GH response to ITT of less than 3 µg/L [8,12]. Many of the studies documenting the beneficial effects of GH replacement therapy in adults have included subject with a peak GH response less than 5 µg/L, however. Other tests have their specific cutoff values for defining severe GHD ranging between 16.5 and 20.3 µg/L for Arg-GHRH, depending on BMI, and less than 10 µg/L for the GHRH-GHRP-6 stimulation test. The performance of the GH assay should also be considered in the interference with GH-binding proteins; the recombinant international reference preparation (IRP) 88/624 has been recommended for use in the GH assays [8]. Moreover, in patients who have a more limited degree of hypopituitarism, in particular those who have two or fewer pituitary hormone deficiencies, two independent stimulation test are recommended to diagnose severe GHD in adults [7,20].

Growth hormone replacement therapy

The aim of GH replacement is to achieve normalization of metabolic, functional, and psychologic consequences of adult hypopituitarism and severe GHD. Before commencement of GH replacement, it is necessary to ensure appropriate replacement of other pituitary hormone deficiencies, to determine the presence and severity of clinical features associated with the GHD, and to predict sensitivity to GH.

Appropriate evaluation of the hypothalamic-pituitary-adrenal axis is mandatory because there are indications that GH may attenuate the peripheral action of glucocorticoids by reducing the activity if the enzyme 11β hydroxysteroid dehydrogenase type 1 [21]. This enzyme preferably converts inactive cortisone to active cortisol, and commencing GH therapy may therefore push an individual who has incipient adrenal insufficiency into overt adrenal insufficiency. Moreover, administration of GH probably enhances the peripheral conversion of T4 to T3 and may thereby influence the thyroxin replacement therapy [22]. Finally, there are well-known interactions between sex steroids and GH action. It is well established that oral estrogen administration markedly reduces GH responsiveness [23] and there are also indications that testosterone potentiates the IGF-I response and the antinatriuretic action of GH [24]. A careful evaluation of other pituitary functions and an adequate replacement therapy, if needed, are therefore mandatory before initiating GH replacement therapy.

To monitor the clinical response to GH, it is necessary to determine the severity of the features associated with GHD in patients being considered for replacement therapy. Baseline assessment should include measurement of body composition, bone mineral density (BMD) testing, glucose and lipid measurement, and assessment of psychologic well being. Baseline variables showing the greatest abnormalities can become important treatment targets for the individual patient. As a minimum, assessment of body composition should include measurement of height and weight, calculation of BMI, and measurement of waist and hip circumferences. Bio-impedance assessment (BIA) is a simple and relatively inexpensive technique that can be used in the clinic setting. Dual energy x-ray absorptiometry (DEXA) scanning performed at baseline can assess BMD and provide an estimate of lean body mass and total and regional fat distribution.

Dose selection

Knowledge of individual sensitivity to GH is helpful for deciding the initial dose of GH and the pace of dose titration in the individual patient. The first attempt to understand which patients develop side effects, and can therefore be considered more GH sensitive, was made by Holmes and Shalet [25]. They observed that in patients receiving a fixed dose of GH/kg body weight, side effects occurred more frequently in those who were older and more obese, and in whom the onset of GHD was during adult life. Furthermore, other studies demonstrated the effects of GH on body composition, lipid profile, and markers of bone turnover to be more marked in men [26–28]. In particular, women receiving oral estrogen replacement therapy were more resistant to GH in the serum IGF-I response [23]. There are also some indications that women receiving oral estrogen replacement therapy may be less prone to developing peripheral edema and other fluid-related side effects of GH treatment [29]. Other factors determining individual responsiveness to GH are not as obvious in the clinical setting and they include baseline adiposity and serum levels of GH-binding proteins [30]. The decision of final dose of GH and guiding during dose titration is therefore dependent on monitoring of clinical and biologic response markers [31,32]. An example of poor responsiveness is a young woman with childhood-onset disease and oral estrogen replacement therapy who is likely to need high doses of GH. She can be started with a dosage of 0.1 mg/d or higher and expected to need fast dose titration and in larger steps than a middle-aged man, who should commence treatment with 0.1 mg/d and will be dose-titrated in few and small steps toward an estimated target dosage of between 0.2 and 0.4 mg/d.

Initial dose monitoring

There is a significant overlap of serum IGF-I levels between normal and GHD individuals, with up to half of all individuals who have confirmed GHD having IGF-I levels in the low-normal age-dependent range. This overlap reflects the multiple influences on IGF-I other than age, including genetic factors, nutritional status, and sex steroids [33]. The dose titration and monitoring of subjects who have normal baseline serum IGF-I levels usually allows a small dose of GH, whereas patients who have below-normal serum levels of IGF-I at baseline can tolerate larger doses before IGF-I exceeds the upper limit of normal. In the latter case, dose titration toward clinically meaningful endpoints is easier. By using serum IGF-I as a dose titration monitor, therefore, those who have the lowest serum IGF-I at baseline most likely receive the highest maintenance dose of GH [34].

Serum IGF-I is still the most useful biologic marker for dose titration of GH in adults. It is a more sensitive serum marker than other well-known GH-dependent markers, such as serum IGF binding protein (IGFBP)–3 and acid-labile subunit (ALS), which respond less markedly to the same dose of GH [35]. In a comparative study, serum IGF-I, IGFBP3, and ALS concentrations in GHD adults were measured before and after commencement of three different doses of GH [35]. Compared with IGF-I, a higher dose of GH was necessary to increase IGFBP3 and ALS into their respective normal ranges, and throughout the study IGFBP3 and ALS proved less sensitive to the effects of GH than IGF-I. In another dose titration study, measurement of serum IGFBP3 and ALS was of no additional value over measurement of serum IGF-I as a marker for monitoring dose titration [32].

The serum IGF-I response to GH administration mainly reflects the hepatic effect of GH, because more than 70% of the circulating IGF-I is produced in the liver [36]. The overall effect of GH depends both on GH and IGF-I and it is likely that many of the anabolic and metabolic effects of GH are primarily mediated through IGF-I [37].

Another important consideration is that the relationship between serum IGF-I response during GH treatment and other treatment effects, such as metabolic endpoints and body composition, is poor [27,38]. The serum IGF-I response and the achieved serum IGF-I levels therefore cannot be used as a surrogate marker for other efficacy variables. Although serum IGF-I can no longer be assumed to reflect the effect of GH in all tissues, it remains a useful and important marker to detect over-replacement with GH. Recent studies have demonstrated that serum levels of IGF-I in the upper normal range in normal subjects are predictive of increased risk for breast, colon, and prostate cancer. Although these findings should not be extrapolated to physiologic GH replacement in adults who have GHD, it is considered prudent to maintain serum IGF-I levels in the normal range during GH replacement therapy [39].

Because IGF-I only reflects one aspect of efficacy of GH replacement, other simple measures to monitor GH dose titration have been evaluated, such as changes in extracellular water (ECW) measured by BIA [40] and normalization of other compartments of the body composition, such as body fat and body cell mass [31]. The effect of GH to increase ECW through its antinatriuretic action occurs within 3 to 5 days of starting GH replacement [41]. Because increased ECW is one of the measurements that changes most consistently during GH treatment, it may be a more useful endpoint for monitoring GH replacement than other aspects of body composition. This concept was first explored by using BIA, a marker of tissue hydration [40]. Electrical impedance was increased in GHD subjects, indicating reduced body hydration, and exhibited a dose-dependent decrease during GH replacement. The dose that normalized tissue hydration was similar to the dose that normalized serum IGF-I in most of the study subjects [35].

Growth hormone dose titration

The observation that a GH dose that is inadequate in one subject can lead to side effects of overdosage in another has prompted the development of methods of individual dose titration. In this paradigm, the dose of GH is titrated against both clinical features of GHD and evidence of overtreatment, determined by serum IGF-I and clinical evidence of side effects, in particular symptoms and signs related to fluid retention. In a study of an individualized dose regimen of GH replacement in which the dose of GH was titrated against serum IGF-I, body composition and clinical response comparison was made with a conventional weight-based regimen [31]. Following 1 year of GH replacement, a mean dosage of 0.45 mg/d was obtained during individualized dose titration and a mean dosage of 0.55 mg/d was obtained during the weight-based regimen after some dose adjustments because of side effects. Although the efficacy of GH treatment was similar in the two groups, side effects occurred in 30% of cases during individualized dose titration compared with 70% in the weight-based group (Fig. 1). Also, within the group receiving dose titration those with the lowest starting dose of GH experienced the fewest side effects of treatment. The interpretation is therefore to commence treatment with a low daily dose of GH independent of body weight and thereafter dose-titrate toward adequate clinical effects and normalization of serum IGF-I.

In a subsequent study, the GH replacement dose was titrated against serum IGF-I, aiming for levels in the upper half of the age-related reference range, and results were compared with retrospective data from subjects treated with a weight-based regimen (Fig. 2). Median maintenance dosages in this study were 0.27 mg/d in men and 0.4 mg/d in women, significantly lower than maintenance dosages of 0.5 mg/d used in the weight-based regimen. An important observation was the longer duration of dose titration in the women if initial dose and dose titration steps were the same in men and women (see Fig. 2). There was no difference in the degree of improvement of QOL observed in the different regimens. Taken together, these two studies provide important indications that using an individualized dose titration allows similar average beneficial effects with fewer side effects and a lower maintenance GH dose than using a weight-based regimen. Some

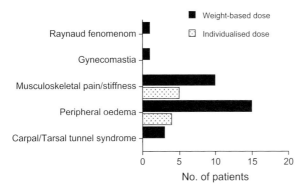

Fig. 1. Side effects in 60 adults receiving 6 months of GH replacement therapy. Thirty patients received a weight-based dosage regimen with 12 μg/kg/d and 30 patients received an individually titrated dosage regimen. (*Adapted from* Johannsson G, Rosén T, Bengtsson B-Å. Individualized dose titration of growth hormone (GH) during GH replacement in hypopituitary adults. Clin Endocrinol [Oxf] 1997;47:575; with permission.)

variations on the initial dose and pace and steps during dose titration should be considered based on the estimated final dose that can be anticipated from published data on individual sensitivity [26,30,42,43]. An older patient may need a lower initial dose of GH with slow dose titration (0.05–0.1 mg at a time), whereas a young woman can be commenced on a larger dose (0.1–0.3 mg) with faster dose titration in larger steps.

Another important observation from postmarketing surveillance studies reveals that subjects initially treated with a high dose of GH are more likely to remain on a high dose in the long term and frequently exhibit supraphysiologic serum levels of IGF-I [44]. Presently there are no data available to

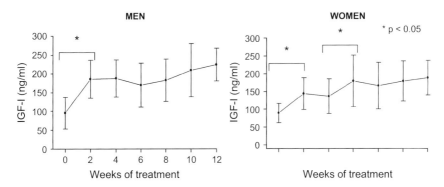

Fig. 2. Serum levels of IGF-I versus weeks of treatment during dose titration in men and women. Serum IGF-I increased significantly between 0 and 2 weeks in males and remained stable thereafter. In contrast, serum IGF-I increased between 0 and 2 weeks and 4 and 6 weeks in females. (*From* Drake WM, Coyte D, Camacho-Hübner C, et al. Opimizing growth hormone replacement therapy by dose titration in hypopituitary adults. J Clin Endocrinol Metab 1998;83:3915; with permission.)

demonstrated harmful effects of slightly high serum levels of IGF-I during GH replacement in adults. An indication, however, comes from a small open study on subjects who received GH replacement early, when high weight-based doses were used for years, demonstrating that their left ventricular mass increased to levels greater than those to be expected from normative values [45].

Evidence is accumulating that the optimum approach to GH replacement is to commence treatment at a low dose and then titrate upward depending on the response of GH-dependent variables, avoiding serum IGF-I levels greater than the upper limit of age-adjusted normative values. Evidence from several studies suggests that variables most sensitive to change are those that are the most abnormal at baseline. A baseline evaluation is therefore helpful to determine which endpoint is most important for each individual patient to follow.

Quality of life and well-being

The beneficial effect of GH replacement on psychologic well-being and QOL in adults who have GHD is an important endpoint for replacement therapy. Many of the patients being considered for GH treatment have high expectations for improved well-being; this aspect of treatment should therefore be monitored clinically or by using standardized questionnaires, or preferably both.

To date, no studies have specifically used QOL as an endpoint against which to titrate GH. In a study that only included GHD patients who had a subjectively low QOL on clinical interview [46], GH was replaced using a dose-titration regimen, aiming to increase IGF-I into the upper half of the normal range. As expected, QOL was seen to improve using two different questionnaires, but the improvement was most marked in those in whom QOL was most abnormal at baseline. This finding also included patients who had childhood-onset GHD who have as a group normal baseline QOL and do not on average respond to treatment [43]. The findings of that study support the strategy to quantify the degree of psychologic impairment at baseline and use it in the monitoring of treatment.

A study of long-term effects on QOL found that improvements that occur shortly after the initiation of GH therapy are sustained in the longer term [47]. Moreover, it seems that some aspects of QOL continue to improve during long-term therapy. Another aspect to consider during monitoring of treatment is that almost one third of the patients who experienced beneficial effects of GH therapy stated that such effects had become noticeable only after GH had been administered for at least 6 months [47]. This finding has clear implications for clinical practice: once the decision has been made to initiate GH therapy, the therapy should be continued for an adequate period of time before judgments are made regarding its efficacy in improving QOL and psychologic well-being. The onset of the effects of GH may be further delayed in some patients who receive individual dose titration after starting with a low dose of GH.

Glucose and lipid metabolism

Using several techniques, it has been demonstrated that insulin sensitivity is reduced in adults who have GHD [48]. Insulin sensitivity is further reduced following short-term GH replacement but returns later to baseline values; no further change in insulin sensitivity has been seen in observational studies lasting up to 7 years using the hyperinsulinemic glucose clamp for assessment [49]. In an open 5-year study comprising more than 100 patients, glycosylated hemoglobin and serum triglyceride levels tended to decrease after more than 3 years of treatment [50]. This long-term neutral or slightly beneficial effect probably reflects a balance between the counterregulatory effect of GH and the indirect insulin sensitizing effect of the other metabolic effects of treatment, such as reduced visceral fat. A large postmarketing surveillance study did not demonstrate an increase in the incidence of diabetes during GH replacement more than what is expected in the background population [51].

GH also exerts complex effects on plasma lipids. The net effect of these changes during GH replacement is consistent reduction in total and LDL cholesterol and apoprotein B, and in some studies increases in HDL cholesterol concentration [52]. Like the improvements in body composition and QOL, the most marked improvements occur in individuals in whom the most abnormal baseline values exist.

Plasma lipids should be measured before treatment is commenced as an overall assessment of cardiovascular risk profile, and should be monitored on a regular basis, particularly in patients who have baseline abnormalities or other cardiovascular risk factors. Assessment of glucose metabolism can be performed by using the oral glucose tolerance test, performed in selected patients who have other strong risk factors for diabetes, or more simply by measuring fasting levels of glucose to monitor long-term trends.

Long-term monitoring and safety

The beneficial effects of GH replacement on body composition, lipid profile, and psychologic well-being that are early response variables are sustained in small studies of up to 10 years' duration [53]. Endpoints and efficacy variables that have only been detected in controlled and open longitudinal studies after more prolonged duration of treatment are changes in bone mineral content (BMC) and BMD. More than 12 to 18 months of treatment have been needed to detect a progressive increase in these endpoints in both childhood-onset [54,55] and adult-onset patients [27,56]. The magnitude of increase of BMC in lumbar spine is approximately 9%, with women responding less than men [50,56] (Fig. 3). The increase in the

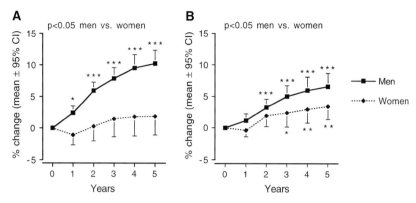

Fig. 3. Lumbar spine BMC (*A*) and BMD (*B*) in men and women during 5 years of GH replacement in 118 patients who had adult-onset GHD. The vertical bars indicate the 95% CI for the mean values shown. Between-group P values are based on an analysis of the percent change from baseline, whereas within-group P values in men and women are based on an analysis of the absolute values. *, $P<.05$; **, $P<.01$; ***, $P<.001$ versus baseline. (*From* Gotherstrom G, Svensson J, Koranyi J, et al. A prospective study of 5 years of GH replacement therapy in GH-deficient adults: sustained effects on body composition, bone mass, and metabolic indices. J Clin Endocrinol Metab 2001;86(10):4663; with permission.)

amount of bone induces an increase in BMC and bone area resulting in less marked effect on BMD [57]. This reduced effect is because of increased endosteal and periosteal bone formation in cortical bone with less marked effect on trabecular bone as shown in histomorphometric data from men who had childhood onset disease who were treated with GH for 5 years [54]. BMD is normalized in most patients [50].

The effect on muscle strength and muscle function is delayed for 12 to 18 months of treatment [27], except in one study demonstrating increased muscle strength in one muscle group after 6 months of treatment [58]. With a treatment duration of 2 to 5 years in a large group of patients mean muscle strength is normalized [59] (Fig. 4).

Although there is no evidence to date that GH increases the risk for diabetes mellitus in adults [51], data suggest that it may do so in children. Reported data from an international surveillance study including more than 20,000 subjects indicated that GH treatment was associated with a sixfold increase in new cases of type 2 diabetes [60]. Although long-term data in adults are encouraging, ongoing monitoring of glucose metabolism in patients receiving long-term GH replacement is recommended [39].

No study has directly demonstrated reduced vascular morbidity and mortality in response to long-term GH replacement in adults. One of the most useful surrogate markers for atherosclerosis is intima-media thickness (IMT). IMT is increased in adults who have GHD and some short-term studies have demonstrated that GH replacement reduces IMT [61,62]. Moreover, recent preliminary data suggest that mortality during long-term

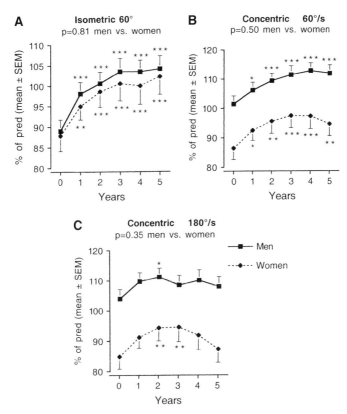

Fig. 4. Observed/predicted value ratios for isometric 60 degrees (*A*), concentric 60 degrees/s (*B*), and concentric 180 degrees/s (*C*) knee flexor strength in men and women during 5 years of GH replacement in 109 patients who had GHD of adult onset. The vertical bars indicate the SE for the mean values shown. Between-group P values are based on an analysis of the percentage change from baseline, whereas within-group P values in men and women are based on an analysis of the absolute values. *, $P < .05$; **, $P < .01$; ***, $P < .001$ versus baseline. (*From* Svensson J, Stibrant Sunnerhagen K, Johannsson G. Five years of growth hormone replacement therapy in adults: age- and gender-related changes in isometric and isokinetic muscle strength. J Clin Endocrinol Metab 2003;88(5):2066; with permission.)

GH replacement in adults is similar to the mortality in the background population [63]. Although the data are reassuring they are preliminary and with a patient population heavily burdened by cardiovascular risk factors and increased premature vascular mortality, cardiovascular risk factors, such as lipids, blood pressure, smoking, and glucose metabolism, should be monitored and effectively treated in accordance with available guidelines.

Neoplasia

The mitogenic and growth-promoting effects of GH and IGF-I and their impact on neoplasia provide a theoretic basis by which GH treatment could increase cancer risk and promote tumor growth. When assessing cancer rates, it must be taken into consideration that subjects who have hypopituitarism secondary to genetic or tumor-related causes might have an inherently higher risk for neoplasia compared with normal subjects, and furthermore, that some treatment modalities, including radiotherapy, might also increase the risk for secondary neoplasia. Two retrospective studies have supported such notions. These studies found increased rate of and mortality from neoplasia in patients who had pituitary adenomas, suggesting an association that may be inherent or attributable to increased surveillance of this patient population [64,65].

Three recent prospective studies have demonstrated that serum IGF-I levels within the upper normal range are predictive of cancer. In a meta-analysis of hormonal predictors of prostate cancer, it was found that men who had either serum testosterone or IGF-I levels in the upper quartile of the population had an approximately twofold higher risk for developing prostate cancer [66]. In other prospective trials among premenopausal women there was a 4.5-fold relative risk for breast cancer in the highest quartile of serum IGF-I compared with the lowest quartile [67]. Similar results were also found for colorectal cancer in men [68]. These studies have also demonstrated an inverse association between risk for cancer and serum IGFBP-3 concentration (ie, the highest risk is among subjects who have high serum IGF-I and low serum IGFBP-3 levels). The significance of these studies, both for the hypopituitary and the general population, is unknown. In the context of GH replacement it provides a clear rationale for maintaining IGF-I within the age-matched normal range.

For several other reasons, long-term monitoring of subjects receiving GH replacement remains essential. Monitoring is necessary to detect changes in GH dose requirements. There is some evidence that sensitivity to GH increases during long-term GH replacement [69] and GH replacement doses are lower in the elderly; any change in requirement is likely to be a reduction. Additionally, changes in replacement of other hormones, in particular commencement or discontinuation of oral estrogen, may alter GH requirement [23]. The consensus is that clinical screening for neoplasia should be based on general recommendations for adults and that current data do not support a more extensive screening program for patients who have hypopituitarism and GH replacement [39].

Neuroimaging surveillance

Another important aspect to consider during long-term monitoring and surveillance of GH replacement is the risk for regrowth or recurrence of the hypothalamic-pituitary or brain tumor responsible for most of the hypopituitarism seen in patients who have adult-onset disease. The progression-free survival in patients who have pituitary tumor has been reported as between 12% and 69% after surgery, and between 72% and 92% following surgery and radiotherapy [70]. These data are from the time before GH became part of the replacement therapy of hypopituitarism.

In a prospective study of 100 patients who had pituitary and peripituitary tumors receiving GH replacement therapy only one case of slight intracellular enlargement was noted after 1 to 4 years of follow-up [71]. Ninety-one percent of the patients had received external radiotherapy. Another retrospective study of patients who had craniopharyngioma compared 32 patients who received GH replacement for a mean period of 6.3 years with 53 patients who had not received GH who had similar tumor and treatment characteristics and follow-up period [72]. During the period of observation, 4 patients treated with GH and 22 not receiving GH developed tumor recurrence. Finally a prospective study of 60 childhood brain tumor survivors receiving adult GH replacement therapy detected only one incurable ependymoma and one residual meningioma that progressed in size over a mean period of 6.7 years [73]. Secondary neoplasia was detected both before and after commencing GH treatment. The results from these studies do not indicate that GH replacement increases tumor recurrence or tumor regrowth, but the studies are small and the follow-up periods are limited. Vigilance and long-term surveillance are therefore indicated in patients previously treated for tumors, in particular those who have increased risk for a secondary tumor, as was the case in childhood brain tumor survivors.

Specific conditions, including pregnancy

The labeled contraindication to GH treatment in adults is uncontrolled malignancy. Relative contraindications are proliferative diabetic retinopathy and benign intracranial hypertension. Patients who have hypopituitarism and diabetes mellitus can be treated, but initial dose should be low and dose titration slow, with close monitoring of glucose metabolism. In patients who have type I diabetes the response in serum IGF-I is markedly attenuated because of reduced availability of insulin to the liver [74]. Normalization of serum IGF-I cannot always be obtained, therefore, and dose titration and monitoring should be guided by other measures.

During pregnancy, the pulsatile release of pituitary GH is progressively suppressed and replaced by a continuous secretion of placental GH into the maternal circulation [75]. After delivery, there is a gradual normalization of the pituitary GH secretion [76]. Placental GH is a major regulator of maternal serum IGF-I levels during pregnancy [77]. GH does not cross the placenta and therefore its effects on the fetus are probably indirect and mediated by maternal IGF-I production and by actions on substrate supply to the fetus [78]. Normal pregnancies have been reported in GH-deficient women [79].

Currently, because of the lack of safety data, GH therapy is not licensed for use during pregnancy. If one takes into account the normal GH physiology during pregnancy, it would be advisable and justifiable to maintain the replacement therapy in women who have GHD, at least until the time enough placental production of GH is achieved. In an attempt to reproduce the gestational GH physiology, eight GH-deficient women during 12 distinct pregnancies received their ordinary pregestational GH dose during the first 3 months of gestation, and the dose was gradually tapered off during the subsequent 3 months [80]. This treatment strategy was demonstrated to be safe to the mother and the fetus. No important adverse events or major obstetric complications were observed. Moreover, no women complained of excess fatigue during the first 3 months of gestation, as was the experience in some cases when GH replacement was discontinued when pregnancy was confirmed [80].

Summary

GH replacement therapy in adults, as documented in published work with correct diagnosis of severe GHD, using dose titration and monitoring of treatment by serum IGF-I and clinical efficacy, is a safe therapy. In patients who have hypothalamic-pituitary tumor as underlying cause of hypopituitarism neuroimaging is recommended before commencement of GH replacement and thereafter according to ordinary surveillance routines. Although the long-term experience is limited in patients who have remnant tumor, all present data indicate that the treatment is safe in such patients. Before GH replacement is started, baseline assessment of clinical features of GHD, optimal route, and dose of other hormone replacement therapy should be ensured and the patient's sensitivity to GH should be estimated. This evaluation should then guide the decision of the starting dose of GH and the pace of dose titration along with what clinical variables are most useful to monitor. Serum IGF-I should be monitored and used as a safety variable, avoiding being greater than the upper limit of age-related normative values. Long-term clinical efficacy and safety data are growing fast. Although all data support the safety and efficacy of GH replacement in adults, continued monitoring during long-term treatment should include safety assessments with respect to this powerful metabolic hormone.

References

[1] Rosén T, Bengtsson B-Å. Premature mortality due to cardiovascular diseases in hypopituitarism. Lancet 1990;336:285–8.
[2] Tomlinson JW, Holden N, Hills RK, et al. Association between premature mortality and hypopituitarism. West Midlands Prospective Hypopituitary Study Group. Lancet 2001; 357:425–31.
[3] Bengtsson B-Å, Edén S, Lönn L, et al. Treatment of adults with growth hormone (GH) deficiency with recombinant human GH. J Clin Endocrinol Metab 1993;76:309–17.
[4] Salomon F, Cuneo RC, Hesp R, et al. The effects of treatment with recombinant human growth hormone on body composition and metabolism in adults with growth hormone deficiency. N Engl J Med 1989;321:1797–803.

[5] Jørgensen JOL, Pedersen SA, Thuesen L, et al. Beneficial effect of growth hormone treatment in GH-deficient adults. Lancet 1989;i:1221–5.
[6] Toogood AA, Bearwell CG, Shalet SM. The severity of growth hormone deficiency in adults with pituitary disease is related to the degree of hypopituitarism. Clin Endocrinol [Oxf] 1994; 41:511–6.
[7] Shalet SM, Toogood A, Rahim A, et al. The diagnosis of growth hormone deficiency in children and adults. Endocr Rev 1998;19:203–23.
[8] Invited report of a workshop. Consensus guidelines for the diagnosis of growth hormone deficiency: summary statement of the Growth Hormone Research Society workshop on adult growth hormone deficiency. J Clin Endocrinol Metab 1998;83:379–81.
[9] Svensson J, Johannsson G, Bengtsson B-Å. Insulin-like growth factor-I in growth hormone-deficient adults: relationship to population-based normal values, body composition and insulin tolerance test. Clin Endocrinol [Oxf] 1997;46:579–86.
[10] Biller BM, Samuels MH, Zagar A, et al. Sensitivity and specificity of six tests for the diagnosis of adult GH deficiency. J Clin Endocrinol Metab 2002;87(5):2067–79.
[11] Roth J, Glick SM, Yalow RS, et al. Hypoglycemia: potent stimulus to secretion of growth hormone. Science 1963;140:987.
[12] Hoffman DM, O'Sullivan AJ, Baxter RC, et al. Diagnosis of growth hormone deficiency in adults. Lancet 1994;343:1064–8.
[13] Cordido F, Dieguez C, Casanueva FF. Effect of central cholinergic neurotransmission enhancement by pyridostigmine on the growth hormone secretion elicited by clonidine, arginine, or hypoglycemia in normal and obese subjects. J Clin Endocrinol Metab 1990;70(5): 1361–70.
[14] Hoeck HC, Vestergaard P, Jakobsen PE, et al. Test of growth hormone secretion in adults: poor reproducibility of the insulin tolerance test. Eur J Endocrinol 1995;133(3):305–12.
[15] Jones SL, Trainer PJ, Perry L, et al. An audit of the insulin tolerance test in adult subjects in an acute investigation unit over one year. Clin Endocrinol [Oxf] 1994;41(1):123–8.
[16] Ghigo E, Aimaretti G, Gianotti L, et al. New approach to the diagnosis of growth hormone deficiency in adults. Eur J Endocrinol 1996;134(3):352–6.
[17] Corneli G, Di Somma C, Baldelli R, et al. The cut-off limits of the GH response to GH-releasing hormone-arginine test related to body mass index. Eur J Endocrinol 2005;153(2): 257–64.
[18] Darzy KH, Aimaretti G, Wieringa G, et al. The usefulness of the combined growth hormone (GH)-releasing hormone and arginine stimulation test in the diagnosis of radiation-induced GH deficiency is dependent on the post-irradiation time interval. J Clin Endocrinol Metab 2003;88(1):95–102.
[19] Popovic V, Leal A, Micic D, et al. GH-releasing hormone and GH-releasing peptide-6 for diagnostic testing in GH-deficient adults. Lancet 2000;356(9236):1137–42.
[20] Consensus guidelines for the diagnosis and the treatment of growth hormone (GH) deficiency in childhood and adolescence: summary statement of the GH Research Society. J Clin Endocrinol Metab 2000;85:3990–3.
[21] Weaver JU, Thaventhiran L, Noonan K, et al. The effect of growth hormone replacement on cortisol metabolism and glucocorticoid sensitivity in hypopituitary adults. Clin Endocrinol [Oxf] 1994;41:639–48.
[22] Jørgensen JOL, Pedersen SA, Laurberg P, et al. Effects of growth hormone therapy on thyroid function of growth hormone-deficient adults with and without concomitant thyroxine-substituted central hypothyroidism. J Clin Endocrinol Metab 1989;69:1127–32.
[23] Wolthers T, Hoffman DM, Nugent AG, et al. Oral estrogen antagonizes the metabolic actions of growth hormone in growth hormone-deficient women. Am J Physiol Endocrinol Metab 2001;281:E1191–6.
[24] Johannsson G, Gibney J, Wolthers T, et al. Independent and combined effects of testosterone and growth hormone on extracellular water in hypopituitary men. J Clin Endocrinol Metab 2005;90(7):3989–94.

[25] Holmes SJ, Shalet SM. Which adults develop side-effects of growth hormone replacement? Clin Endocrinol [Oxf] 1995;43:143–9.
[26] Burman P, Johansson AG, Siegbahn A, et al. Growth hormone (GH)-deficient men are more responsive to GH replacement therapy than women. J Clin Endocrinol Metab 1997;82: 550–5.
[27] Johannsson G, Grimby G, Stibrant Sunnerhagen K, et al. Two years of growth hormone (GH) treatment increases isometric and isokinetic muscle strength in GH-deficient adults. J Clin Endocrinol Metab 1997;82:2877–84.
[28] Johannsson G, Oscarsson J, Rosén T, et al. Effects of 1 year of growth hormone therapy on serum lipoprotein levels in growth hormone-deficient adults: influence of gender and apo(a) and apoE phenotypes. Arterioscler Thromb Vasc Biol 1995;15:2142–50.
[29] Janssen YJ, Helmerhorst F, Frölich M, et al. A switch from oral (2 mg/day) to transdermal (50 microg/day) 17beta-estradiol therapy increases serum insulin-like growth factor-I levels in recombinant human growth hormone (GH)-substituted women with GH deficiency J Clin Endocrinol Metab 2000;85:464–7.
[30] Johannsson G, Bjarnason R, Bramnert M, et al. The individual responsiveness to growth hormone (GH) treatment in GH-deficient adults is dependent on the level of GH binding protein, body mass index, age and gender. J Clin Endocrinol Metab 1996;81:1575–81.
[31] Johannsson G, Rosén T, Bengtsson B-Å. Individualized dose titration of growth hormone (GH) during GH replacement in hypopituitary adults. Clin Endocrinol [Oxf] 1997;47: 571–81.
[32] Drake WM, Coyte D, Camacho-Hübner C, et al. Opimizing growth hormone replacement therapy by dose titration in hypopituitary adults. J Clin Endocrinol Metab 1998;83:3913–9.
[33] Landin-Wilhelmsen K, Wilhelmsen L, Lappas G, et al. Serum insulin-like growth factor I in a random population sample of men and women: relationship to age, sex, smoking habits, coffee consumption and physical activity, blood pressure and concentrations of plasma lipids, fibrinogen, parathyroid hormone and osteocalcin. Clin Endocrinol [Oxf] 1994;41: 351–7.
[34] Murray RD, Howell SJ, Lissett CA, et al. Pre-treatment IGF-I level is the major determinant of GH dosage in adult GH deficiency. Clin Endocrinol [Oxf] 2000;52(5):537–42.
[35] de Boer H, Blok GJ, Popp-Snijders C, et al. Monitoring of growth hormone replacement therapy in adults, based on measurements of serum markers. J Clin Endocrinol Metab 1996;81:1371–7.
[36] Sjögren K, Liu JL, Blad K, et al. Liver-derived insulin-like growth factor I (IGF-I) is the principal source of IGF-I in blood but is not required for postnatal body growth in mice. Proc Natl Acad Sci USA 1999;96:7088–92.
[37] Sjogren K, Wallenius K, Liu JL, et al. Liver-derived IGF-I is of importance for normal carbohydrate and lipid metabolism. Diabetes 2001;50(7):1539–45.
[38] Johannsson G, Sverrisdottir YB, Ellegard L, et al. GH increases extracellular volume by stimulating sodium reabsorption in the distal nephron and preventing pressure natriuresis. J Clin Endocrinol Metab 2002;87(4):1743–9.
[39] Consensus. Critical evaluation of the safety of recombinant human growth hormone administration: statement from the Growth Hormone Research Society. J Clin Endocrinol Metab 2001;86:1868–70.
[40] de Boer H, Blok GJ, Voerman B, et al. The optimal growth hormone replacement dose in adults, derived from bioimpedance analysis. J Clin Endocrinol Metab 1995;80:2069–76.
[41] Valk NK, vd Lely AJ, de Herder WW, et al. The effects of human growth hormone (GH) administration in GH-deficient adults: a 20-day metabolic ward study. J Clin Endocrinol Metab 1994;79:1070–6.
[42] Attanasio AF, Howell S, Bates PC, et al. Body composition, IGF-I and IGFBP-3 concentrations as outcome measures in severely GH-deficient (GHD) patients after childhood GH treatment: a comparison with adult onset GHD patients. J Clin Endocrinol Metab 2002; 87:3368–72.

[43] Attanasio AF, Lamberts SWJ, Matranga AMC, et al. Adult growth hormone (GH)-deficient patients demonstrate heterogeneity between childhood onset and adult onset before and during human GH treatment. J Clin Endocrinol Metab 1997;82:82–8.
[44] Abs R, Bengtsson B-Å, Hernberg-Stahl E, et al. GH replacement in 1034 growth hormone deficient hypopituitary adults: demographic and clinical characteristics, dosing and safety. Clin Endocrinol [Oxf] 1999;50:703–13.
[45] Johannsson G, Bengtsson B-Å, Andersson B, et al. Long-term cardiovascular effects of growth hormone treatment in GH-deficient adults: preliminary data from a small group of patients. Clin Endocrinol [Oxf] 1996;45:305–14.
[46] Murray RD, Skillicorn CJ, Howell SJ, et al. Influences on quality of life in GH deficient adults and their effect on response to treatment. Clin Endocrinol [Oxf] 1999;51:565–73.
[47] Wirén L, Bengtsson B-Å, Johannsson G. Beneficial effects of long-term GH replacement therapy on quality of life in adults with GH deficiency. Clin Endocrinol [Oxf] 1998;48:613–20.
[48] Johansson J-O, Fowelin J, Landin K, et al. Growth hormone-deficient adults are insulin-resistant. Metabolism 1995;44:1126–9.
[49] Svensson J, Fowelin J, Landin K, et al. Effects of seven years of GH-replacement therapy on insulin sensitivity in GH-deficient adults. J Clin Endocrinol Metab 2002;87(5):2121–7.
[50] Gotherstrom G, Svensson J, Koranyi J, et al. A prospective study of 5 years of GH replacement therapy in GH-deficient adults: sustained effects on body composition, bone mass, and metabolic indices. J Clin Endocrinol Metab 2001;86(10):4657–65.
[51] Monson JP, Bengtsson BA, Abs R, et al. Can growth hormone therapy cause diabetes? KIMS Strategic Committee. Lancet 2000;355(9216):1728–9.
[52] Maison P, Griffin S, Nicoue-Beglah M, et al. Impact of growth hormone (GH) treatment on cardiovascular risk factors in GH-deficient adults: a metaanalysis of blinded, randomized, placebo-controlled trials. J Clin Endocrinol Metab 2004;89(5):2192–9.
[53] Gibney J, Wallace JD, Spinks T, et al. The effects of 10 years of recombinant human growth hormone (GH) in adult GH-deficient patients. J Clin Endocrinol Metab 1999;84:2596–602.
[54] Bravenboer N, Holzmann PJ, ter Maaten JC, et al. Effect of long-term growth hormone treatment on bone mass and bone metabolism in growth hormone-deficient men. J Bone Miner Res 2005;20(10):1778–84.
[55] Koranyi J, Svensson J, Gotherstrom G, et al. Baseline characteristics and the effects of five years of GH replacement therapy in adults with GH deficiency of childhood or adulthood onset: a comparative, prospective study. J Clin Endocrinol Metab 2001;86(10):4693–9.
[56] Bex M, Abs R, Maiter D, et al. The effects of growth hormone replacement therapy on bone metabolism in adult-onset growth hormone deficiency: a 2-year open randomized controlled multicenter trial. J Bone Miner Res 2002;17(6):1081–94.
[57] Johannsson G, Rosén T, Bosaeus I, et al. Two years of growth hormone (GH) treatment increases bone mineral content and density in hypopituitary patients with adult-onset GH deficiency. J Clin Endocrinol Metab 1996;81:2865–73.
[58] Cuneo RC, Salomon F, Wiles CM, et al. Growth hormone treatment in growth hormone-deficient adults. I. Effects on muscle mass and strength. J Appl Physiol 1991;70:688–94.
[59] Svensson J, Stibrant Sunnerhagen K, Johannsson G. Five years of growth hormone replacement therapy in adults: age- and gender-related changes in isometric and isokinetic muscle strength. J Clin Endocrinol Metab 2003;88(5):2061–9.
[60] Cutfield WS, Wilton P, Bennmarker H, et al. Incidence of diabetes mellitus and impaired glucose tolerance in children and adolescents receiving growth-hormone treatment. Lancet 2000;355:610–3.
[61] Pfeifer M, Verhovec M, Zi Zek B, et al. Growth hormone (GH) treatment reverses early atherosclerotic changes in GH-deficient adults. J Clin Endocrinol Metab 1999;84:453–7.
[62] Chrisoulidou A, Beshyah SA, Rutherford O, et al. Effects of 7 years of growth hormone replacement therapy in hypopituitary adults. J Clin Endocrinol Metab 2000;85(10):3762–9.

[63] Svensson J, Bengtsson BA, Rosen T, et al. Malignant disease and cardiovascular morbidity in hypopituitary adults with or without growth hormone replacement therapy. J Clin Endocrinol Metab 2004;89(7):3306–12.
[64] Popovic V, Damjanovic S, Micic D, et al. Increased incidence of neoplasia in patients with pituitary adenomas. The Pituitary Study Group. Clin Endocrinol (Oxf) 1998;49:441–5.
[65] Nilsson B, Gustavsson-Kadaka E, Bengtsson BA, et al. Pituitary adenomas in Sweden between 1958 and 1991: incidence, survival, and mortality. J Clin Endocrinol Metab 2000; 85:1420–5.
[66] Shaneyfelt T, Husein R, Bubley G, et al. Hormonal predictors of prostate cancer: A meta-analysis. J Clin Oncol 2000;18:847–53.
[67] Hankinson SE, Willett WC, Colditz GA, et al. Circulating concentrations of insulin-like growth factor-I and risk of breast cancer. Lancet 1998;351:1393–6.
[68] Ma J, Pollak MN, Giovannucci E, et al. Prospective study of colorectal cancer risk in men and plasma levels of insulin-like growth factor (IGF)-I and IGF-binding protein-3. J Natl Cancer Inst 1999;91:620–5.
[69] Johannsson G, Rosén T, Lindstedt G, et al. Effects of 2 years of growth hormone treatment on body composition and cardiovascular risk factors in adults with growth hormone deficiency. Endocrinol Metab 1996;3(Suppl A):3–12.
[70] Brada M, Rajan B, Traish D, et al. The long-term efficacy of conservative surgery and radiotherapy in the control of pituitary adenomas. Clin Endocrinol [Oxf] 1993;38:571–8.
[71] Frajese G, Drake WM, Loureiro RA, et al. Hypothalamo-pituitary surveillance imaging in hypopituitary patients receiving long-term GH replacement therapy. J Clin Endocrinol Metab 2001;86(11):5172–5.
[72] Karavitaki N, Warner JT, Marland A, et al. GH replacement does not increase the risk of recurrence in patients with craniopharyngioma. Clin Endocrinol [Oxf] 2006;64(5):556–60.
[73] Jostel A, Mukherjee A, Hulse PA, et al. Adult growth hormone replacement therapy and neuroimaging surveillance in brain tumour survivors. Clin Endocrinol [Oxf] 2005;62(6): 698–705.
[74] Brismar K, Fernqvist-Forbes E, Wahren J, et al. Effect of insulin on the hepatic production of insulin-like growth factor-binding protein-1 (IGFBP-1), IGFBP-3, and IGF-I in insulin-dependent diabetes. J Clin Endocrinol Metab 1994;79(3):872–8.
[75] Hennen G, Frankenne F, Closset J, et al. A human placental GH: increasing levels during second half of pregnancy with pituitary GH suppression as revealed by monoclonal antibody radioimmunoassays. Int J Fertil 1985;30:27–33.
[76] Boguszewski CL, Boguszewski MCS, de Zegher F, et al. Growth hormone isoforms in newborns and postpartum women. Eur J Endocrinol 2000;142:353–8.
[77] Caufriez A, Frankenne F, Englert Y, et al. Placental growth hormone as a potential regulator of maternal IGF-I during human pregnancy. Am J Physiol 1990;258:E1014–9.
[78] Bauer MK, Harding JE, Bassett NS, et al. Fetal growth and placental function. Mol Cell Endocrinol 1998;140:115–20.
[79] Hall K, Enberg G, Hellem E, et al. Somatomedin levels in pregnancy: longitudinal study in healthy subjects and patients with growth hormone deficiency. J Clin Endocrinol Metab 1984;59:587–94.
[80] Wiren L, Boguszewski CL, Johannsson G. Growth hormone (GH) replacement therapy in GH-deficient women during pregnancy. Clin Endocrinol [Oxf] 2002;57:235–59.

Quality of Life in Growth Hormone Deficiency and Acromegaly

Susan M. Webb, MD, PhD[a],*, Xavier Badia, MD, PhD[b]

[a]Department of Endocrinology, Hospital de Sant Pau, Autonomous University of Barcelona, Pare Claret 167, 08025 Barcelona, Spain
[b]Health Economics and Outcome Research, IMS Health, Avda. Diagonal 618, 1°C, 08021-Barcelona, Spain

Quality of life (QoL) has emerged in recent years as an important indicator of health; it reflects the overall effects of a disease and its treatment and is the sum on the patient of the effects of symptoms, function, and well-being, the effects of which are variable and complicated by a process of adaptation; this means that with a disability the patient maintains good QoL by taking up activities or challenges that are within his or her capabilities. In any case, the patient is the best judge of his or her QoL, because it considers the patient's perspective, and relates to how an individual feels, functions, and responds in daily life; it is influenced by the patient's goals, expectations, standards, and concerns, and the patient's cultural context.

Any medical condition is capable of restricting a patient's life, involving physical, emotional, or social aspects. The increasing interest in measures reflecting the personal point of view of patients' health and his or her active participation in decision making regarding treatment options has led to an extended demand for reliable and valid standardized questionnaires for evaluating health-related QoL (HRQoL) [1]. These questionnaires are used to evaluate the health status of both individuals and populations; they differ in aims, content, language, and scoring among other features. Depending on item content, scope, and target population, instruments can basically be classified as generic or disease-generated or -specific [2].

Generic questionnaires may be applied to a general population or different groups of patients, which can then be compared as to how much their specific disease or disability affects their QoL; examples are the Nottingham Health Profile (NHP), the Psychological General Well Being Schedule (PGWS), the 5 Dimensions EuroQoL, and the SF-36.

---

* Corresponding author.
 E-mail address: swebb@santpau.es (S.M. Webb).

The NHP measures perceived health problems and the extent to which they affect daily activity [3]. The PGWS is designed to provide an index of self-perception of affective or emotional states, reflecting a sense of subjective well-being or distress [4]. The 5 Dimensions EuroQoL was designed for self-completion by respondents and is ideally suited for use in postal surveys, in clinics, and in face-to-face interviews; it has been used for both clinical and economic appraisal [5–7]. Originally designed to complement other generic instruments or disease-specific questionnaires, it is now increasingly used as a stand-alone measure. Each of the five dimensions (mobility; self-care; pain and discomfort; anxiety and depression; usual activities [work, study, housework, family, leisure]) has three levels (no problem, some problems, and major problems), with a total of 245 possible health states including "unconscious" and "dead" and "perfect health state" defined as patients having no problems in any of the five dimensions (ie, who scored 11,111). It also includes a visual analog scale (EQ-VAS), which respondents use to rate their own health, where 0 is the "worst" and 100 is the "best" imaginable health state. The SF-36 questionnaire was designed for use in clinical practice and research, health policy evaluations, and general population surveys. Originally developed in American English, it has since been translated into many different languages, and includes 36 items across eight health dimensions, summarized into a physical and mental component [8].

Disease-specific questionnaires are designed to evaluate those areas that are mainly affected in a determined disease and are more sensitive to detect impairments and changes related to the disease; examples include the Adult GH Deficiency Assessment (AGHDA) score [9] and the Questions on Life Satisfaction-Hypopituitarism (QLS-H) questionnaire for adult patients with growth hormone deficiency [10,11], and Acromegaly Quality of Life (AcroQoL) questionnaire for patients with acromegaly [12].

HRQoL questionnaires, especially disease-generated ones, favor further in-depth clinical knowledge of the specific condition, both in clinical settings and therapy research; they can be used for evaluating different types of interventions and gaining greater knowledge of the clinical impact of the disease. In the area of health policies and financing, they can provide valid information for the manager or health care policy maker to make decisions regarding assigning resources, and may be used as a tool in pharmacoeconomic studies. From the patient's point of view, they can contribute to improve the patient's perception of high-quality management and is an opportunity to improve the physician-patient relationship.

Adult patients with growth hormone deficiency and quality of life

Impairment of QoL in AGHD was revealed in the first randomized, placebo-controlled clinical trials performed in the late 1980s and early 1990s [13–15], even though it was suggested, but subsequently overlooked, three decades before by Raben [16]. A number of generic questionnaires designed to cover domains related to perception of health problems (NHP), psychiatric morbidity (PWBS and the General Health Questionnaire GHQ), and life events in the preceding month that might bias current mood (life events inventor) were initially included in clinical trials [14,17]. At entry, AGHD patients were found to have worse scores than matched controls for both the NHP and PGWS, indicating that these adult patients with AGHD were psychologically compromised, to an extent that 4 out of 10 would have been severe enough to justify a psychiatric referral. Subsequent studies have shown that improvements in energy and emotional reactions tend to occur within the first 6 months (but may even occur later) and are still present after several years, so that treated patients do not differ from matched controls anymore, whereas untreated patients showed no improvement in impaired QoL [11,18–24]. From a practical point of view, it is noteworthy that most adult patients who have experienced the benefits of recombinant human growth hormone (rhGH) choose to continue on daily subcutaneous injections, which is in sharp contrast with diabetic subjects who often ask if they could not change insulin for oral antidiabetic drugs. Altogether, these facts argue strongly against the idea that improvement of QoL constitutes a placebo effect of rhGH replacement therapy, and supports the concepts that GH replacement should be considered standard practice, as with other hormone deficiencies of hypopituitarism [25,26].

An increasing interest in measures reflecting the personal point of view of these AGHD patients led to the development of questionnaires designed to address their particular issues of concern and which the clinicians wished to monitor; they were more likely to be relevant and sensitive for areas requiring clinical monitoring in these patients. The AGHDA score has subsequently been validated and includes dimensions related to dislike of body image, low energy, poor concentration and memory, and increased irritability. Normative data are available for the Spanish general population where AGHD patients were found to score worse (except for patients aged over 60 years) [27], and Swedish control population [28]. Its initial definition as "disease-specific" has been criticized, because it was also found to identify impaired QoL in acromegaly [29]; a more appropriate denomination is "disease-generated" or "disease-orientated" questionnaire, even though it reflects the patient's global situation (ie, health dimensions, including symptoms, deficiencies, treatments, and so forth), but also professional, hobby, humor, libido, and other related areas. It should also be remembered that extreme situations like GHD and acromegaly can both share common symptoms (ie, lack of energy, dislike of body image, and so forth), so no QoL instrument provides absolute discrimination between different patients or control groups, making individual clinical assessment paramount.

Another questionnaire for AGHD is the QLS-H, which has the advantage of weighting each item by the individual patient; patients are first asked how important each item is to them and then how satisfied they are with each item; translated into several languages, normative data are available in six European countries and the United States [10]. In 66 patients with severe childhood-onset GHD in the transition phase to adulthood, QLS-H scores for ability to become sexually aroused, ability to tolerate stress, body shape, concentration, initiative and drive, physical stamina, and self-confidence were lower than the normal average, even though overall baseline QoL was not compromised, probably because of the adaptation phenomenon previously mentioned; after treatment dimensions related to age-specific psychologic problems (ie, sexual arousal and body shape) responded positively to GH replacement [30].

The mechanism by which AGHD produces bad QoL is probably multifactorial, in parallel to the widespread physiologic actions of GH (which include normalization of body composition; increased exercise capacity and muscle strength; recovery of decreased body water, which has been related to fatigue; normalization of metabolic disturbances; and possible neuroendocrine effects in the central nervous system). This is further supported by the lack of definite correlations between improvements in QoL and biochemical, metabolic, or body composition parameters in treated AGHD.

Since the initial clinical trials in the early 1990s, the required dose of GH in adults has been reduced to around half or a third of that initially prescribed, with no loss of improved QoL or clinical benefit, but significantly less side effects like edema and arthralgia [31,32].

In the untreated state, AGHD patients do not always score badly on QoL questionnaires; this has led more skeptical physicians to doubt the relevance of QoL as an end point for the therapeutic benefits of GH replacement. Whenever AGHD cohorts have been compared with general populations, however, the former have always scored significantly worse [24], and GH-replaced AGHD patients maintain improved QoL, whereas untreated subjects show a mild gradual decline over the subsequent years, especially for vitality [27,33,34]. Childhood-onset GHD patients tend to show less impairment of QoL, probably explained by an adaptation mechanism whereby they aspire to less demanding challenges and aims in life, more in accordance to their limitations.

The degree of improvement in QoL with GH replacement has been shown to be proportional to the baseline deviation from normality. This explains that in some randomized, double-blind, placebo-controlled studies unselected at study entry in terms of QoL, no changes in QoL or cognitive function were observed after 18 months of GH therapy [35]; it should be pointed out, however, that only generic questionnaires were used in those studies that showed no improvement [36]. Because improvement of QoL after GH replacement is proportional to basal impairment, this is not really surprising because patients with normal QoL at outset can hardly be expected to achieve any improvement during substitution therapy. When instruments specifically developed for AGHD patients have been used to assess QoL, consistent improvement with GH replacement therapy has been reported, mainly in the dimension of vitality [24].

A practical reflection of improved QoL is the fact that days of sick leave, number of doctor visits, and days in hospital were reduced after starting replacement therapy with GH and leisure time physical activity and satisfaction with physical activity increased [37].

Acromegaly and quality of life

Acromegaly, a chronic disease that implies physical limitations (low libido, sex drive, and energy, together with headaches and joint pains) and morphologic changes (in body image, excessive sweating, change in the subject's voice) often not completely reversible, may go undiagnosed for years despite the presence of signs and symptoms; the impact of the disease and its treatment on the patients' QoL is conceivably great. Successful surgery or response to medical treatment may be followed by marked improvement in the patient's general condition, often, but not always, accompanied by improvement or normalization of biochemical parameters, such as GH and insulin-like growth factor (IGF)-I. Whether QoL is also restored to normal was unknown until recently, when interest in its evaluation emerged.

QoL was evaluated with the SF-36 questionnaire in 168 North American patients with pituitary disease, including 36 patients with acromegaly [38]; all pituitary disease patients had a lower perceived QoL than normal population in physical and mental measures; patients with acromegaly showed impaired physical function but no difference in the mental measures.

In 1999 the authors set out to develop a disease-generated QoL questionnaire for acromegaly. After reviewing Medline between 1982 and 2000 using acromegaly and QoL as headers, they came across a few reports dealing with several aspects of physical and psychologic restrictions in acromegaly (Table 1) [39–51]. These were heterogeneous, and described case reports [39], personal experiences [40], comparisons of acromegaly with other endocrine diseases or pituitary tumors [41,42], and response to surgical [42,43] or medical treatments [44]. The only relatively systematic attempt to study the psychopathologic concomitants of acromegaly until now dated back to 1951 [45] and reported in 28 patients that acromegaly was not associated with any psychoses, but was commonly accompanied by mood liability and personality change. After this literature search, a panel of expert endocrinologists identified the areas they believed were important for HRQoL in acromegaly. These dimensions were physical, psychologic, social, capacity to perform daily activities, symptoms, cognitive ability, health perception, sleeping abnormalities, sexual function, energy and vitality, pain, and body image. Semistructured interviews performed in 10 acromegalic patients asked to describe how the disease had affected their QoL were taped and transcribed; after a qualitative reduction, remaining items in question form were rated by the panel of experts; items were selected for final inclusion if importance, clarity, and frequency were rated as high. The resulting 38-item

Table 1
Physical and psychologic problems reported in acromegaly that are involved in the patient's health-related quality of life

Reference	Depression	Social withdrawal	Low energy/ fatigability	Anxiety	Mood lability	Amnesia	Low libido
Abed et al, 1987 [46]	No						
Avery, 1973 [47]	Yes						
Beaumont, 1972 [48]						Yes	
Bleuler, 1951 [45]					Yes		
Ezzat, 1992 [49]		Yes			Yes		Yes
Flitsch et al, 2000 [42]			Yes				
Landolt, 1990 [50]	Yes		Yes		Yes		
Margo, 1981 [39]	Yes						
Sablowski et al, 1986 [43]				Yes			
Smith et al, 1972 [51]			Yes			Yes	Yes
Sonino et al, 1999 [44]		Yes					

(*Adapted from* Webb SM, Prieto L, Badia X, et al. Quality of life in acromegaly. In: Lamberts SWJ, Ghigo E, editors. The expanding role of octreotide. A forum in Endocrinology and Eye Disease. Bioscientifica; Bristol (UK): Bioscientifica; 2002. p. 19–25.)

questionnaire used frequency of occurrence (always, most of the time, sometimes, rarely, or never) and degree of agreement with the statement (completely agree, moderately agree, neither agree nor disagree, moderately disagree, or completely disagree) as response choices in a five-point Likert-type scale. This initial questionnaire was administered to 72 patients with acromegaly throughout Spain and the answers were analyzed by Rasch analysis [52], a scaling approach, and reduction procedure model, which allowed a further reduction to obtain the final 22-item AcroQoL questionnaire [12]. AcroQoL provides clinicians with a cost-effective means to assess patient self-perceived status, can be used to screen patients who require appropriate further evaluation, and in longitudinal research of QoL assessments can increase the knowledge of the impact of the disease on patient perception of well-being and functioning. This is particularly useful in the evaluation of intervention or treatment effects.

The AcroQoL questionnaire is currently available in 12 languages (Spanish, English, French, German, Italian, Greek, Dutch, Swedish, Hungarian,

Polish, Turkish, and Portuguese). Translations were all performed and validated by standard procedures, by professional translators. Available data allow a comparison of results between different countries (Table 2) [53–58]. When patients in remission are compared, results are very similar globally (mean scores range from 64–68) and for each dimension; patients scored lowest (ie, most impact on HRQoL) for appearance (between 51 and 65) and highest (least impact on HRQoL) for personal relations. Furthermore, patients with active disease scored lower than patients in remission in The Netherlands [53,54], Switzerland [56], Spain [58], the United Kingdom [55], and Turkey [57] (Table 3).

The use of AcroQoL in clinical practice in the United Kingdom has confirmed severe impairment HRQoL in patients with acromegaly [55]. Eighty patients, 42 males, with a mean age of 54 years, completed two generic questionnaires (PGWS, EuroQoL-5D) and a symptom and signs score and AcroQoL; GH and IGF-I were also measured. The PGWS showed that acromegalic patients had significantly worse HRQoL as compared with the general population (70 versus 82, $P < .05$), especially in the domains of general health and vitality. These patients also scored worse on the PGWS than patients who had been treated for a nonfunctioning pituitary adenoma, and as bad as or worse than adults with GH deficiency. Significant correlations were observed between the Visual Analog Scale of the 5 Dimensions EuroQoL, PGWS, and AcroQoL, indicative that the dimensions evaluated by AcroQoL are those that impact HRQoL.

QoL was shown to be severely impaired in 118 acromegalic patients in long-term (on average 12 years) remission (defined as having basal GH <1.9 ng/mL or <0.38 ng/mL after an oral glucose tolerance test and normal sex- and age-matched IGF-I) from The Netherlands as demonstrated with the SF-36 and NHP generic questionnaires, for which normative data from Dutch general population are available [53]. Disease duration, age, treatment with radiotherapy, and presence of joint problems were negatively correlated with the AcroQoL scores [53,54]; only the appearance dimension (which was the most impaired) was weakly but significantly correlated with circulating GH levels ($P < .05$). Endocrine control is not associated with normalization of QoL in acromegaly and seems to imply that an earlier diagnosis before irreversible changes had occurred is desirable. A linear regression analysis performed to investigate the factors determining

Table 2
Acromegaly quality of life dimension scores from acromegalic patients in remission using the Dutch, German, and Spanish versions of the questionnaire

	Global	Physical	Psychologic	Appearance	Personal relations
Biermasz et al [53] n = 118	68 ± 17	64 ± 21	71 ± 17	63 ± 22	78 ± 15
Trepp et al [56] n = 33	68 ± 17	67 ± 20	68 ± 18	51 ± 16	73 ± 20
Webb et al [58] n = 42	64 ± 20	60 ± 27	66 ± 19	56 ± 23	77 ± 19

Table 3
Acromegaly quality of life scores from patients with active and controlled acromegaly in different countries and languages

	Biermasz, NL [53]	Rowles, UK [55]	Trepp, CH-D [56]	Deyneli, TK [57]	Webb, E [58]
Global	68 ± 17	57 (range 18–93)	—	—	61 ± 19
Active	—	—	43 ± 25	48 ± 13	56 ± 20
Remission	68 ± 17	—	68 ± 17	71 ± 14	65 ± 18

Mean ± SD, unless specified otherwise.
Abbreviations: CH, Switzerland–German; E, Spain–Spanish; NL, Netherlands–Dutch; TK, Turkey–Turkish; UK, United Kingdom–English.

QoL confirmed that patients with joint problems scored significantly worse globally and for each dimension ($P < .01$); personal relations and the psychologic dimension were also significantly related to age ($P < .05$) and disease duration ($P < .01$) [54].

The selection of an ideal control group for any QoL questionnaire is difficult because the subjects should have the same underlying diagnosis and have undergone the same treatment, require frequent follow-up, but not have (in this case) acromegaly. Because this ideal control group cannot exist, one should be aware that control subjects from the general population have better QoL and other chronic or pituitary diseases with frequent follow-up are not completely comparable. For validation of the psychometric properties that any such questionnaire should fulfill (reliability or capacity to show the same results when the clinical situation is stable and sensitivity to change [ie, improvement of the QoL score when the disease is successfully treated]) the authors chose obese patients as a control group, because these patients also have problems with appearance and physical dimensions and are accessible in the endocrine clinics. The rationale for this was that if any differences were found between these controls and the acromegalic patients, they could be assumed to be specifically related to acromegaly.

Patients with acromegaly and obese controls showed more problems on the EuroQoL than general Spanish population [58]. Globally, AcroQoL scored worse in the sensitivity to change group (which comprised 42 patients with active disease studied basally and 6 months after treatment) than in the reliability group (formed by 64 acromegalic patients with treated, stable disease, studied twice within 1 month: 56 ± 20 versus 65 ± 18, $P < .05$) but did not discriminate between patients and obese controls. The psychologic domain was worse in the sensitivity-to-change group than obese controls. Appearance was the most affected subscale in acromegaly and significantly worse than in obese controls. Interestingly, the subscale personal relationships of AcroQoL were less affected in the stable group than in obese controls. Longitudinal retesting disclosed no change in the reliability group, demonstrating good test-retest reliability, whereas in the sensitivity-to-change group, after 6 months of treatment there was improvement in

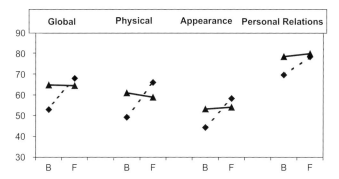

Fig. 1. Longitudinal evaluation of the AcroQoL questionnaire in patients with stable disease (*full lines with triangles*) or active disease treated for 6 months (*dotted lines*). No change was observed at re-evaluation in the stable group, whereas improvement was seen globally and for each subscale of AcroQoL in the active group after treatment ($P < .01$). B, basal; F, follow-up. (*Data from* Webb SM, Badia X, Lara-Surinach N, for the Spanish AcroQoL Study Group. Validity and clinical applicability of the Acromegaly Quality of Life Questionnaire (AcroQoL): a six months prospective study. Eur J Endocrinol 2006;155:269–77.)

AcroQoL score (Fig. 1). No correlation between AcroQoL and GH or IGF-I was observed, although there was a near significant trend between the subscale of appearance (the most affected dimension of AcroQoL) and IGF-I ($P = 0.051$). These findings validated the AcroQoL questionnaire as a valid tool for the assessment of QoL in clinical practice in patients with acromegaly [58].

Summary

This article highlights QoL as an important end point in the clinical and therapeutic evaluation of chronic endocrine diseases, such as AGHD and acromegaly. Despite being extreme conditions of the GH status, both impair QoL significantly and even share some common problems like easy fatigability or dislike of body image. The availability of questionnaires designed to analyze the problems that characterize these patients has shed light on dimensions often ignored by clinicians, which are of concern for the affected patients. Examples are the AGHDA score and the QLS-H for AGHD and the AcroQoL questionnaire for acromegaly. Replacement therapy with GH in AGHD is associated with improvement in QoL, proportionally to the degree of initial impairment, and is maintained over several years of therapy. In acromegaly, despite improvement in QoL in parallel to endocrine control following successful therapy, it still seems to remain below normative scores in the general population; appearance and physical dimensions are the most affected, and probably benefit from an earlier diagnosis and control of the disease, before irreversible changes take place.

References

[1] Patrick DL, Erickson P. Health states and health policy: quality of life in health care evaluation and resource allocation. Oxford: Oxford University Press; 1993.
[2] McDowell I, Newell C. Measuring health: a guide to rating scales and questionnaires. 2nd edition. Oxford: Oxford University Press; 1996.
[3] Hunt SM, McKenna SP, McEwen J, et al. The Nottingham Health profile: subjective health status and medical consultations. Soc Sci Med 1981;15A:221–9.
[4] Gray LC, Goldsmith HF, Livieratos BB, et al. Individual and contextual social-status contributions to psychological well-being. Soc & Soc Res 1983;68:78–95.
[5] Dolan P. Modeling valuations for EuroQol health states. Med Care 1997;35:1095–108.
[6] Badia X, Herdman M, Schiaffino A. A determining correspondence between scores on the EQ-5D thermometer and a 5-point categorical rating scale. Med Care 1999;37:671–7.
[7] Brooks R. EuroQol: the current state of play. Health Policy (New York) 1996;37:53–7.
[8] Ware JE, Snow KK, Kosinski M, et al. SF-36 Health Survey: manual and interpretation guide. Boston: The Health Institute, New England Medical Center; 1993.
[9] McKenna SP, Doward LC, Alonso J, et al. The QoL.AGHDA: an instrument for the assessment of quality of life in adults with growth hormone deficiency. Qual Life Res 1999;8: 373–83.
[10] Blum WF, Shavrikova EP, Edwards DJ, et al. Decreased quality of life in adult patients with growth hormone deficiency compared with general populations using the new, validated, self-weighted questionnaire, questions on life satisfaction hypopituitarism module. J Clin Endocrinol Metab 2003;88:4158–67.
[11] Rosilio M, Blum WF, Edwards DJ, et al. Long-term improvement of quality of life during growth hormone (GH) replacement therapy in adults with GH deficiency, as measured by QLS-H. J Clin Endocrinol Metab 2004;89:1684–93.
[12] Webb SM, Prieto L, Badia X, et al. Acromegaly Quality of Life Questionnaire (ACROQOL) a new health-related quality of life questionnaire for patients with acromegaly: development and psychometric properties. Clin Endocrinol (Oxf) 2002;57:251–8.
[13] Degerblad M, Grunditz R, Hall K, et al. Substitution therapy with recombinant growth hormone (Somatrem) in adults with growth hormone deficiency. Acta Paediatr Scand Suppl 1987;337:170–1.
[14] Salomon F, Cuneo R, Hesp R, et al. The effects of treatment with recombinant human growth hormone on body composition and metabolism in adults with growth hormone deficiency. N Engl J Med 1989;321:1797–803.
[15] Cuneo R, Salomon F, McGauley G, et al. The growth hormone deficiency syndrome in adults. Clin Endocrinol 1992;37:387–97.
[16] Raben MS. Clinical use of human growth hormone. N Engl J Med 1962;266:82–6.
[17] McGauley GA, Cuneo RC, Salomon F, et al. Psychological well-being before and after growth hormone treatment in adults with growth hormone deficiency. Horm Res 1990; 33(Suppl 4):52–4.
[18] Wiren L, Bengtsson BA, Johannsson G. Beneficial effects of long-term GH replacement therapy on quality of life in adults with GH deficiency. Clin Endocrinol (Oxf) 1998;48:613–20.
[19] McMillan CV, Bradley C, Gibney J, et al. Psychological effects of withdrawal of growth hormone therapy from adults with growth hormone deficiency. Clin Endocrinol (Oxf) 2003;59: 467–75.
[20] Malik IA, Foy P, Wallymahmed M, et al. Assessment of quality of life in adults receiving long-term growth hormone replacement compared to control subjects. Clin Endocrinol (Oxf) 2003;59:75–81.
[21] Hakkaart-van Roijen L, Beckers A, Stevenaert A, et al. The burden of illness of hypopituitary adults with growth hormone deficiency. Pharmacoeconomics 1998;14:395–403.
[22] Sonksen PH, McGauley G. Lies, damn lies and statistics. Growth Horm IGF Res 2005;15: 173–6.

[23] Gibney J, Johannsson G. Clinical monitoring of growth hormone replacement in adults. Front Horm Res 2005;33:86–102.
[24] Woodhouse LJ, Mukherjee A, Shalet SM, et al. The influence of growth hormone status on physical impairments, functional limitations, and health-related quality of life in adults. Endocr Rev 2006;27:287–317.
[25] Consensus guidelines for the diagnosis and treatment of adults with growth hormone deficiency: summary statement of the Growth Hormone Research Society workshop on adult GHD. J Clin Endocrinol Metab 1998;83:379–81.
[26] Molitch ME, Clemmons DR, Malozowski S, et al. Evaluation and treatment of adult growth hormone deficiency: an Endocrine Society clinical practice guideline. J Clin Endocrinol Metab 2006;91:1621–34.
[27] Badia X, Lucas A, Sanmarti A, et al. One-year follow-up of quality of life in adults with untreated growth hormone deficiency. Clin Endocrinol (Oxf) 1998;49:765–71.
[28] Wiren L, Whalley D, McKenna S, et al. Application of a disease-specific, quality-of-life measure (QoL-AGHDA) in growth hormone-deficient adults and a random population sample in Sweden: validation of the measure by Rasch analysis. Clin Endocrinol (Oxf) 2000;52:143–52.
[29] Barkan AL. The Quality of Life-Assessment of Growth Hormone Deficiency in Adults questionnaire: can it be used to assess quality of life in hypopituitarism? J Clin Endocrinol Metab 2001;86:1905–7.
[30] Attanasio AF, Shavrikova EP, Blum WF, et al. Quality of life in childhood onset growth hormone-deficient patients in the transition phase from childhood to adulthood. J Clin Endocrinol Metab 2005;90:4525–9.
[31] Murray RD, Shalet SM. The use of self-rating questionnaires as a quantitative measure of quality of life in adult growth hormone deficiency. J Endocrinol Invest 1999;22(5 Suppl):118–26.
[32] Webb SM, Mo M, Li Y, et al, on behalf of the HypoCCS International Advisory Board. The Hypopituitary Control and Complication Study (HypoCCS) 1996–2006: a decade of care of adult patients with growth hormone deficiency (GHD) [abstract 3-812]. In: Programs and abstracts of the 88th annual meeting of the Endocrine Society. Boston; 2006.
[33] Gilchrist FJ, Murray RD, Shalet SM. The effect of long-term untreated growth hormone deficiency (GHD) and 9 years of GH replacement on the quality of life (QoL) of GH-deficient adults. Clin Endocrinol (Oxf) 2002;57:363–70.
[34] Abs R, Bengtsson BA, Hernberg-Stahl E, et al. GH replacement in 1034 growth hormone deficient hypopituitary adults: demographic and clinical characteristics, dosing and safety. Clin Endocrinol (Oxf) 1999;50:703–13.
[35] Baum HB, Katznelson L, Sherman JC, et al. Effects of physiological growth hormone (GH) therapy on cognition and quality of life in patients with adult-onset GH deficiency. J Clin Endocrinol Metab 1998;83:3184–9.
[36] Arwert LI, Deijen JB, Drent ML. Effects of growth hormone deficiency and growth hormone treatment on quality of life in growth hormone- deficient adults. Front Horm Res 2005;33:196–208.
[37] Svensson J, Mattsson A, Rosen T, et al. Three-years of growth hormone (GH) replacement therapy in GH-deficient adults: effects on quality of life, patient-reported outcomes and healthcare consumption. Growth Horm IGF Res 2004;14:207–15.
[38] Johnson MD, Woodburn CJ, Vance ML. Quality of life in patients with a pituitary adenoma. Pituitary 2003;6:81–5.
[39] Margo A. Acromegaly and depression. Br J Psychiatry 1981;139:467–8.
[40] Furman K, Ezzat S. Psychological features of acromegaly. Psychother Psychosom 1998;67:147–53.
[41] Fava GA, Sonino N, Morphy MA. Psychosomatic view of endocrine disorders. Psychother Psychosom 1993;59:20–33.

[42] Flitsch J, Spitzner S, Lüdecke DK. Emotional disorders in patients with different types of pituitary adenomas and factors affecting the diagnostic process. Exp Clin Endocrinol Diabetes 2000;108:480–5.
[43] Sablowski N, Pawlik K, Lüdecke DK, et al. Aspects of personality in patients with pituitary adenomas. Acta Neurochir (Wien) 1986;83:8–11.
[44] Sonino N, Scarpa E, Paoletta A, et al. Slow-release lanreotide treatment in acromegaly: effects on quality of life. Psychother Psychosom 1999;68:165–7.
[45] Bleuler M. The psychopathology of acromegaly. J Nerv Ment Dis 1951;113:497–511.
[46] Abed RT, Clark J, Elbadawy MHF, et al. Psychiatric morbidity in acromegaly. Acta Psychiatr Scand 1987;75:635–9.
[47] Avery TL. A case of acromegaly and gigantism with depression. Br J Psychiatr 1973;122:599–600.
[48] Beaumont PJV. Endocrines and psychiatry. Br J Hosp Med 1972;7:561–79.
[49] Ezzat S. Living with acromegaly. Endocrinol Metab Clin North Am 1992;21:753–60.
[50] Landolt AM. Acromegaly. In: Labhart A, editor. Clinical endocrinology. Barcelona: Salvat; 1990. p. 121–2.
[51] Smith CK, Barish J, Correa J, et al. Psychiatric disturbance in endocrine disease. Psychosom Med 1972;34:69–82.
[52] Rasch G. Probabilistic models for some intelligence and attainment tests. Chicago: Mesa Press; 1993.
[53] Biermasz NK, Van Thiel SW, Pereira AM, et al. Decreased quality of life in patients with acromegaly despite long-term cure of growth hormone excess. J Clin Endocrinol Metab 2004;89:5369–76.
[54] Biermasz NK, Pereira AM, Smit JWA, et al. Morbidity after long-term remission for acromegaly: persisting joint-related complaints cause reduced quality of life. J Clin Endocrinol Metab 2005;90:2731–9.
[55] Rowles SV, Prieto L, Badia X, et al. Quality of life (QOL) in patients with acromegaly is severely impaired: use of a novel measure of QOL. Acromegaly quality of life questionnaire. J Clin Endocrinol Metab 2005;90:3337–41.
[56] Trepp R, Everts R, Stettler C, et al. Assessment of quality of life in patients with uncontrolled versus controlled acromegaly using the acromegaly quality of life questionnaire (AcroQoL). Clin Endocrinol (Oxf) 2005;63:103–10.
[57] Deyneli O, Yavuz D, Gozu H, et al. Evaluation of quality of life in Turkish patients with acromegaly [abstract P3–508]. In: Programs and abstracts of the 84th annual meeting of the Endocrine Society. Philadelphia: 2003.
[58] Webb SM, Badia X, Lara-Surinach N, for the Spanish AcroQoL Study Group. Validity and clinical applicability of the Acromegaly Quality of Life Questionnaire (AcroQoL): a six months prospective study. Eur J Endocrinol 2006;155:269–77.

Growth Hormone Supplementation in the Elderly

Ralf Nass, MD[a], Jennifer Park, MD[b], Michael O. Thorner, MB, BS, DSc, FRCP, MACP[a],*

[a]Division of Endocrinology and Metabolism, Department of Medicine, University of Virginia, 450 Ray C. Hunt Drive, Charlottesville, VA 22903, USA
[b]Department of Internal Medicine, University of Virginia, Charlottesville, VA 22903, USA

Aging and its association with the loss of function has become an important issue, partially because of the dramatic increase in life expectancy observed since the end of World War II. In the developed world, people older than 80 years are the fastest growing subset of the population [1]. In the United States alone, the proportion of the population older than 60 years is expected to increase from 35 million (12.4%) in 2000 to 71.6 million (19.6%) in 2030, and the number of persons older than 80 years is projected to increase from 9.3 million in 2000 to 19.5 million in 2030 [2]. Between 2000 and 2030, the worldwide population older than 65 years is projected to increase from 420 million to 973 million. The largest increase in absolute numbers will occur in the developing countries where the population older than 65 years will increase from about 249 million in 2000 to an estimated 690 million in 2030 [3]. During the second half of the twentieth century, a 20-year increase in the average life span was observed, and the average life span worldwide is expected to increase by another 10 years by 2050 [4]. Therefore, the major priority in the future of aging research should be to enhance "healthy aging" and thus reduce the number of years that the elderly spend in a diseased state with disabilities or frailty. Maintaining independence is the number one priority. One of the major causes of dysfunction and disability in the elderly is loss of muscle mass and muscle strength (sarcopenia). This article reviews some of the aspects of sarcopenia in the elderly and the arguments in favor of and against growth hormone (GH) replacement in the elderly.

---

* Corresponding author.
  E-mail address: mot@virginia.edu (M.O. Thorner).

Features of growth hormone deficiency

GH deficiency (GHD) is a clinical syndrome associated with alterations in metabolism, mood, and quality of life. Specifically, there are changes in body composition such as increased total fat mass, principally manifest as central obesity [5,6], and decreased lean body mass [7,8]. Another metabolic characteristic of GHD is decreased bone mass, which is a marker for increased fracture rates [9,10]. Insulin resistance is also increased in adults who have GHD, which may be related to the increased fat mass [11]. Adults who have GHD have a greater number of cardiac risk factors such as abnormal lipid profiles, premature atherosclerosis, and impaired cardiac function that have been associated with an increased risk of death compared with control subjects [12–15]. Adults who have GHD have been reported to have impaired diastolic function [16,17]. Although some studies found preserved systolic function in GHD [17], others have described impaired left ventricular function [9,14]. Colao and colleagues [14] reported decreased left ventricular ejection fraction at peak exercise in elderly patients who had hypopituitarism compared with age-matched control subjects. Patients who have GHD can have impaired quality of life and experience decreased overall well-being, increased social isolation, and decreased energy [18,19], although the validity of the applicability of quality-of-life questionnaires in GHD (quality-of-life assessment of growth hormone deficiency in adults) has been questioned in the past [20].

Changes in body composition in the elderly

Circulating GH levels show a significant decline with aging. In elderly subjects, the 24-hour integrated GH concentration is equal to levels observed in young patients who have GHD. Several investigators have described a 15% to 70% reduction in GH secretory parameters in men and women older than 60 years [21,22]. The age-dependent decrease in GH levels is paralleled by changes in body composition similar to those seen with GHD. There is a change in the distribution of fat mass in the elderly, with an increase in total body fat by about 9% to 18% in men and 12% to 13% in women [23,24]. This age-dependent increase in body fat is associated with an excess of intra-abdominal fat [25–27] rather than subcutaneous abdominal fat [23,28–30]. In addition, a reduction of subcutaneous fat in the upper leg region has been described with aging [31]. Peak bone mass is usually reached by the third decade of life, followed by a significant gradual decline in bone mineral density [24,32–35] that is associated with an increased incidence of fractures. Several studies show an age-dependent decrease in lean body mass (fat-free mass) [23,36,37]. Janssen and colleagues [38] reported that skeletal muscle mass starts to decrease in the third decade when expressed as relative muscle mass, but when reporting absolute muscle mass, it remains stable until age 45 years and starts to decline afterward (Fig. 1). Thus, the relative decrease in muscle mass before age 45 is mainly

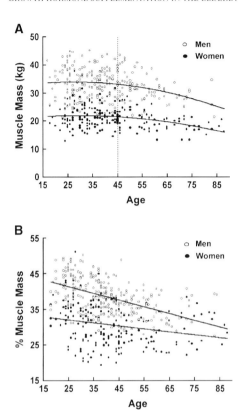

Fig. 1. (*A*) The relationship between whole-body skeletal muscle mass and age in men and women. Solid lines – regression lines. (*B*) The relationship between the relative skeletal muscle mass (body mass/skeletal muscle mass) and age in men and women. (*From* Janssen I, Heymsfield SB, Wang ZM, et al. Skeletal muscle mass and distribution in 468 men and women aged 18–88 yr. J Appl Physiol 2000;89(1):84; with permission.)

the result of an increase in body fat. This finding is in accordance with studies showing that muscle fiber cross-sectional area (ie, contractile muscle), body cell mass, and strength do not change substantially until around age 45 [39–44]. These findings suggest that interventions targeting age-dependent changes in body composition should be started earlier than age 60 years, which is the lower age-limit of most of the GH studies to date.

When reaching the eighth decade, men have lost about 7 kg of absolute muscle mass and women have lost about 3.8 kg of muscle mass [38]. The age-related decrease in lower extremity strength is of the order of 20% to 40% in the seventh and eighth decade and exceeds 50% in the ninth decade [45].

Functional implications of loss of muscle mass

Several studies suggest that the age-dependent loss of muscle mass and function is linked to physical frailty and decreased capacity for independent living. Sarcopenia is recognized as one component of the multisystem decline leading to frailty [46]. The etiology of sarcopenia includes a wide variety of factors such as the loss of alpha-motor neurons [47], the reduction in dietary protein [48], a decreased level of physical activity [49], and an increase in catabolic cytokines such as interleukin (IL)-6 [50] and tumor necrosis factor α, IL-15, and cyliary neutrotrophic factor [51]. Most likely quantitative and qualitative changes in Ca^{2+} and K^+ ion channels are also involved in the age-related decline in muscle force [51,52]. K^+ channels are essential to induce myogenesis and proliferation of muscle cells and are modulated by insulin-like growth factor (IGF)-I [53]. Overexpression of IGF-I in skeletal muscle increases the number of and prevents the age-related decline in voltage-gated L-type Ca^{2+} channels [51]. The decrease in steroid hormones and GH is also possibly involved in the age-dependent decrease in muscle mass [54]. Of interest, sarcopenia and cancer-associated muscle wasting show many similarities [51]. Fig. 2 summarizes the factors contributing to sarcopenia. Several studies have assessed the relationship between sarcopenia and the loss of function in the elderly using different definitions for sarcopenia. Janssen and colleagues [55] studied the relationship between sarcopenia and functional impairment in 4504 adults 60 years and

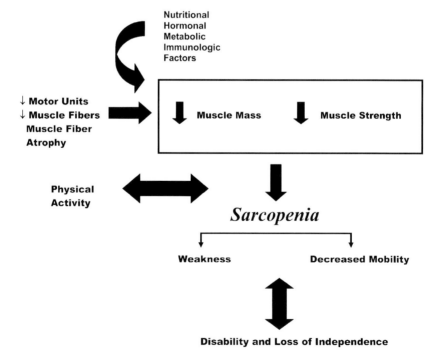

Fig. 2. Factors contributing to sarcopenia. (*From* Doherty TJ. Invited review: aging and sarcopenia. J Appl Physiol 2003;95(4):1721; with permission.)

older using data of the Third National Health and Nutrition Examination Survey. Skeletal muscle mass was estimated using bioelectrical impedance and expressed as skeletal muscle mass index (SMI; skeletal muscle mass/body mass × 100) and compared with young adults. The likelihood of functional impairment and physical disability was two times greater in older men and three times greater in older women who had class II sarcopenia (defined as SMI 2 SD below young adults). Of interest, class I sarcopenia (defined as SMI within -1 to -2 SD of young adults), when adjusted for age, race, health behaviors, and comorbidity, was no longer associated with an increased likelihood of functional impairment and disability. This study suggests that modest reductions in skeletal muscle mass with aging do not cause functional impairment and disability; however, when skeletal muscle mass relative to body weight is 30% below the mean of young adults, an increased risk of functional impairment and disability is found. Baumgartner and collegues [56] studied 808 older non-Hispanic white and Mexican-American men and women. In this study, sarcopenia was independently associated with disability and a history of falling. Melton and colleagues [57] showed an association between sarcopenia and difficulties in walking and an increase in fractures in older men and women.

Effects of growth hormone therapy in the elderly

The rationale for the potential use of GH as an intervention in the elderly is based on the observation that there is an age-dependent decline in GH and IGF-I and that the alterations in body composition and physiology occurring with aging are similar to those observed in adult GHD [21–24,28].

Although several age-dependent changes in body composition have been associated with the age-dependent decline of GH, potential effects on strength, function, and quality of life in the elderly are unclear to date. Some of the studies evaluating the outcome of GH treatment in the elderly are discussed here.

Most data on quality of life and GH have involved patients who have adult-onset or childhood-onset GHD. GH-deficient patients who have adult-onset GHD have lower quality-of-life parameters independent of age or sex [58]. GH replacement in those patients consistently shows an improvement in quality-of-life parameters [58–61]. In addition, withdrawal of GH therapy in GH-deficient patients has been shown to result in detrimental psychologic effects [61]. Similar observations have been published for patients who have childhood-onset GHD [62]. The beneficial effects of GH therapy on quality of life were also observed in patients older than 65 years. Whether the reversal of the reduction of GH action in the elderly results in improvements in quality of life has not been shown to date in placebo-controlled human studies. Animal studies suggest that GH therapy could have positive effects on memory [63]. Donahue and colleagues [63,64] showed not only that GH is expressed in the hippocampus in rodents but also that GH administration to aged rodents decreases oxidative stress in the hippocampus by increasing the endogenous antioxidant glutathione.

Treatment of healthy older adults aged 60 years and older for 1 week using different doses of GH resulted in a dose-dependent increase in IGF-I [65]. Rudman and coworkers [66] showed that 6 months of treatment with a high dose of GH (0.03 mg/kg/wk given subcutaneously 3 times per week) increased lean body mass and spinal bone density and decreased fat mass in men older than 60 years who had low pretreatment IGF-I levels. The GH dose used in this study was significantly higher than the current recommendation for the GH starting dose for GH-deficient adults, which could explain the side effect observed in this study. Similar results were found by Papadakis and colleagues [67] who studied 52 men older than 69 years. After 6 months of GH treatment (0.03 mg/kg given subcutaneously 3 times per week), fat mass decreased by 13.1% and lean body mass increased by 4.3%. These changes, however, did not improve functional ability in this study population. Positive effects on body composition were also found in a 10-week study in 18 healthy elderly men when strength training was combined with GH treatment (0.02 mg/kg/d) [68]. In postmenopausal women, the combination of exercise, diet, and GH resulted in an enhanced loss of truncal fat rather than peripheral fat compared with placebo [69]. In a more recent study, 26 weeks of GH treatment (0.02 mg/kg given subcutaneously 3 times per week) resulted in a decrease in visceral fat (compared with baseline) [70] in men 65 years or older; however, no change was found compared with placebo. In a study of elderly malnourished patients [71], 3 weeks of GH treatment increased midarm muscle circumference. In a 12-week study with a small number of participants (n = 17), treatment with GH of healthy older men increased lean body mass by 3.2 kg [72].

As discussed previously, development of sarcopenia in the elderly is not the result of one single change occurring during aging but is a consequence of multisystem changes [73]. Additional hormones that have been shown to have effects on skeletal muscle in the elderly are sex steroids [74]. Several studies suggest that GH and testosterone have synergistic anabolic actions. Blackman and colleagues [75] found an increase in total body strength in men older than 65 years who received GH and testosterone for 26 weeks. This result was not seen in the group of women in the study who received GH and estrogen. In the same study, Vo_2max increased (not significantly) in the group of men after 26 weeks of GH treatment. When GH and testosterone were administered simultaneously, Vo_2 max increased significantly compared with placebo. In the same study, protein synthesis showed an upward trend in men and women who received low-dose GH treatment. This trend reached statistical significance in the group of men treated with GH and testosterone [76]. Giannoulis and colleagues [77] compared GH, testosterone, and a combination of GH and testosterone treatment in healthy older men between age 65 and 80 years. The GH dose was adjusted according to the IGF-I levels. GH treatment resulted in an increase in lean body mass that was even more pronounced when GH and testosterone where given together. A mild increase in one of six strength measures (knee extension at $90°\sec^{-1}$) was seen when GH and testosterone were given combined. The results of these studies underline the multifactorial nature of the aging process and a possible role for individualized treatment doses. In addition, it should not be forgotten that the changes observed in the seventh, eighth, ninth, and tenth decades have come about from progressive decline of GH secretion from midpuberty onward.

In addition, the question whether the age-related decline in GH is a maladaptive process or an appropriate adaptation to this stage in life is still unanswered and deserves attention, especially given the concern regarding an increased cancer risk [78] with GH treatment in the elderly.

Potential risks of growth hormone replacement in the elderly

The possible side effects of GH therapy in the elderly are similar to the side effects found in young adults; however, concerns have been voiced about the use of GH in the elderly. One concern is whether the risk of cancer is increased with GH therapy. Experimental data suggest that GH/IGF-I provides an antiapoptotic environment, with IGF-I having powerful proliferative effects on almost all tissues. This environment could possibly favor the survival of genetically damaged cells and consequently increase the risk for developing cancer. This concern could be particularly relevant in the older population because there is a greater likelihood of the presence of genetically damaged cells. Long-term data in children and adults treated with GH for 27,000 patient-years, however, have shown no increased overall occurrence of de novo neoplasia or an increased rate of regrowth of primary pituitary tumors [79,80]. A review of the current data suggests that GH replacement therapy in GH-deficient adults is safe and does not lead to tumor formation [81,82]. Patients who have acromegaly with sustained increased IGF-I levels have no significant increase in site-specific cancer rates [81]. Another concern relates to the potential diabetogenic effects of GH because it antagonizes insulin action. Only two cases of reversible diabetes were reported from a combined series of 400 treated GH-deficient adults [83], and insulin sensitivity may even normalize with GH treatment as shown in a 1-year study [84]. After 7 years of GH treatment in GH-deficient adults, no change in insulin sensitivity was found [85]. Nevertheless, the potential effect of GH on glucose metabolism (especially in the elderly) remains a concern and needs to be carefully monitored. Other possible side effects include the development of edema, arthralgia, and carpal tunnel syndrome. These side effects are dose dependent, and reduction of the dose ameliorates them. Another point to consider when discussing the use of GH treatment in the elderly concerns the data showing that GH–knock-out rodents live longer compared with their GH-intact controls [86]. Whether these findings in rodents can be applied to a real-life situation remains questionable [87]. Besson and colleagues [88] showed that patients who had isolated childhood-onset GHD caused by a GH-1 gene deletion and were untreated during childhood and adulthood had a reduction in life span compared with control subjects. The outcome of this study further questions a role for GHD in extending life span in humans as suggested by animal studies with rodent mutant strains.

Growth hormone secretagogues as a potential therapeutic in the elderly

Since the discovery of GH releasing peptides in 1976, several types of GH secretagogues (GHSs) have been developed, including nonpeptidyl GHSs [89,90]. Only a few groups have studied the effects of GHSs in the elderly. Optimal GHS therapy requires the functional integrity of somatotroph cells and an intact hypothalamic-pituitary axis [91]. Both are present in the elderly. In addition, the pituitary GH-releasable pool in the elderly is comparable to young adults [92]. The use of GHSs has the advantage that GH release is physiologic (ie, pulsatile). Because the GH–IGF-I feedback loop remains intact when using a GHS, supraphysiologic GH levels are unlikely to occur, which is not the case with GH treatment in which high doses are associated with supraphysiologic IGF-I levels. Studies regarding the effects of GHSs on circulating IGF-I levels have shown conflicting results, which may be due to varying pharmacodynamic and pharmacokinetic properties of different GHSs. Rahim and colleagues [93] studied the effects of the nonorally active peptidergic GHS hexarelin on body composition in the elderly and could not show any effects after 16 weeks of treatment. Studies performed with an orally active nonpeptidergic GHS (MK-0677) resulted in a significant increase in IGF-I in the elderly (≥ 60 years old) compared with the placebo group after 4 weeks of treatment [94]. In a double-blind placebo-controlled trial, 1-year treatment with MK-0677 led to an increase in fat-free mass (measured by four-compartment model and dual energy X-ray absorptiometry) by 1.6 kg compared with placebo. This gain in fat-free mass was maintained in the second year of the study [95]. In the same study, IGF-I levels were significantly increased for 2 years. In a randomized placebo-controlled study of 292 postmenopausal women, the treatment with MK-0677 alone or together with the antiresorptive agent alendronate resulted in an increase in bone mineral density at the femoral neck [96]. Another study reported the effects of 6 months' treatment with MK-0677 in healthy older adults recovering from hip fracture that showed improvements of some lower extremity functional measures. Overall, GHSs were well tolerated and their use was safe [97].

Summary

Current data suggest that improved muscle function or a decrease in the age-dependent muscle loss in the elderly may result in more independent living and in improved quality of life in this population group. The fact that increased musculoskeletal impairment as a result of muscle wasting in the elderly is not a well-recognized syndrome complicates studies that try to show clinically meaningful efficacy. Overall, the use of GH or GHSs represents potential treatment or options for prevention of musculoskeletal impairment associated with aging. A considerable number of questions remain unanswered relating to the use of GH in the elderly, however, such as whether the age-related decline in GH should be considered a maladaptive change or an appropriate adaptation and what the safety concerns are regarding potential side effects of GH therapy. Based on preliminary study results, the use of GHSs may be more suitable for use in this age group than GH itself, especially because during GHS treatment, feedback mechanisms are intact that prevent overtreatment. Future well-controlled studies of adequate duration are required to answer these questions.

References

[1] Butler RN. Population aging and health. BMJ 1997;315(7115):1082–4.
[2] US Census Bureau international database. Table 094. Midyear population. Available at: http://www.census.gov/cgi-bin/ipc/idbagg.
[3] CDC Public Health and Aging. Trends in aging–United States and worldwide. JAMA 2003; 289:1371–3.
[4] United Nations Report of the Second World Assembly on Aging. Madrid, Spain, April 8–12, 2002. Available at: http://www.on.org/ageing/documents.htm. Accessed December 6, 2006.
[5] Toogood AA, Adams JE, O'Neill PA, et al. Body composition in growth hormone deficient adults over the age of 60 years. Clin Endocrinol (Oxf) 1996;45(4):399–405.
[6] Barreto-Filho JA, Alcantara MR, Salvatori R, et al. Familial isolated growth hormone deficiency is associated with increased systolic blood pressure, central obesity, and dyslipidemia. J Clin Endocrinol Metab 2002;87(5):2018–23.
[7] Snel YE, Brummer RJ, Doerga ME, et al. Adipose tissue assessed by magnetic resonance imaging in growth hormone-deficient adults: the effect of growth hormone replacement and a comparison with control subjects. Am J Clin Nutr 1995;61(6):1290–4.
[8] Lonn L, Kvist H, Grangard U, et al. CT-determined body composition changes with recombinant human growth hormone treatment to adults with growth hormone deficiency. Basic Life Sci 1993;60:229–31.
[9] Amato G, Carella C, Fazio S, et al. Body composition, bone metabolism, and heart structure and function in growth hormone (GH)-deficient adults before and after GH replacement therapy at low doses. J Clin Endocrinol Metab 1993;77(6):1671–6.
[10] Bex M, Abs R, Maiter D, et al. The effects of growth hormone replacement therapy on bone metabolism in adult-onset growth hormone deficiency: a 2-year open randomized controlled multicenter trial. J Bone Miner Res 2002;17(6):1081–94.
[11] Johansson JO, Fowelin J, Landin K, et al. Growth hormone-deficient adults are insulin-resistant. Metabolism 1995;44(9):1126–9.
[12] Rosen T, Bengtsson BA. Premature mortality due to cardiovascular disease in hypopituitarism. Lancet 1990;336(8710):285–8.
[13] Bates AS, Van't Hoff W, Jones PJ, et al. The effect of hypopituitarism on life expectancy. J Clin Endocrinol Metab 1996;81(3):1169–72.
[14] Colao A, Cuocolo A, Di Somma C, et al. Impaired cardiac performance in elderly patients with growth hormone deficiency. J Clin Endocrinol Metab 1999;84(11):3950–5.
[15] Attanasio AF, Bates PC, Ho KK, et al. Human growth hormone replacement in adult hypopituitary patients: long-term effects on body composition and lipid status—3-year results from the HypoCCS Database. J Clin Endocrinol Metab 2002;87(4):1600–6.

[16] Sneppen SB, Steensgaard-Hansen F, Feldt-Rasmussen U. Cardiac effects of low-dose growth hormone replacement therapy in growth hormone-deficient adults. An 18-month randomised, placebo-controlled, double-blind study. Horm Res 2002;58(1):21–9.
[17] Shahi M, Beshyah SA, Hackett D, et al. Myocardial dysfunction in treated adult hypopituitarism: a possible explanation for increased cardiovascular mortality. Br Heart J 1992;67(1): 92–6.
[18] Rosen T, Johannson G, Hallgren P, et al. Beneficial effects of 12 months replacement therapy with recombinant human growth hormone to growth hormone deficient adults. J Clin Endocrinol Metab 1994;1:55–66.
[19] Mahajan T, Crown A, Checkley S, et al. Atypical depression in growth hormone deficient adults, and the beneficial effects of growth hormone treatment on depression and quality of life. Eur J Endocrinol 2004;151(3):325–32.
[20] Barkan AL. The "quality of life-assessment of growth hormone deficiency in adults" questionnaire: can it be used to assess quality of life in hypopituitarism? J Clin Endocrinol Metab 2001;86(5):1905–7.
[21] Finkelstein JW, Roffwarg HP, Boyar RM, et al. Age-related change in the twenty-four-hour spontaneous secretion of growth hormone. J Clin Endocrinol Metab 1972;35(5):665–70.
[22] Zadik Z, Chalew SA, McCarter RJ Jr, et al. The influence of age on the 24-hour integrated concentration of growth hormone in normal individuals. J Clin Endocrinol Metab 1985; 60(3):513–6.
[23] Novack L. Aging, total body potassium, fat-free mass, and cell mass in males and females between ages 18 and 85 years. J Gerontol 1972;27:438–43.
[24] Rudman D. Growth hormone, body composition, and aging. J Am Geriatr Soc 1985;33(11): 800–7.
[25] Haarbo J, Marslew U, Gotfredsen A, et al. Postmenopausal hormone replacement therapy prevents central distribution of body fat after menopause. Metabolism 1991;40(12):1323–6.
[26] Ashwell M, Cole TJ, Dixon AK. Obesity: new insight into the anthropometric classification of fat distribution shown by computed tomography. Br Med J (Clin Res Ed) 1985;290(6483): 1692–4.
[27] DeNino WF, Tchernof A, Dionne IJ, et al. Contribution of abdominal adiposity to age-related differences in insulin sensitivity and plasma lipids in healthy nonobese women. Diabetes Care 2001;24(5):925–32.
[28] Clasey JL, Weltman A, Patrie J, et al. Abdominal visceral fat and fasting insulin are important predictors of 24-hour GH release independent of age, gender, and other physiological factors. J Clin Endocrinol Metab 2001;86(8):3845–52.
[29] Enzi G, Gasparo M, Biondetti P, et al. Subcutaneous and visceral fat distribution according to sex, age and overweight, evaluated by computed tomography. Am J Clin Nutr 1986;44: 739–46.
[30] Shimokata H, Tobin J, Muller D, et al. Studies in the distribution of body fat. I. Effects of age, sex, and obesity. J Gerontol 1989;44:67–73.
[31] Borkan GA, Hults DE, Gerzof SG, et al. Age changes in body composition revealed by computed tomography. J Gerontol 1983;38(6):673–7.
[32] Marcus R. Skeletal aging—understanding the functional and structural basis of osteoporosis. Trends Endocrinol Metab 1991;2:53–8.
[33] Hannan M, Felson D, Anderson J. Bone mineral density in elderly men and women: results from the Framingham Osteoporosis Study. J Bone Miner Res 1992;7:547–53.
[34] Stiegler C, Leb G. One year of replacement therapy in adults with growth hormone deficiency. Endocrinol Metab 1994;1(Suppl A):37–42.
[35] de Boer H, Blok GJ, van Lingen A, et al. Consequences of childhood-onset growth hormone deficiency for adult bone mass. J Bone Miner Res 1994;9(8):1319–26.
[36] Thompson JL, Butterfield GE, Marcus R, et al. The effects of recombinant human insulin-like growth factor-I and growth hormone on body composition in elderly women. J Clin Endocrinol Metab 1995;80(6):1845–52.

[37] Forbes G, Reina J. Adult lean body mass declines with age: some longitudinal observations. Metabolism 1970;19:653–63.
[38] Janssen I, Heymsfield SB, Wang ZM, et al. Skeletal muscle mass and distribution in 468 men and women aged 18–88 yr. J Appl Physiol 2000;89(1):81–8.
[39] Tseng BS, Marsh DR, Hamilton MT, et al. Strength and aerobic training attenuate muscle wasting and improve resistance to the development of disability with aging. J Gerontol A Biol Sci Med Sci 1995;50:113–9.
[40] Bemben MG, Massey BH, Bemben DA, et al. Isometric muscle force production as a function of age in healthy 20- to 74-yr-old men. Med Sci Sports Exerc 1991;23(11):1302–10.
[41] Clement FJ. Longitudinal and cross-sectional assessments of age changes in physical strength as related to sex, social class, and mental ability. J Gerontol 1974;29(4):423–9.
[42] Hurley BF. Age, gender, and muscular strength. J Gerontol A Biol Sci Med Sci 1995;50:41–4.
[43] Kehayias JJ, Fiatarone MA, Zhuang H, et al. Total body potassium and body fat: relevance to aging. Am J Clin Nutr 1997;66(4):904–10.
[44] Lexell J, Downham D, Sjostrom M. Distribution of different fibre types in human skeletal muscles. Fibre type arrangement in m. vastus lateralis from three groups of healthy men between 15 and 83 years. J Neurol Sci 1986;72(2–3):211–22.
[45] Doherty TJ. Invited review: aging and sarcopenia. J Appl Physiol 2003;95(4):1717–27.
[46] Walston J, Hadley EC, Ferrucci L, et al. Research agenda for frailty in older adults: toward a better understanding of physiology and etiology: summary from the American Geriatrics Society/National Institute on Aging Research Conference on Frailty in Older Adults. J Am Geriatr Soc 2006;54(6):991–1001.
[47] Brown WF. A method for estimating the number of motor units in thenar muscles and the changes in motor unit count with ageing. J Neurol Neurosurg Psych 1972;35(6):845–52.
[48] Young VR. Amino acids and proteins in relation to the nutrition of elderly people. Age Ageing 1990;19(4):S10–24.
[49] Westerterp KR. Daily physical activity, aging and body composition. J Nutr Health Aging 2000;4(4):239–42.
[50] Roubenoff R, Harris TB, Abad LW, et al. Monocyte cytokine production in an elderly population: effect of age and inflammation. J Gerontol A Biol Sci Med Sci 1998;53(1):M20–6.
[51] Argiles JM, Busquets S, Felipe A, et al. Molecular mechanisms involved in muscle wasting in cancer and ageing: cachexia versus sarcopenia. Int J Biochem Cell Biol 2005;37(5):1084–104.
[52] Delbono O. Molecular mechanisms and therapeutics of the deficit in specific force in ageing skeletal muscle. Biogerontology 2002;3(5):265–70.
[53] Gamper N, Fillon S, Huber SM, et al. IGF-1 up-regulates K+ channels via PI3-kinase, PDK1 and SGK1. Pflugers Arch 2002;443(4):625–34.
[54] Labrie F, Belanger A, Luu-The V, et al. DHEA and the intracrine formation of androgens and estrogens in peripheral target tissues: its role during aging. Steroids 1998;63(5–6):322–8.
[55] Janssen I, Heymsfield SB, Ross R. Low relative skeletal muscle mass (sarcopenia) in older persons is associated with functional impairment and physical disability. J Am Geriatr Soc 2002;50(5):889–96.
[56] Baumgartner RN, Koehler KM, Gallagher D, et al. Epidemiology of sarcopenia among the elderly in New Mexico. Am J Epidemiol 1998;147(8):755–63.
[57] Melton LJ III, Khosla S, Crowson CS, et al. Epidemiology of sarcopenia. J Am Geriatr Soc 2000;48(6):625–30.
[58] Blum WF, Shavrikova EP, Edwards DJ, et al. Decreased quality of life in adult patients with growth hormone deficiency compared with general populations using the new, validated, self-weighted questionnaire, questions on life satisfaction hypopituitarism module. J Clin Endocrinol Metab 2003;88(9):4158–67.

[59] Salomon F, Cuneo R, Hesp R, et al. The effects of treatment with recombinant human growth hormone on body composition and metabolism in adults with growth hormone. N Engl J Med 1989;321:1797–803.
[60] Monson JP, Jonsson P. Aspects of growth hormone (GH) replacement in elderly patients with GH deficiency: data from KIMS. Horm Res 2003;60(Suppl 1):112–20.
[61] McMillan CV, Bradley C, Gibney J, et al. Psychological effects of withdrawal of growth hormone therapy from adults with growth hormone deficiency. Clin Endocrinol (Oxf) 2003; 59(4):467–75.
[62] Attanasio AF, Lamberts SW, Matranga AM, et al. Adult growth hormone (GH)-deficient patients demonstrate heterogeneity between childhood onset and adult onset before and during human GH treatment. Adult Growth Hormone Deficiency Study Group. J Clin Endocrinol Metab 1997;82(1):82–8.
[63] Donahue CP, Kosik KS, Shors TJ. Growth hormone is produced within the hippocampus where it responds to age, sex, and stress. Proc Natl Acad Sci U S A 2006;103(15):6031–6.
[64] Donahue AN, Aschner M, Lash LH, et al. Growth hormone administration to aged animals reduces disulfide glutathione levels in hippocampus. Mech Ageing Dev 2006;127(1):57–63.
[65] Marcus R, Butterfield G, Holloway L, et al. Effects of short term administration of recombinant human growth hormone to elderly people. J Clin Endocrinol Metab 1990;70:519–27.
[66] Rudman D, Feller A, Nagraj H, et al. Effects of human growth hormone in men over 60 years old. N Engl J Med 1990;323:1–6.
[67] Papadakis MA, Grady D, Black D, et al. Growth hormone replacement in healthy older men improves body composition but not functional ability. Ann Intern Med 1996;124(8): 708–16.
[68] Taaffe DR, Pruitt L, Reim J, et al. Effect of recombinant human growth hormone on the muscle strength response to resistance exercise in elderly men. J Clin Endocrinol Metab 1994;79(5):1361–6.
[69] Taaffe DR, Thompson JL, Butterfield GE, et al. Recombinant human growth hormone, but not insulin-like growth factor-I, enhances central fat loss in postmenopausal women undergoing a diet and exercise program. Horm Metab Res 2001;33(3):156–62.
[70] Muenzer T, Harman S, Hees P, et al. Effects of GH and/or sex steroid administration on abdominal subcutaneous and visceral fat in healthy aged women and men. J Clin Endocrinol Metab 2001;86:3604–10.
[71] Kaiser FE, Silver AJ, Morley JE. The effect of recombinant human growth hormone on malnourished older individuals. J Am Geriatr Soc 1991;39(3):235–40.
[72] Lange KH, Isaksson F, Rasmussen MH, et al. GH administration and discontinuation in healthy elderly men: effects on body composition, GH-related serum markers, resting heart rate and resting oxygen uptake. Clin Endocrinol (Oxf) 2001;55(1):77–86.
[73] Mauras N. Growth hormone and sex steroids. Interactions in puberty. Endocrinol Metab Clin North Am 2001;30(3):529–44.
[74] Snyder PJ, Peachey H, Hannoush P, et al. Effect of testosterone treatment on body composition and muscle strength in men over 65 years of age. J Clin Endocrinol Metab 1999;84(8): 2647–53.
[75] Blackman MR, Sorkin JD, Munzer T, et al. Growth hormone and sex steroid administration in healthy aged women and men: a randomized controlled trial. JAMA 2002;288(18): 2282–92.
[76] Huang X, Blackman MR, Herreman K, et al. Effects of growth hormone and/or sex steroid administration on whole-body protein turnover in healthy aged women and men. Metabolism 2005;54(9):1162–7.
[77] Giannoulis MG, Sonksen PH, Umpleby M, et al. The effects of growth hormone and/or testosterone in healthy elderly men: a randomized controlled trial. J Clin Endocrinol Metab 2006;91(2):477–84.
[78] Toogood AA. The somatopause: an indication for growth hormone therapy? Treat Endocrinol 2004;3(4):201–9.

[79] Monson JP. Long-term experience with GH replacement therapy: efficacy and safety. Eur J Endocrinol 2003;148(Suppl 2):S9–14.
[80] Jenkins PJ, Mukherjee A, Shalet SM. Does growth hormone cause cancer? Clin Endocrinol (Oxf) 2006;64(2):115–21.
[81] Orme SM, McNally RJ, Cartwright RA, et al. Mortality and cancer incidence in acromegaly: a retrospective cohort study. United Kingdom Acromegaly Study Group. J Clin Endocrinol Metab 1998;83(8):2730–4.
[82] GHR Society Consensus. Critical evaluation of the safety of recombinant human growth hormone administration: statement from the Growth Hormone Research Society. J Clin Endocrinol Metab 2001;86:1868–70.
[83] Chipman JJ, Attanasio AF, Birkett MA, et al. The safety profile of GH replacement therapy in adults. Clin Endocrinol (Oxf) 1997;46(4):473–81.
[84] Hwu CM, Kwok CF, Lai TY, et al. Growth hormone (GH) replacement reduces total body fat and normalizes insulin sensitivity in GH-deficient adults: a report of one-year clinical experience. J Clin Endocrinol Metab 1997;82(10):3285–92.
[85] Svensson J, Fowelin J, Landin K, et al. Effects of seven years of GH-replacement therapy on insulin sensitivity in GH-deficient adults. J Clin Endocrinol Metab 2002;87(5):2121–7.
[86] Bartke A. Long-lived Klotho mice: new insights into the roles of IGF-1 and insulin in aging. Trends Endocrinol Metab 2006;17(2):33–5.
[87] Nass R, Thorner MO. Life extension versus improving quality of life. Best Pract Res Clin Endocrinol Metab 2004;18(3):381–91.
[88] Besson A, Salemi S, Gallati S, et al. Reduced longevity in untreated patients with isolated growth hormone deficiency. J Clin Endocrinol Metab 2003;88(8):3664–7.
[89] Bowers CY, Chang JK, Fong TTW. A synthetic pentapeptide which specifically releases GH, in vitro. Proceedings of the 59th Meeting of the Endocrine Society. Chicago; 1977. p. 2332.
[90] Smith RG. Development of growth hormone secretagogues. Endocr Rev 2005;26(3):346–60.
[91] Ghigo E, Arvat E, Aimaretti G, et al. Diagnostic and therapeutic uses of growth hormone-releasing substances in adult and elderly subjects. Baillieres Clin Endocrinol Metab 1998;12(2):341–58.
[92] Ghigo E, Arvat E, Gianotti L, et al. Hypothalamic growth hormone-insulin-like growth factor-I axis across the human life span. J Pediatr Endocrinol Metab 2000;13(Suppl 6):1493–502.
[93] Rahim A, O'Neill PA, Shalet SM. Growth hormone status during long-term hexarelin therapy. J Clin Endocrinol Metab 1998;83(5):1644–9.
[94] Chapman IM, Bach MA, Van Cauter E, et al. Stimulation of the growth hormone (GH)-insulin-like growth factor I axis by daily oral administration of a GH secretogogue (MK-677) in healthy elderly subjects. J Clin Endocrinol Metab 1996;81(12):4249–57.
[95] Thorner MO, Nass R, Pezzoli S, et al. Orally active ghrelin mimetic (MK-677) prevents and partially reverses sarcopenia in healthy older men and women: a double-blind, placebo-controlled, crossover study. Presented at the 88th Annual Meeting of the Endocrine Society. June 24–26, 2006 [OR5–5].
[96] Murphy MG, Weiss S, McClung M, et al. Effect of alendronate and MK-677 (a growth hormone secretagogue), individually and in combination, on markers of bone turnover and bone mineral density in postmenopausal osteoporotic women. J Clin Endocrinol Metab 2001;86(3):1116–25.
[97] Bach MA, Rockwood K, Zetterberg C, et al. The effects of MK-0677, an oral growth hormone secretagogue, in patients with hip fracture. J Am Geriatr Soc 2004;52(4):516–23.

Growth Hormone Treatment and Cancer Risk

Indraneel Banerjee, MBBS, MD, MRCPCH[a],
Peter E. Clayton, MBChB, BSc, MD, MRCP, FRCPCH[b],*

[a]Department of Pediatric Endocrinology, Royal Manchester Children's Hospital, Hospital Road, Pendlebury, Swinton, Manchester M27 4HA, UK
[b]Endocrine Science Research Group, Division of Human Development, School of Medicine, University of Manchester, Room 3.705, Stopford Building, Oxford Road, Manchester M13 9PT, UK

There are significant benefits to the clinical use of growth hormone (GH) in children and adults, with indications for treatment in children extending beyond those of growth hormone deficiency (GHD). In addition, evidence is accumulating that for persons who have childhood-onset GHD, GH treatment should extend beyond the childhood phase of growth into the mid-twenties to complete full somatic development and possibly should continue throughout life [1–5]. It is recognized, however, that GH and insulin-like growth factor-I (IGF-I) do have mitogenic and anti-apoptotic activity [6] and that there is a theoretical risk that GH treatment may be associated with tumor development, particularly in persons previously treated for a malignant lesion. It therefore is important to undertake regular reviews of the incidence of cancer in those who receive GH treatment at any stage of life.

The evidence that GH or IGF-I may have a role in cancer occurrence or recurrence originally comes from in vitro cell and in vivo animal models. This evidence has been supported by epidemiologic studies in large populations relating circulating levels of IGF-I and its main binding protein (IGFBP-3) to a relative risk of certain cancers in adults. Surveillance studies targeted at specific populations receiving GH treatment also have provided data on the relative risk of tumors. These data are discussed to provide an overall picture of the safety profile of GH with regard to cancer risk.

* Corresponding author.
E-mail address: peter.clayton@manchester.ac.uk (P.E. Clayton).

The growth hormone–insulin-like growth factor axis in tumorigenesis and tumor progression

In vitro experiments

It is widely held that the GH–IGF-I system plays a role in the development and progression of cancers [7, 8], although the exact pathophysiology of neoplastic transformation is not clear. Most but not all the actions of GH are mediated by IGF-I, and it therefore is likely that both GH and IGF-I have roles in the pathogenesis of tumors [7,9–11]. GH is blastogenic in lymphocyte cell cultures [12], and IGF-I in high doses has the potential to induce proliferation of human leukemic cells in vitro [13]. Such cells often express the GH receptor (GHR) and transduce signals by activation of signal transducer and activator of transcription–5 [14]; the inference is that GH is likely to augment proliferation of these cells. Not all activity of GH is anti-apoptotic, however: GH may enhance natural killer cell activity against glioma, suggesting an immune-mediated protective role against malignancy [15].

IGF-I has been noted to have mitogenic activity in breast cancer cell lines, an action mediated through the IGF-I receptor (IGF-IR) [16], whereas an IGF-IR kinase antagonist inhibited tumor xenograft proliferation [17]. In rat hepatoma cells, IGF-I increased DNA replication, chiefly through the phosphoinositol-3′ kinase pathway [18]. IGF-IR was expressed in a significant proportion of primary low-passage craniopharyngioma cells lines [19]. Treatment with an IGF-IR inhibitor reduced Akt phosphorylation and led to growth arrest in these cells, raising the possibility that selective disruption of the IGF–I axis by anti IGF-IR agents might be a cancer treatment modality [20].

IGF-I is bound in the circulation and tissues to various binding proteins. IGFBP-3 limits the bioactivity of IGF-I and thus would be expected to have inhibitory effects on tumor growth, as shown by its induction of apoptosis in *p53*-negative prostate cancer cell lines [21]. In contrast, IGFBP-2 bound to the extracellular matrix enhanced proliferation and metastatic behavior of neuroblastoma cells [22]. IGFBP-4 has anti-angiogenic and anti-tumorigenic properties in human glioblastoma cells [23] and therefore could be a possible targeted therapy [24].

In vivo animal studies

The possibility that GH may have a role in cancer tumorigenesis came from observations that hypophysectomy suppressed metastasis and progression of specific malignancies [25], whereas the administration of GH to hypophysectomized animals induced tumors in various organs [26–28]. In hypophysectomized rats, GH but not IGF-I induced expression of the proto-oncogene c-myc, suggesting that GH itself is mitogenic [29]. GH-releasing hormone antagonists inhibited growth of mammary, renal, lymphoma, and other cell lines xenografted into nude mice [30]. Mice transgenic for human GH developed mammary tumors [31], whereas the introduction of lipopeptide GH-releasing hormone antagonists inhibited prostate cancer xenografts in nude mice [32]. In rectal cancer biopsy specimens, those negative for GHR mRNA were associated with higher radiosensitivity, suggesting a potential role for GHR antagonists before radiotherapy and a role for GHR in tumor progression [33]. Within the GH signaling pathway, suppressor of cytokine signaling-2 (SOCS2) normally acts to limit GH activity. In GH transgenic mice with no, one, or two functional *SOCS2* genes, loss of SOCS2 was associated with increased local IGF-I production and multiple hyperplastic and lymphoid polyps [34], indicating a link between GH signaling and neoplasia.

If an upregulated GH–IGF axis seems to be favorable to tumorigenesis, then a downregulated GH–IGF state may be beneficial in slowing tumor growth, a view in keeping with epidemiologic observations. Tumor xenografts grew slowly if IGF-I concentrations were low because of either GH deficiency or pharmacologic intervention [35–37]. Mice with low circulating IGF-I levels but normal tissue expression had a lower risk of developing colon and breast cancers [38]. Those transgenic for a constitutively active IGF-IR developed tumors [17] that were inhibited by an IGF-IR antagonist [39].

IGFBP-3 also modulates tumor growth as suggested by experiments in which upregulation of IGFBP-3 by apigenin inhibited the growth of human prostate cancer xenografted in nude mice [40]. Experiments in long probasin promoter (LPB-Tag) mice that develop prostate tumors because of increased expression of T-cell antigen have demonstrated that expression of IGFBP-3 inhibits neoplastic growth by IGF-I–dependent and –independent mechanisms [41].

These studies have indicated clearly that GH, GH signaling molecules, IGF-I, IGFBP-3, IGF-IR, and IGF-I signaling molecules have a role in tumor growth. These direct experimental observations need to be translated into clinical implications by assessment of cancer risk in both normal and at-risk populations.

Epidemiologic observations

A number of large population-based studies have identified a link between cancer risk, tall stature, and IGF-I levels at the upper end of the normal range. Review of anthropometric data in relation to cancer incidence in large cohorts [42,43] has shown that relatively rapid growth in childhood and adolescence was associated with the risk of developing cancers, particularly of the breast, prostate, and colon [44,45]. In cohorts in whom a measurement of circulating levels of IGF-I has been available, longitudinal follow-up has revealed that those who had high-normal levels of IGF-I (in the upper quartile of the study population) had increased rates of breast [46] and prostate [47] cancers, although inconsistencies remain across different studies [48,49]. The overall relative risk (RR) as defined by meta-analysis is 1.7. In some of these studies, risk was highest for those who had a high IGF-I but low IGFBP-3, and IGFBP-3 alone had a protective effect. In addition, reduced IGFBP-3 expression was noted to be associated with disease progression and death from epithelial ovarian cancer [50]. On review and meta-analysis, however, the RR of cancer in relation to IGFBP-3 levels was not significant [51,52].

The other IGFBPs also may modulate the risk posed by IGF-I. In children who had acute myeloid leukemia, there was a greater risk of relapse after hematopoietic stem cell transplantation if serum IGFBP-2 levels were high [53]. In contrast with laboratory observations, serum IGFBP-4 levels were low at diagnosis of acute lymphoblastic leukemia but were not related to risk category for the leukemia [54].

There may be genetic variants in the IGF system that determine relative tumor risk. Polymorphisms in the *IGF-I* gene were shown to be associated with increased risk of prostate cancer [55]. In contrast, polymorphisms in the *IGF-I* gene [56] or in the GH synthesis pathway [57] were not significantly associated with breast cancer. For IGFBP-3, a strong association was noted between a specific promoter polymorphism and mammographic density, a known risk factor for breast cancer development [58], although no direct association with breast cancer was observed in systematic review [59] or in a large case-control study [56]. Although IGFBP-3 polymorphic alleles were modestly associated with risk of colorectal cancer [60,61], there was no significant association between IGFBP-3 polymorphisms and the risk of prostate cancer [62,63]. It remains to be seen if these genetic variations could determine a higher cancer risk in susceptible individuals receiving GH treatment.

An important disease state for assessment of cancer risk in relation to persistently high levels of GH and IGF-I over many years is acromegaly. Case series including more than 1000 patients have indicated an increased incidence of colonic polyps and increased mortality from colon cancer in a study from the United Kingdom and an increased standardized incidence rate (2.5) for colon cancer and also for bone, kidney, thyroid, and brain tumors from a Scandinavian registry [64–66]. Another perspective has been provided by a study in adult men in whom increasing serum basal GH concentrations were inversely related to the risk of prostate cancer. It was proposed that this finding indicated a negative feedback effect of tumor-produced IGF-I on pituitary GH secretion [67].

These observations add to the experimental evidence that the GH–IGF axis plays a role in tumor development, but the RRs are low. The importance of assessment of risk in cohorts to whom GH has been given exogenously is reinforced, however.

Growth hormone treatment: de novo malignancies

Major studies of GH treatment and risk of de novo cancers are shown in Table 1.

Table 1
Major studies of growth hormone treatment and risk of de novo cancers

Study	N	Patient-years of growth hormone treatment	Principal malignancy type(s)	Risk estimate
Swerdlow et al [81]	1848	39,178	colorectal, Hodgkin lymphoma	SMR 2.8
Mehls et al [86]	2422	–	renal carcinoma	4 recorded
Fradkin et al [71]	6284[a]	59,736[a]	leukemia/lymphoma	RR 2.6[a]
Tuffli et al [80]	12,209	51,000	nonleukemic neoplasms	SMR = 1.0
Allen et al [75]	24,417[a]	119,846	leukemia	SIR 0–1.5
Nishi et al [74]	32,000	–	leukemia	SIR 0.7–1.0

Abbreviations: N, number; RR, relative risk; SIR, standardized incidence ratio; SMR, standardized mortality ratio.

[a] Indicates values for which pre-existing risk factors were not excluded.

Leukemia/lymphoma

Case reports of children treated with GH developing acute leukemia and other malignancies [68–70] generated considerable anxiety about the safety profile of human GH. One such report was in a Japanese cohort of children who had GHD treated with GH, in which five children developed acute lymphoblastic leukemia, five acute myeloid leukemia, one chronic myeloid leukemia, and one malignant histiocytosis [70]. These findings contrasted with those observed in 6284 recipients of pituitary GH under the National Hormone Pituitary Program in the United States between 1963 and 1985 [71]. There were three cases of leukemia in 59,736 patient-years of GH treatment, a rate that was not significantly higher than the 1.66 cases expected for an age-, ethnicity-, and sex-matched population. In extended follow-up, an additional three cases were found (for a total of six cases), and this total was significantly greater than the 2.26 cases expected for the population. As in the Japanese cohort, five of the six subjects developing leukemia had had a brain tumor, and four had received radiotherapy before GH treatment. There was no increase in the rate of leukemia during the follow-up of children who had idiopathic GHD.

Since these initial reports, case reports of de novo leukemia following GH treatment have been published. That the occurrence of leukemia in conjunction with GH treatment is not related causally is suggested also by reports that leukemia has occurred in GH-deficient children not treated with GH; it is speculated that GHD per se may induce a risk for leukemia [72,73].

A decade after the initial report of cancer secondary to GH administration, the Foundation for Growth Science in Japan investigated 32,000 individuals treated with GH to reassess the safety profile of GH therapy [74]. The occurrence of de novo leukemia was noted in 15 patients (9 during GH treatment and 6 after treatment stopped), of whom 6 had known risk factors such as Fanconi anemia, radiation, and previous chemotherapy predisposing them to tumor development. The incidence of leukemia in those without risk factors was estimated at 3.0 per 100,000 patient-years, similar to that in the general population in the age range from birth to 15 years. The report concluded that the risk of inducing tumors with GH treatment was negligible in the absence of known risk factors.

The safety of GH treatment was reconfirmed in data from the National Cooperative Growth Study (NCGS), a large 20-year follow-up surveillance program in North America involving 435 centers and approximately 47,000 patients [75]. Three cases of leukemia were recorded in comparison with an expected 3.42 cases in 119,846 patient-years of GH treatment.

Brain tumors

There is no evidence of an increase in the incidence of de novo intracranial tumors in children treated with GH in on-going surveillance studies such as the NCGS [76]. Similarly, there is no evidence of an increased incidence of type 1 neurofibromatosis (NF-1), which itself is associated with an enhanced risk of developing tumors, particularly optic nerve gliomas [77]. Although neurofibromata in NF-1 may stain positively for the GHR [78], it seems unlikely that GH treatment will exacerbate tumor risk in children who have NF-1: a retrospective analysis of 102 children who had NF-1 treated with GH showed that the rate of tumor occurrence in those treated with GH was similar to that in those not treated [79].

Other tumors

In line with the concerns about GH and leukemia, data from 12,209 patients treated with GH and enrolled in the NCGS were analyzed for a possible increase in nonleukemic neoplasms [80]. Ten new cases of malignancies were noted in the cohort representing 51,000 patient-years at risk. The number of new cases was not greater than that expected, indicating that GH is not implicated in the occurrence of solid tumors.

Safety concerns for GH treatment were raised, however, in a retrospective long-term cohort study in the United Kingdom that reported a nearly threefold increase in cancer mortality associated with the use of human pituitary GH in 1848 individuals between 1959 and 1985 [81]. In comparison with the general population, controlled for age, sex, and calendar period, there was a significant increase in mortality from Hodgkin lymphoma (two cases; standardized mortality ratio [SMR], 11.4) and colorectal carcinoma (two cases; SMR, 10.8). The GH treatment regimens used during this period were very different from those used currently. All patients received standard GH doses given two or three times per week, and serum IGF-I levels were not monitored. It therefore may be argued that these findings are unlikely to be replicated using current GH treatment regimens. Nevertheless, this study emphasizes that the safety profile of GH needs to be kept constantly under review, and vigilance must be maintained for the occurrence of infrequent or unusual malignancies in large surveillance programs.

In addition, case reports continue to implicate GH in sporadic tumor development [82,83], but the risk of a true association seems to be minimal; examination of 86,000 patients receiving GH, representing 250,000 GH-years, revealed only one 15-year-old previously treated with radiotherapy who was affected with intestinal adenocarcinoma and another who had Turner syndrome who developed spontaneous colon cancer after stopping GH treatment [84].

An important indication for GH use is poor growth associated with chronic renal failure before renal transplantation. Children and young persons undergoing renal transplantation are at increased risk of malignant change in the transplanted kidneys related to immunosuppressive therapy. A recent case report raised the possibility of developing renal cysts and later renal carcinoma in the posttransplantation period, possibly attributable to GH treatment [85]. Therefore the Kabi International Growth Study (KIGS) and NCGS databases were examined to establish the likelihood of this development in large populations of patients [86]. A total of three individuals (including those reported previously [85]) who had graft malignancy and one who had native kidney cancer were found in the KIGS cohort of 938 children treated before and 314 children treated after kidney transplantation. In the NCGS cohort, 1170 children who had renal failure were treated with GH, and no case of renal cell carcinoma was reported. In contrast, in those who did not have renal disease, two cases of renal cell carcinoma were found in the 43,000 patients enrolled in the NCGS and in 42,000 children enrolled in the KIGS registries and treated with GH in childhood [86]. The number of malignancies seemed disproportionately high for the relatively small number of children who had chronic renal failure and who were receiving GH treatment. It is possible that the age of the graft recipient and the donor as well as the cumulative dose of immunosuppressive agents may have played a greater part than GH in increasing tumor risk. Nevertheless, GH treatment should be used with caution in the posttransplantation period, with routine renal ultrasound scanning to aid early tumor detection.

Growth hormone treatment: tumor recurrences

Major studies of GH treatment and risk of tumor recurrences are shown in Table 2.

Leukemia/lymphoma

Survivors of childhood leukemias are at risk of developing a range of endocrine late effects [87–90], including growth failure, which may require treatment with GH. The incidence of GH deficiency in children treated

Table 2
Major studies of growth hormone treatment and risk of tumor recurrences

Study	N	Patient-years on growth hormone treatment	Principal malignancy type(s)	Risk estimate
Ogilvy-Stuart et al [105]	53	–	brain tumors	RR 0.8
Sklar et al [98]	172	–	brain tumors	RR 0.8
Swerdlow et al [104]	180	–	brain tumors	RR 0.6
Blethen et al [96]	19,000	47,000	leukemia	no increased risk
Maneatis et al [97]	33,161	113,000	leukemia	SMR 0.7
			nonleukemic neoplasms	SMR 0.4
Wyatt [95]	~33,000	135,431	nonleukemic neoplasms	SIR 0–1.6

Abbreviations: N, number; RR, relative risk; SIR, standardized incidence ratio; SMR, standardized mortality ratio.

for leukemia has decreased with the reduction in the use of prophylactic cranial irradiation. GHD associated with total-body irradiation remains a risk, however, and GHD also has been reported with regimens using only chemotherapy [91]. Evidence also is accumulating that GH treatment may be beneficial in the management of the consequences of chemotherapy noted in young adults, such as cardiotoxicity [92], metabolic syndrome [93], and poor bone mineralization [94]. Thus in the future GH may be used more often to treat adult leukemia survivors.

Although a link between the de novo occurrence of leukemia and GH treatment is unproven, clinical concerns remain that children completing therapy for leukemia/lymphomas may be at a greater risk of recurrence if given GH. Reassuringly, data from the NCGS surveillance program [95] suggest that GH is safe and without an increased risk of leukemia recurrences [96,97]. An independent study on a smaller scale, involving 47 children who received GH treatment after leukemia and 544 controls who did not receive GH, examined leukemia recurrence rates at 7 years and 11 years after continuous hematologic remission [90]. There was no increase in the disease recurrence rate in the GH-treated cohort. In 122 acute leukemia survivors in the Childhood Cancer Survivor Study (CCSS) treated with GH, the RR was not increased in comparison with 4545 children not treated with GH [98]. Another study was undertaken in 72 children who had leukemia in the United Kingdom, accounting for a total of 149 patient-years of GH treatment [6,99]. The analysis took into account factors such as previous number of relapses, chemotherapy protocol, and prognostic factors for leukemia cure and relapse. As observed in other studies, there was no increased risk in the GH-treated children. Caution must be used in interpretation, however, when controlling for multiple variables in a small study.

Although chemotherapy regimens have changed during the last 2 decades, and intracranial irradiation is not standard treatment in children who have leukemia, it still is important to continue sustained and careful vigilance in large cohorts over long periods of time to identify leukemia relapses in GH recipients. Surveillance studies are the only practical means of evaluating tumor risks, but it must be recognized that such studies may be incomplete [90] and therefore prone to underreporting.

Brain tumors

GHD is a common late effect in patients who have tumors in the hypothalamic–pituitary axis, caused by local compression, surgery, or irradiation to the hypothalamus and pituitary [9]. GHD also is common in those who have tumors distant from the hypothalamic–pituitary axis who receive radiation to the whole head as part of their treatment. Patients who have proven GHD are treated with GH during childhood, primarily to optimize growth but also to aid bone mineralization, maintain normal body composition and exercise tolerance, and potentially improve quality of life. It also is proposed that those who have persistent GHD when retested at the end of growth should continue GH therapy until at least their mid-twenties to optimize somatic development [5].

As with leukemia, there has been clinical concern that GH may play a part in tumor recurrence; thus single- and multicenter studies have been reported defining the RR of disease recurrence in those treated with GH. In one of the first reports, follow-up on 34 children who had various brain tumors [100] found that 8 of 24 subjects who received GH for GHD had tumor recurrences, in comparison with 3 of 11 who did not receive GH. Similarly, the rate of tumor recurrence in 24 children who had brain tumors distant from the hypothalamus and pituitary was not increased in those treated with GH as compared with those not treated with GH [101].

Craniopharyngioma, a benign slow-growing tumor with significant local effects, is one of the commonest types of brain tumor for which children receive GH treatment. In the NCGS experience in the United States, children receiving GH had a recurrence rate of 6.4% [76], considerably lower than the estimates of 20% to 25% reported elsewhere [102]. Other studies have concurred with these findings and show no evidence that GH has a role in craniopharyngioma recurrence [97,103].

In a follow-up study of 180 children who had brain tumors, the RR of first tumor recurrence in the group treated with GH was actually reduced (0.6; 95% confidence interval [CI], 0.4–0.9), after adjusting for potentially confounding variables [104]. The results were similar to those of a previous finding in 53 children who had brain tumor treated with GH, reporting a low RR (0.8), albeit with a wide 95% CI (0.2–2.3) [105].

More recently, investigation of 172 young persons who had brain tumors surviving beyond 5 years and enrolled in the CCSS in the United States reiterated the safety record of GH treatment: the RR for disease recurrence was 0.8 (95% CI, 0.3–1.8) [98]. There remains, however, some concern about selection bias and the lack of matched controls that limit the validity of conclusions drawn from these studies [9]. Nevertheless, evidence to date is reassuring, indicating that GH does not induce an increase in brain tumor recurrence in children and young persons.

The risk of tumor recurrence has been assessed in adults treated with GH. There was no recurrence of malignancy in a group of 24 adult survivors of childhood-onset brain tumors who were treated with GH in adult life for severe GHD [106]. Similar safety profiles were found in 1034 GH-treated hypopituitary adults, totaling 818 patient-years of treatment [107]. There were six recurrences of brain tumors, a rate comparable with that in the control population. In 100 patients who had adult-onset GHD secondary to hypothalamo-pituitary tumors, there was no progression in tumor size in imaging surveillance at 6 and 12 months following treatment with GH [108].

Because GH now is being considered as a lifelong treatment for those who have severe radiation-induced GHD [109], adequate surveillance programs remain very important. This need is emphasized further by the trend for an increase in the RR of death seen on longer-term follow-up of these patients [104].

Other tumors

Targeted radiation therapy for solid tumors in extracranial sites, such as nasal rhabdomyosarcoma [110], often is complicated by endocrine sequelae necessitating the use of GH. These tumors are relatively rare, and there are no large studies reporting relapse rates in those receiving GH. Patients who have Langerhans histiocytosis may develop GHD and be treated with GH. Recurrence with GH has not been an issue [111]. For rare tumor groups, it is very important that surveillance be undertaken to establish a cohort of sufficient size to assess RR of relapse with and without GH.

Growth hormone treatment: second malignant neoplasms

Childhood cancer survivors often receive GH in relatively high doses to maximize final adult height [112]. It is possible that GH could induce mitogenic activity in cells already predisposed to neoplastic change and hence increase the theoretical risk of developing second neoplasms. Data from the CCSS, which includes 13,581 children who had various cancers treated across North America, were analyzed for incidence of second malignant neoplasms (SMNs) by standardized incidence ratios [98,113]. In 298 individuals, 314 SMNs were identified with a high standardized incidence rate of 6.3 (95% CI, 5.6–7.1), the highest being for bone and breast cancers [113]. In further analysis, 361 GH-treated cancer survivors were identified, and a time-dependent Cox regression demonstrated a RR for SMNs of 3.2 (95% CI, 1.8–5.4) over those not treated with GH [98]. This increased RR resulted from three osteogenic sarcomas in the GH-treated group. In contrast, there were only two SMNs in the 4500 leukemia survivors not treated with GH. The most recent assessment of this cohort has shown that the RR for SMNs has fallen but remains significantly elevated at 2.1 (95% CI, 1.3–3.5) at extended follow-up assessment [114].

The risk of SMNs after GH treatment needs to be monitored carefully, and the findings need to be validated in larger cohorts. A number of SMNs have been reported in isolation [115–117]; although small in number, these adverse events do generate some concern and indicate that GH should be used with caution in individuals who have other risk factors for SMNs.

Balance of risks and implications for management

The occurrence or recurrence of tumors can be related to many factors, either intrinsic, such as genetic predisposition and diet, or extrinsic, such as environment or drug/toxin exposure. With regard to the GH debate, the environment may play a greater role in neoplastic transformation than previously envisaged. Insulin was found to sensitize cells to GH activation of induced extracellular signal-regulated kinase [118], whereas obesity in pregnant rats induced mammary neoplastic change [119]. It is possible that nutrition and insulin [120] may affect the GH–IGF-I axis and signaling pathways to influence background cancer risk in an individual child. Such factors have not been taken into account in surveillance studies to date.

Reassuringly, in studies from around the world that include considerable follow-up data on thousands of children, there is no clear evidence from clinical practice that GH treatment has a causal relationship with tumor occurrence or recurrence. The one area that requires further follow-up and assessment is that of second tumors, because there is some evidence that childhood cancer survivors who receive GH treatment have an increased risk of developing a second neoplasm.

The benefits of GH treatment for growth and metabolism are well recognized, and for children who have tumors associated with GHD, replacement treatment with GH is recommended. For those who have nonmalignant tumors such as craniopharyngioma, initiation of GH treatement after tumor treatment is complete is appropriate. In those who have malignant lesions, an interval of at least 1 year after completed tumor treatment should be considered. Monitoring of serum IGF-I levels (at least annually) during GH treatment should be undertaken, aiming to maintain IGF-I levels within the normal range. Most importantly all children who had malignancies and who received GH or continue to receive GH into adulthood should be under lifelong surveillance for any late-onset tumor.

References

[1] Molitch ME, Clemmons DR, Malozowski S, et al. Evaluation and treatment of adult growth hormone deficiency: an Endocrine Society Clinical Practice Guideline. J Clin Endocrinol Metab 2006;91(5):1621–34.
[2] Salerno M, Esposito V, Farina V, et al. Improvement of cardiac performance and cardiovascular risk factors in children with GH deficiency after two years of GH replacement therapy: an observational, open, prospective, case-control study. J Clin Endocrinol Metab 2006;91(4):1288–95.
[3] Attanasio AF, Shavrikova E, Blum WF, et al. Continued growth hormone (GH) treatment after final height is necessary to complete somatic development in childhood-onset GH-deficient patients. J Clin Endocrinol Metab 2004;89(10):4857–62.
[4] Hoffman AR, Kuntze JE, Baptista J, et al. Growth hormone (GH) replacement therapy in adult-onset GH deficiency: effects on body composition in men and women in a double-blind, randomized, placebo-controlled trial. J Clin Endocrinol Metab 2004;89(5): 2048–56.
[5] Clayton PE, Cuneo RC, Juul A, et al. Consensus statement on the management of the GH-treated adolescent in the transition to adult care. Eur J Endocrinol 2005;152(2): 165–70.
[6] Ogilvy-Stuart AL, Gleeson H. Cancer risk following growth hormone use in childhood: implications for current practice. Drug Saf 2004;27(6):369–82.
[7] Yakar S, LeRoith D, Brodt P. The role of the growth hormone/insulin-like growth factor axis in tumor growth and progression: lessons from animal models. Cytokine Growth Factor Rev 2005;16(4–5):407–20.
[8] Yakar S, Pennisi P, Kim CH, et al. Studies involving the GH-IGF axis: lessons from IGF-I and IGF-I receptor gene targeting mouse models. J Endocrinol Invest 2005;28(5 Suppl): 19–22.
[9] Sklar CA. Growth hormone treatment: cancer risk. Horm Res 2004;62(Suppl 3):30–4.
[10] Furstenberger G, Senn HJ. Insulin-like growth factors and cancer. Lancet Oncol 2002;3(5): 298–302.
[11] LeRoith D, Roberts CT Jr. The insulin-like growth factor system and cancer. Cancer Lett 2003;195(2):127–37.
[12] Mercola KE, Cline MJ, Golde DW. Growth hormone stimulation of normal and leukemic human T-lymphocyte proliferation in vitro. Blood 1981;58(2):337–40.
[13] Estrov Z, Meir R, Barak Y, et al. Human growth hormone and insulin-like growth factor-1 enhance the proliferation of human leukemic blasts. J Clin Oncol 1991;9(3): 394–9.
[14] Manabe N, Kubota Y, Kitanaka A, et al. Src transduces signaling via growth hormone (GH)-activated GH receptor (GHR) tyrosine-phosphorylating GHR and STAT5 in human leukemia cells. Leuk Res 2006;30(11):1391–8.
[15] Shimizu K, Adachi K, Teramoto A. Growth hormone enhances natural killer cell activity against glioma. J Nippon Med Sch 2005;72(6):335–40.
[16] Cullen KJ, Yee D, Sly WS, et al. Insulin-like growth factor receptor expression and function in human breast cancer. Cancer Res 1990;50(1):48–53.
[17] Carboni JM, Lee AV, Hadsell DL, et al. Tumor development by transgenic expression of a constitutively active insulin-like growth factor I receptor. Cancer Res 2005;65(9): 3781–7.
[18] Alexia C, Fourmatgeat P, Delautier D, et al. Insulin-like growth factor-I stimulates H4II rat hepatoma cell proliferation: dominant role of PI-3′K/Akt signaling. Exp Cell Res 2006; 312(7):1142–52.
[19] Ulfarsson E, Karstrom A, Yin S, et al. Expression and growth dependency of the insulin-like growth factor I receptor in craniopharyngioma cells: a novel therapeutic approach. Clin Cancer Res 2005;11(13):4674–80.

[20] Haddad T, Yee D. Targeting the insulin-like growth factor axis as a cancer therapy. Future Oncol 2006;2(1):101–10.
[21] Rajah R, Valentinis B, Cohen P. Insulin-like growth factor (IGF)-binding protein-3 induces apoptosis and mediates the effects of transforming growth factor-beta1 on programmed cell death through a p53- and IGF-independent mechanism. J Biol Chem 1997;272(18): 12181–8.
[22] Russo VC, Schutt BS, Andaloro E, et al. Insulin-like growth factor binding protein-2 binding to extracellular matrix plays a critical role in neuroblastoma cell proliferation, migration, and invasion. Endocrinology 2005;146(10):4445–55.
[23] Moreno MJ, Ball M, Andrade MF, et al. Insulin-like growth factor binding protein-4 (IGFBP-4) is a novel anti-angiogenic and anti-tumorigenic mediator secreted by dibutyryl cyclic AMP (dB-cAMP)-differentiated glioblastoma cells. Glia 2006;53(8):845–57.
[24] Durai R, Davies M, Yang W, et al. Biology of insulin-like growth factor binding protein-4 and its role in cancer. Int J Oncol 2006;28(6):1317–25.
[25] Luft R, Olivecrona H. Hypophysectomy in the treatment of malignant tumors. Cancer 1957;10(4):789–94.
[26] Moon HD, Simpson ME, Li CH, et al. Neoplasms in rats treated with pituitary growth hormone; pulmonary and lymphatic tissues. Cancer Res 1950;10(5):297–308.
[27] Moon HD, Simpson ME, Li CH, et al. Neoplasms in rats treated with pituitary growth hormone; adrenal glands. Cancer Res 1950;10(6):364–70.
[28] Moon HD, Simpson ME, Li CH, et al. Neoplasms in rats treated with pituitary growth hormone. III. Reproductive organs. Cancer Res 1950;10(9):549–56.
[29] Murphy LJ, Bell GI, Friesen HG. Growth hormone stimulates sequential induction of c-myc and insulin-like growth factor I expression in vivo. Endocrinology 1987;120(5): 1806–12.
[30] Schally AV, Varga JL. Antagonists of growth hormone-releasing hormone in oncology. Comb Chem High Throughput Screen 2006;9(3):163–70.
[31] Tornell J, Carlsson B, Pohjanen P, et al. High frequency of mammary adenocarcinomas in metallothionein promoter-human growth hormone transgenic mice created from two different strains of mice. J Steroid Biochem Mol Biol 1992;43(1–3):237–42.
[32] Zarandi M, Varga JL, Schally AV, et al. Lipopeptide antagonists of growth hormone-releasing hormone with improved antitumor activities. Proc Natl Acad Sci U S A 2006; 103(12):4610–5.
[33] Wu X, Wan M, Li G, et al. Growth hormone receptor overexpression predicts response of rectal cancers to pre-operative radiotherapy. Eur J Cancer 2006;42(7):888–94.
[34] Michaylira CZ, Ramocki NM, Simmons JG, et al. Haplotype insufficiency for suppressor of cytokine signaling-2 enhances intestinal growth and promotes polyp formation in growth hormone-transgenic mice. Endocrinology 2006;147(4):1632–41.
[35] Kineman RD. Antitumorigenic actions of growth hormone-releasing hormone antagonists. Proc Natl Acad Sci U S A 2000;97(2):532–4.
[36] Yang XF, Beamer WG, Huynh H, et al. Reduced growth of human breast cancer xenografts in hosts homozygous for the lit mutation. Cancer Res 1996;56(7):1509–11.
[37] Kahan Z, Nagy A, Schally AV, et al. Inhibition of growth of MX-1, MCF-7-MIII and MDA-MB-231 human breast cancer xenografts after administration of a targeted cytotoxic analog of somatostatin, AN-238. Int J Cancer 1999;82(4):592–8.
[38] Yakar S, Pennisi P, Zhao H, et al. Circulating IGF-1 and its role in cancer: lessons from the IGF-1 gene deletion (LID) mouse. Novartis Found Symp 2004;262:3–9.
[39] Haluska P, Carboni JM, Loegering DA, et al. In vitro and in vivo antitumor effects of the dual insulin-like growth factor-I/insulin receptor inhibitor, BMS-554417. Cancer Res 2006; 66(1):362–71.
[40] Shukla S, Mishra A, Fu P, et al. Up-regulation of insulin-like growth factor binding protein-3 by apigenin leads to growth inhibition and apoptosis of 22Rv1 xenograft in athymic nude mice. FASEB J 2005;19(14):2042–4.

[41] Silha JV, Sheppard PC, Mishra S, et al. Insulin-like growth factor (IGF) binding protein-3 attenuates prostate tumor growth by IGF-dependent and IGF-independent mechanisms. Endocrinology 2006;147(5):2112–21.
[42] Gunnell DJ, Smith GD, Frankel SJ, et al. Socio-economic and dietary influences on leg length and trunk length in childhood: a reanalysis of the Carnegie (Boyd Orr) survey of diet and health in prewar Britain (1937–39). Paediatr Perinat Epidemiol 1998;12(Suppl 1): 96–113.
[43] Ahlgren M, Melbye M, Wohlfahrt J, et al. Growth patterns and the risk of breast cancer in women. N Engl J Med 2004;351(16):1619–26.
[44] Gunnell D, Okasha M, Smith GD, et al. Height, leg length, and cancer risk: a systematic review. Epidemiol Rev 2001;23(2):313–42.
[45] Jenkins PJ, Mukherjee A, Shalet SM. Does growth hormone cause cancer? Clin Endocrinol (Oxf) 2006;64(2):115–21.
[46] Hankinson SE, Willett WC, Colditz GA, et al. Circulating concentrations of insulin-like growth factor-I and risk of breast cancer. Lancet 1998;351(9113):1393–6.
[47] Wolk A, Mantzoros CS, Andersson SO, et al. Insulin-like growth factor 1 and prostate cancer risk: a population-based, case-control study. J Natl Cancer Inst 1998;90(12):911–5.
[48] Jernstrom H, Barrett-Connor E. Obesity, weight change, fasting insulin, proinsulin, C-peptide, and insulin-like growth factor-1 levels in women with and without breast cancer: the Rancho Bernardo Study. J Womens Health Gend Based Med 1999;8(10):1265–72.
[49] Kurek R, Tunn UW, Eckart O, et al. The significance of serum levels of insulin-like growth factor-1 in patients with prostate cancer. BJU Int 2000;85(1):125–9.
[50] Wiley A, Katsaros D, Fracchioli S, et al. Methylation of the insulin-like growth factor binding protein-3 gene and prognosis of epithelial ovarian cancer. Int J Gynecol Cancer 2006; 16(1):210–8.
[51] Ali O, Cohen P, Lee KW. Epidemiology and biology of insulin-like growth factor binding protein-3 (IGFBP-3) as an anti-cancer molecule. Horm Metab Res 2003;35(11–12):726–33.
[52] Renehan AG, Zwahlen M, Minder C, et al. Insulin-like growth factor (IGF)-I, IGF binding protein-3, and cancer risk: systematic review and meta-regression analysis. Lancet 2004; 363(9418):1346–53.
[53] Dawczynski K, Kauf E, Schlenvoigt D, et al. Elevated serum insulin-like growth factor binding protein-2 is associated with a high relapse risk after hematopoietic stem cell transplantation in childhood AML. Bone Marrow Transplant 2006;37(6):589–94.
[54] Wex H, Ahrens D, Hohmann B, et al. Insulin-like growth factor-binding protein 4 in children with acute lymphoblastic leukemia. Int J Hematol 2005;82(2):137–42.
[55] Cheng I, Stram DO, Penney KL, et al. Common genetic variation in IGF1 and prostate cancer risk in the multiethnic cohort. J Natl Cancer Inst 2006;98(2):123–34.
[56] Canzian F, McKay JD, Cleveland RJ, et al. Polymorphisms of genes coding for insulin-like growth factor 1 and its major binding proteins, circulating levels of IGF-I and IGFBP-3 and breast cancer risk: results from the EPIC study. Br J Cancer 2006;94(2):299–307.
[57] Canzian F, McKay JD, Cleveland RJ, et al. Genetic variation in the growth hormone synthesis pathway in relation to circulating insulin-like growth factor-I, insulin-like growth factor binding protein-3, and breast cancer risk: results from the European prospective investigation into cancer and nutrition study. Cancer Epidemiol Biomarkers Prev 2005; 14(10):2316–25.
[58] Lai JH, Vesprini D, Zhang W, et al. A polymorphic locus in the promoter region of the IGFBP3 gene is related to mammographic breast density. Cancer Epidemiol Biomarkers Prev 2004;13(4):573–82.
[59] Fletcher O, Gibson L, Johnson N, et al. Polymorphisms and circulating levels in the insulin-like growth factor system and risk of breast cancer: a systematic review. Cancer Epidemiol Biomarkers Prev 2005;14(1):2–19.
[60] Morimoto LM, Newcomb PA, White E, et al. Insulin-like growth factor polymorphisms and colorectal cancer risk. Cancer Epidemiol Biomarkers Prev 2005;14(5):1204–11.

[61] Le Marchand L, Kolonel LN, Henderson BE, et al. Association of an exon 1 polymorphism in the IGFBP3 gene with circulating IGFBP-3 levels and colorectal cancer risk: the multi-ethnic cohort study. Cancer Epidemiol Biomarkers Prev 2005;14(5):1319–21.
[62] Schildkraut JM, Demark-Wahnefried W, Wenham RM, et al. IGF1 (CA)19 repeat and IGFBP3 -202 A/C genotypes and the risk of prostate cancer in black and white men. Cancer Epidemiol Biomarkers Prev 2005;14(2):403–8.
[63] Friedrichsen DM, Hawley S, Shu J, et al. IGF-I and IGFBP-3 polymorphisms and risk of prostate cancer. Prostate 2005;65(1):44–51.
[64] Bogazzi F, Cosci C, Sardella C, et al. Identification of acromegalic patients at risk of developing colonic adenomas. J Clin Endocrinol Metab 2006;91(4):1351–6.
[65] Orme SM, McNally RJ, Cartwright RA, et al. Mortality and cancer incidence in acromegaly: a retrospective cohort study. United Kingdom Acromegaly Study Group. J Clin Endocrinol Metab 1998;83(8):2730–4.
[66] Baris D, Gridley G, Ron E, et al. Acromegaly and cancer risk: a cohort study in Sweden and Denmark. Cancer Causes Control 2002;13(5):395–400.
[67] Fuhrman B, Barba M, Schunemann HJ, et al. Basal growth hormone concentrations in blood and the risk for prostate cancer: a case-control study. Prostate 2005;64(2):109–15.
[68] Watanabe S, Tsunematsu Y, Fujimoto J, et al. Leukaemia in patients treated with growth hormone. Lancet 1988;1(8595):1159–60.
[69] Wada E, Murata M, Watanabe S. Acute lymphoblastic leukemia following treatment with human growth hormone in a boy with possible preanemic Fanconi's anemia. Jpn J Clin Oncol 1989;19(1):36–9.
[70] Watanabe S, Mizuno S, Oshima LH, et al. Leukemia and other malignancies among GH users. J Pediatr Endocrinol 1993;6(1):99–108.
[71] Fradkin JE, Mills JL, Schonberger LB, et al. Risk of leukemia after treatment with pituitary growth hormone. JAMA 1993;270(23):2829–32.
[72] Kubota M, Fujii K, Yamanaka C, et al. Leukaemia in children with growth hormone deficiency not treated with growth hormone. Eur J Pediatr 1995;154(5):418–9.
[73] Rapaport R, Oberfield SE, Robison L, et al. Relationship of growth hormone deficiency and leukemia. J Pediatr 1995;126(5 Pt 1):759–61.
[74] Nishi Y, Tanaka T, Takano K, et al. Recent status in the occurrence of leukemia in growth hormone-treated patients in Japan. GH Treatment Study Committee of the Foundation for Growth Science, Japan. J Clin Endocrinol Metab 1999;84(6):1961–5.
[75] Allen DB, Rundle AC, Graves DA, et al. Risk of leukemia in children treated with human growth hormone: review and reanalysis. J Pediatr 1997;131(1 Pt 2):S32–6.
[76] Moshang T JR, Rundle AC, Graves DA, et al. Brain tumor recurrence in children treated with growth hormone: the National Cooperative Growth Study experience. J Pediatr 1996;128(5 Pt 2):S4–7.
[77] Sharif S, Ferner R, Birch JM, et al. Second primary tumors in neurofibromatosis 1 patients treated for optic glioma: substantial risks after radiotherapy. J Clin Oncol 2006;24(16):2570–5.
[78] Cunha KS, Barboza EP, Da Fonseca EC. Identification of growth hormone receptor in localised neurofibromas of patients with neurofibromatosis type 1. J Clin Pathol 2003;56(10):758–63.
[79] Howell SJ, Wilton P, Lindberg A, et al. Growth hormone replacement and the risk of malignancy in children with neurofibromatosis. J Pediatr 1998;133(2):201–5.
[80] Tuffli GA, Johanson A, Rundle AC, et al. Lack of increased risk for extracranial, nonleukemic neoplasms in recipients of recombinant deoxyribonucleic acid growth hormone. J Clin Endocrinol Metab 1995;80(4):1416–22.
[81] Swerdlow AJ, Higgins CD, Adlard P, et al. Risk of cancer in patients treated with human pituitary growth hormone in the UK, 1959–85: a cohort study. Lancet 2002;360(9329):273–7.

[82] Cabanas P, Garcia-Caballero T, Barreiro J, et al. Papillary thyroid carcinoma after recombinant GH therapy for Turner syndrome. Eur J Endocrinol 2005;153(4):499–502.
[83] Freedman RJ, Malkovska V, LeRoith D, et al. Hodgkin lymphoma in temporal association with growth hormone replacement. Endocr J 2005;52(5):571–5.
[84] Sperling MA, Saenger PH, Ray H, et al. Growth hormone treatment and neoplasia-coincidence or consequence? J Clin Endocrinol Metab 2002;87(12):5351–2.
[85] Tyden G, Wernersson A, Sandberg J, et al. Development of renal cell carcinoma in living donor kidney grafts. Transplantation 2000;70(11):1650–6.
[86] Mehls O, Wilton P, Lilien M, et al. Does growth hormone treatment affect the risk of post-transplant renal cancer? Pediatr Nephrol 2002;17(12):984–9.
[87] Kirk JA, Raghupathy P, Stevens MM, et al. Growth failure and growth-hormone deficiency after treatment for acute lymphoblastic leukaemia. Lancet 1987;1(8526):190–3.
[88] Leung W, Hudson MM, Strickland DK, et al. Late effects of treatment in survivors of childhood acute myeloid leukemia. J Clin Oncol 2000;18(18):3273–9.
[89] Katz JA, Pollock BH, Jacaruso D, et al. Final attained height in patients successfully treated for childhood acute lymphoblastic leukemia. J Pediatr 1993;123(4):546–52.
[90] Leung W, Rose SR, Zhou Y, et al. Outcomes of growth hormone replacement therapy in survivors of childhood acute lymphoblastic leukemia. J Clin Oncol 2002;20(13):2959–64.
[91] Haddy TB, Mosher RB, Nunez SB, et al. Growth hormone deficiency after chemotherapy for acute lymphoblastic leukemia in children who have not received cranial radiation. Pediatr Blood Cancer 2006;46(2):258–61.
[92] Lipshultz SE, Vlach SA, Lipsitz SR, et al. Cardiac changes associated with growth hormone therapy among children treated with anthracyclines. Pediatrics 2005;115(6):1613–22.
[93] Follin C, Thilen U, Ahren B, et al. Improvement in cardiac systolic function and reduced prevalence of metabolic syndrome after two years of growth hormone (GH) treatment in GH-deficient adult survivors of childhood-onset acute lymphoblastic leukemia. J Clin Endocrinol Metab 2006;91(5):1872–5.
[94] Jarfelt M, Fors H, Lannering B, et al. Bone mineral density and bone turnover in young adult survivors of childhood acute lymphoblastic leukaemia. Eur J Endocrinol 2006;154(2):303–9.
[95] Wyatt D. Lessons from the national cooperative growth study. Eur J Endocrinol 2004;151(Suppl 1):S55–9.
[96] Blethen SL, Allen DB, Graves D, et al. Safety of recombinant deoxyribonucleic acid-derived growth hormone: The National Cooperative Growth Study experience. J Clin Endocrinol Metab 1996;81(5):1704–10.
[97] Maneatis T, Baptista J, Connelly K, et al. Growth hormone safety update from the National Cooperative Growth Study. J Pediatr Endocrinol Metab 2000;13(Suppl 2):1035–44.
[98] Sklar CA, Mertens AC, Mitby P, et al. Risk of disease recurrence and second neoplasms in survivors of childhood cancer treated with growth hormone: a report from the Childhood Cancer Survivor Study. J Clin Endocrinol Metab 2002;87(7):3136–41.
[99] Leiper A. Growth hormone deficiency in children treated for leukaemia. Acta Paediatr Suppl 1995;411:41–4.
[100] Arslanian SA, Becker DJ, Lee PA, et al. Growth hormone therapy and tumor recurrence. Findings in children with brain neoplasms and hypopituitarism. Am J Dis Child 1985;139(4):347–50.
[101] Clayton PE, Shalet SM, Gattamaneni HR, et al. Does growth hormone cause relapse of brain tumours? Lancet 1987;1(8535):711–3.
[102] Weiss M, Sutton L, Marcial V, et al. The role of radiation therapy in the management of childhood craniopharyngioma. Int J Radiat Oncol Biol Phys 1989;17(6):1313–21.
[103] Karavitaki N, Warner JT, Marland A, et al. GH replacement does not increase the risk of recurrence in patients with craniopharyngioma. Clin Endocrinol (Oxf) 2006;64(5):556–60.

[104] Swerdlow AJ, Reddingius RE, Higgins CD, et al. Growth hormone treatment of children with brain tumors and risk of tumor recurrence. J Clin Endocrinol Metab 2000;85(12): 4444–9.
[105] Ogilvy-Stuart AL, Ryder WD, Gattamaneni HR, et al. Growth hormone and tumour recurrence. BMJ 1992;304(6842):1601–5.
[106] Murray RD, Darzy KH, Gleeson HK, et al. GH-deficient survivors of childhood cancer: GH replacement during adult life. J Clin Endocrinol Metab 2002;87(1):129–35.
[107] Abs R, Bengtsson BA, Hernberg-Stahl E, Monson JP, et al. GH replacement in 1034 growth hormone deficient hypopituitary adults: demographic and clinical characteristics, dosing and safety. Clin Endocrinol (Oxf) 1999;50(6):703–13.
[108] Frajese G, Drake WM, Loureiro RA, et al. Hypothalamo-pituitary surveillance imaging in hypopituitary patients receiving long-term GH replacement therapy. J Clin Endocrinol Metab 2001;86(11):5172–5.
[109] Darzy KH, Shalet SM. Pathophysiology of radiation-induced growth hormone deficiency: efficacy and safety of GH replacement. Growth Horm IGF Res 2006;16(Suppl):30–40.
[110] Punyko JA, Mertens AC, Gurney JG, et al. Long-term medical effects of childhood and adolescent rhabdomyosarcoma: a report from the childhood cancer survivor study. Pediatr Blood Cancer 2005;44(7):643–53.
[111] Donadieu J, Rolon MA, Pion I, et al. Incidence of growth hormone deficiency in pediatric-onset Langerhans cell histiocytosis: efficacy and safety of growth hormone treatment. J Clin Endocrinol Metab 2004;89(2):604–9.
[112] Brownstein CM, Mertens AC, Mitby PA, et al. Factors that affect final height and change in height standard deviation scores in survivors of childhood cancer treated with growth hormone: a report from the childhood cancer survivor study. J Clin Endocrinol Metab 2004; 89(9):4422–7.
[113] Neglia JP, Friedman DL, Yasui Y, et al. Second malignant neoplasms in five-year survivors of childhood cancer: childhood cancer survivor study. J Natl Cancer Inst 2001;93(8): 618–29.
[114] Ergun-Longmire B, Mertens AC, Mitby P, et al. Growth hormone treatment and risk of second neoplasms in the childhood cancer survivor. J Clin Endocrinol Metab 2006;91(9): 3494–8.
[115] Forbes GM, Cohen AK. Primary cerebral lymphoma: an association with craniopharyngioma or cadaveric growth hormone therapy? Med J Aust 1992;157(1):27–8.
[116] Bordigoni P, Turello R, Clement L, et al. Osteochondroma after pediatric hematopoietic stem cell transplantation: report of eight cases. Bone Marrow Transplant 2002;29(7):611–4.
[117] Kranzinger M, Jones N, Rittinger O, et al. Malignant glioma as a secondary malignant neoplasm after radiation therapy for craniopharyngioma: report of a case and review of reported cases. Onkologie 2001;24(1):66–72.
[118] Xu J, Keeton AB, Franklin JL, et al. Insulin enhances growth hormone induction of the MEK/ERK signaling pathway. J Biol Chem 2006;281(2):982–92.
[119] de Assis S, Wang M, Goel S, et al. Excessive weight gain during pregnancy increases carcinogen-induced mammary tumorigenesis in Sprague-Dawley and lean and obese Zucker rats. J Nutr 2006;136(4):998–1004.
[120] Holly J, Perks C. Growth hormone and cancer: are we asking the right questions? Clin Endocrinol (Oxf) 2006;64(2):122–4.

Index

Note: Page numbers of article titles are in **boldface** type.

A

Acid-labile subunit (ALS), efficacy in growth hormone deficiency diagnosis, 115–116
  IGF-I regulation by, 110–111, 114–115
    growth hormone therapy and, 207
    idiopathic short stature and, 161

Acquired diseases, idiopathic short stature related to, 161

Acromegaly, growth hormone and, 93–94
    glucose/fat metabolism effects of, 76–77, 82–83
    measurements of, 101–105
    insulin-like growth factor system markers of, 120–122
    quality of life in, 222, 225–229

Acromegaly Quality of Life (AcroQoL), 222, 226
    languages available/results of, 226–228
    longitudinal evaluation of, 228–229
    Netherlands results of, 227–228
    United Kingdom results of, 227

ACTH deficiency, growth hormone deficiency and, 20, 30, 32

Activation failure, in growth hormone receptor mutations, 6

Adipocytokines, growth hormone impact on, 84

Adipokines, growth hormone impact on, 84

Adolescence. See *Puberty to adulthood transition.*

Adrenal gland, deficit of, in growth hormone deficiency, 20, 30, 32, 205

Adult GH Deficiency Assessment (AGHDA), of quality of life, 222–224

Adult height, 134

Adult onset growth hormone deficiency (AOGHD), diagnosis of, 203–205
    IGF-I as marker of, 116–117
    growth hormone therapy management of, **203–220**
      baseline assessment for, 205–206
      contraindications to, 215
      dose titration for, 208–210
      during pregnancy, 215–216
      estrogen replacement therapy and, 205–206
      glucose metabolism with, 211
      goal of, 205
      HPA axis evaluation for, 205
      initial dose monitoring of, 206–208
      initial dose selection for, 206
      lipid metabolism with, 211
      long-term monitoring and safety of, 211–215
      overview of, 203, 216
      quality of life and well-being with, 210–211
      side effects of, 208–209
    pathophysiology of, 78, 188–190, 197
    quality of life and, 110–113, 203

Adulthood, puberty transition to, growth hormone and, **187–201**
    deficiency diagnosis of, glucose metabolism and, 77–79
    re-establishing, 191–192
    overview of, 187, 198
    physiology of, 187–188
    replacement therapy during, 191–198
      discontinuation at final height, 188–191
      dosing of, 193
      effect on body and fat mass, 193–194
      effect on bone, 195–196
      effect on glucose, 197
      effect on lipids, 196

A

Adulthood (*continued*)
  quality of life aspects, 197–198
  sex steroids modulation of, 58. See also *Gonadal steroids*.

Aging, body composition changes with, 234–235
  global statistics on, 233
  growth hormone deficiency with, 234. See also *Adult onset growth hormone deficiency (AOGHD)*.
  growth hormone secretion regulation and, 39

Alanine, in growth hormone receptor mutations, 7

Alpha-motor neurons, loss of, sarcopenia associated with, 236

American Association of Clinical Endocrinologists, on growth hormone for non-GHD growth disorders, 133, 167

Amino acid oxidation, growth hormone effects on, 89–100
  animal studies of, 90
  during prolonged fasting, 91–92
  glucose and, 94–95
  in vitro studies, 90–91
  obesity and, 92–93
  overview of, 89–90, 95–96
  postabsorptive, 90–91

Anabolism, growth hormone effects on, 89–100
  animal studies of, 90
  during prolonged fasting, 91–92
  gonadal steroids modulation of, 65–67
  in vitro studies, 90–91
  mechanisms of, 93–95
  obesity and, 92–93
  overview of, 89–90, 95–96
  postabsorptive, 90–91
  hormones associated with, 238

Androgen replacement therapy, for growth hormone deficiency, 58

Androgens, growth hormone action regulation by, linear growth and, 64
  neurosecretory modulation, 58–59
  peripheral biologic effects modulation, 60–61
  substrate metabolism and, 66–67
  growth hormone secretion regulation by, 47–48

IGF-I secretion regulation by, 112
IGFBP-3 regulation by, 114

Animal studies, of growth hormone effects, on protein metabolism, 90
  of growth hormone secretion regulation, 38, 40–42
  gonadal steroids and, 47–48
  nutritional influences, 49
  of growth hormone—insulin-like growth factor axis in tumorigenesis, 248–249
  of isolated growth hormone deficiency type II, 27–28

Anti-apoptotic activity, of IGF-I, 247

Antibodies, in idiopathic short stature, 161
  polyclonal, in growth hormone assays, 103–104

Aortic aneurysm, growth hormone therapy risk for, 146

Arginine test, for growth hormone deficiency, 192, 204–205

Aspartate, in growth hormone receptor mutations, 6

Assays, of growth hormone, design of, 103–104
  growth hormone binding protein impact on, 104
  heterogeneity of, 101–102
  standard assay preparations and units, 104–105

Atherosclerosis, growth hormone therapy effects on, 196–197

Autoimmune thyroid disease, idiopathic short stature related to, 161

B

Bio-impedance assessment (BIA), for growth hormone deficiency, 206, 208

Body composition/mass, changes with aging, 234–235
  growth hormone therapy effects on, 237–239
  maturation during puberty, 188
  growth hormone deficiency effects on, 188–189, 234
  diagnostic applications of, 204–205
  therapeutic applications of, 205–206
  growth hormone effects on, gonadal steroids regulation of, 64–68
  growth hormone therapy effects on, 193–194, 224–225
  protein turnover and, 91

Bone mass, growth hormone deficiency impact on, 206, 234
    growth hormone therapy effects on, 195–196
Bone maturation, during puberty, 187–188
Bone mineral content (BMC), growth hormone therapy effects on, 195–196
    long-term monitoring of, 211–212
Bone mineral density (BMD), growth hormone deficiency impact on, 206, 234
    growth hormone therapy effects on, 195–196
        long-term monitoring and safety of, 211–212
Brain tumors, growth hormone treatment related to, 251–252
        neuroimaging surveillance for, 214–215
        recurrences of, 255–256

C

C-terminal domain, of growth hormone receptor, 2
Ca^{2+} ion channels, in sarcopenia, 236
Calibration, of growth hormone assays, 104–105
Canadian Pediatric Endocrine Group, on growth hormone for Turner Syndrome, 143–144
Cancer(s), growth hormone treatment risks for, **247–263**
    brain tumors in, 251–252
        recurrence of, 255–256
    clinical benefits versus, 236, 247
    colorectal tumors in, 251–253
    de novo malignancies in, 250–253
    epidemiologic observations of, 249–250
    evidence for, 197–198, 239, 247
    growth hormone—insulin-like growth factor axis in, 248–249
    Hodgkin lymphoma in, 251–253
    in children, 166
    in vitro experiments on, 248
    in vivo animal studies on, 248–249
    leukemia/lymphoma in, 251–252
        recurrence of, 253–255
    management implications of, 257
    renal tumors in, 251, 253

    second malignant neoplasms in, 256–257
    tumor recurrences in, 253–256
Candidate genes, in isolated growth hormone deficiency type IB, 23, 25
Carbohydrate metabolism, growth hormone effects on, gonadal steroids modulation of, 65–66
    in non—GHD growth disorders, 140, 145, 166–167
Cardiac disease, growth hormone deficiency risk for, 234
Cardiac function, growth hormone deficiency impact on, 234
Catabolism, sarcopenia associated with, 236
Cellular analysis, of growth hormone deficiency, 31–32
Cellular mechanisms, of growth hormone therapy effect on glucose/fat metabolism, 79–82
    FFA role in, 75–77, 83–84
Childhood onset growth hormone deficiency, **187–201**
    diagnosis of, glucose metabolism and, 189–191
        IGF-I as marker of, 116–117
        re-establishing, 191–192
    overview of, 187, 198
    physiology of, 187–188
    replacement therapy for, 191–198
        discontinuation at final height, 188–191
        dosing of, 193
Chronic renal insufficiency (CRI), growth disturbance related to, 136–140
    etiology of, 137
    growth hormone treatment of, efficacy of, 137–139
        mechanism of, 135–136
        overview of, 132
        safety of, 139–140
    growth pattern in, 137–138
    pathology of, 136
    untreated adult height in, 137
CIS domain, of growth hormone receptor signaling, 3–4
Clinical trials, on growth hormone, for non—GHD growth disorders, 138–139, 142–145, 147–148, 150, 168
    idiopathic, 162–165
Clonidine test, for growth hormone deficiency, 204

Colorectal tumors, growth hormone
    treatment related to, 214,
    251–253
Consensus, on growth hormone treatment,
    during puberty, 187, 192, 198
    of non—GHD growth disorders,
        167–168
Cortisol, acromegaly and, 120
    growth hormone secretion regulation
        by, 45–46
        therapeutic implications for
            adults, 205
Cranial irradiation, growth hormone
    deficiency related to, 204–205
CSH gene cluster, 21–22
Cytokine receptors, growth hormone
    receptor as, 1–2
    signaling suppressors of, 3
Cytokines, catabolic, sarcopenia associated
    with, 236
    in growth hormone status assessment,
        113
    in tumorigenesis, 249
Cytoplasmic domains, of growth hormone
    receptor, 1–3

D

d3GHR allele. See *Exon 3 deletion (d3)*.
De novo malignancies, growth hormone
    treatment related to, 250–253
DEXA scan, for growth hormone deficiency
    assessment, 206, 234
    for growth hormone therapy effect
        assessment, 195–196
Diabetes, FFA inhibition of insulin in,
    81
    growth hormone associated with, 94
    growth hormone therapy risk for, 212,
        239
    in children, 140, 145, 151,
        157–158
Diabetic retinopathy, as growth hormone
    therapy contraindication, 215
Dietary proteins, reduction of. See *Protein metabolism*.
Disability, in the elderly, with growth
    hormone deficiency, 233,
    235–237
    with sarcopenia, 236–237
L-Dopa test, for growth hormone
    deficiency, 204

E

Elderly patients, growth hormone deficiency
    in, 234
    functional impairment with, 233,
        235–237, 240
    treatment of, 237–240
    growth hormone therapy for, **233–245**
        body composition changes and,
            234–235
        clinical indications for, 234
        effects of, 237–239
        functional indications for, 233,
            235–237
        muscle mass loss implications,
            235–237
        overview of, 233, 240–241
        potential risks of, 239–240
        secretagogues for, 240
Endocrinology, growth hormone receptor
    polymorphisms in, 9, 11
    health-related quality of life and,
        221–222
    of isolated growth hormone deficiency
        type II, 29–30
Energy, growth hormone therapy effects on,
    224–226
Epiphysis, growth hormone effect on, in
    non-GHD growth disorders, 135–136
ESPE Consensus Workshop, on growth
    hormone deficiency management
    during puberty, 187, 192, 198
Estradiol, growth hormone action
    modulation by, 58
    linear growth and, 64
Estrogen replacement therapy, growth
    hormone deficiency and, 58, 65–66
    growth hormone therapy with,
        205–206, 214
Estrogens, growth hormone action
    regulation by, body composition and,
        65–66
        linear growth and, 64
        neurosecretory modulation of,
            59–60
        peripheral biologic effects
            modulation, 61–61
        substrate metabolism and, 65–66
    growth hormone secretion regulation
        by, 47–48
    in growth hormone status assessment,
        114
Exon 3 deletion (d3), in growth hormone
    receptor polymorphisms, 8
    fl genotype associations with,
        11–12

growth hormone therapy for, 10–13
in vitro activity of, 9
in idiopathic short stature, 165

Exon 9, in growth hormone receptor mutations, 6–7

Expression failure, in growth hormone receptor mutations, 6

Extracellular water (ECW), growth hormone therapy impact on, 207–208

F

Familial isolated growth hormone deficiency, 22–27

Fasting, growth hormone therapy during, glucose/fat metabolism effects of, 79, 83–84
    prolonged, IGFBP-3 changes with, 115
    insulin resistance induced by, 93
    protein metabolism during, 91–92

Fat mass, changes with aging, 234–235
    growth hormone deficiency impact on, 234
    growth hormone secretagogues impact on, 240
    growth hormone therapy effects on, 193–194
    truncal, growth hormone therapy discontinuation and, 189–190

Fat metabolism, growth hormone effects on, **75–87**
    acromegaly and, 76–77, 82–83
    cellular mechanisms of, 79–82
        FFA role in, 75–77, 83–84
    discussion on, 82–84
    during fasting, 79, 83–84
    gonadal steroids modulation of, 65
    in growth hormone-deficient patients, 77–79, 83
    in healthy subjects, 75–77
    insulin sensitivity and, 76, 78–79, 83–84
    insulin signaling crosstalk in, 79–81
    lipolysis and, 76, 81–83
    overview of, 75, 84–85
    substrate metabolism and, 76, 79, 83–84
    therapeutic strategies for, 239

Final adult height, 134

Final height, 134

5 Dimensions EuroQoL, 221–222, 227–228

Fluid retention, growth hormone regulation of, gonadal steroids modulation of, 67
    therapeutic implications of, 207–208

Food and Drug Administration (FDA), growth hormone therapy approval by, for non-GHD growth disorders, 136, 138, 150, 158, 167–168
    idiopathic, 162, 164

Free fatty acids (FFAs), growth hormone secretion regulation and, 48–49
    growth hormone therapy effect on, 75–77, 83–84
    during fasting, 79
    protein metabolism role of, 93–94

Full-length growth hormone receptor (fl), in growth hormone receptor polymorphisms, 9
    exon 3 deletion associations with, 11–12
    growth hormone therapy for, 10–13

Functional impairment, in the elderly, with growth hormone deficiency, 233, 235–237, 240
    with sarcopenia, 236–237

G

Gene mutations, in growth hormone deficiency, cellular analysis of, 31–32
    spectrum of, 22–27
    splice site, 23, 31
        endocrine data on, 29–30
    type IA, 22–23
    type IB, 23–26
    type II, 23, 26–27
        clinical implications of, 29, 32
        missense, 30
        splice site, 29–30
    type III, 23, 27
    variable phenotypes of, 31
    in idiopathic short stature, 159–161
    in *SHOX* gene deficiency. See *Short stature homeobox-containing (SHOX) gene deficiency.*
    of growth hormone receptor, activation failure, 6
        expression failure, 6
        growth disorders related to, 5–8
        post-growth hormone receptor defects, 7–8
        signaling failure, 6–7

General Health Questionnaire (GHQ), 223

Genetics, of growth hormone deficiency, **17–36**
  cellular analysis of, 31–32
  classification based on, 21–22
  familial isolated types in, 22–27
  isolated type II, as evolving pituitary deficit, 28–30
    clinical impact of, 29, 32
    gene mutations in, 23, 26–27
    in transgenic mice, 27–28
    overview of, 17–19
    splice site mutation in, 23, 31
      endocrine data on, 29–30
    transcription factors in, 18–21
    variable phenotypes in, 31
  of growth hormone-related tumorigenesis, 250
  of IGF-I deficiency, 17–18
  of IGF-I secretion regulation, 112
  of IGFBP-3 regulation, 115

Genomics, of human growth hormone therapy, 9–13

GH-1 gene, 21–22
  mutations of, cellular analysis of, 31–32
    deletions, 23–24
    frameshift- and nonsense, 23–24
    in idiopathic short stature, 159–160
    in isolated growth hormone deficiency type II, 28–29
    specific trans-acting factor to, 23, 25–26
    variable phenotypes of, 31

GH gene cluster, 21–22

GH-IGF growth plate axis, 135–136
  in non—GHD growth disorders, 149, 159–161

GH-N gene, 102

GH provocation tests, for growth hormone deficiency, 135, 161, 192, 204–205

GH-suppression testing, for acromegaly, 120–122

GH-V gene, 102

GHR gene, polymorphisms of, 8–9
  in idiopathic short stature, 160

Ghrelin, growth hormone secretion regulation by, 42–43
  gonadal steroids modulation of, 57–58
  in idiopathic short stature, 160

GHRH gene, in isolated growth hormone deficiency type IB, 23, 25

GHRH-GH releasing peptide (GHRP)-6, in IGF-I deficiency versus growth hormone deficiency, 17–18, 204–205

GHRH-pyridostigmine test, for growth hormone deficiency, 204

GHRH-receptor gene, in isolated growth hormone deficiency type IB, 23, 25

Glucagon test, for growth hormone deficiency, 192

Glucocorticoids, in growth hormone status assessment, 113, 205

Gluconeogenesis, fasting effects on, 83–84

Glucose, growth hormone secretion regulation and, 48
  growth hormone therapy effects on, 197
  protein metabolism role of, 94–95

Glucose clamp technique, for glucose/fat metabolism detection, 76–78, 81

Glucose intolerance, OGTT for, in growth hormone-deficient patients, 77–79
  with growth hormone treatment, in children, 140, 145, 151, 157–158, 166–167

Glucose metabolism, growth hormone effects on, **75–87**
  acromegaly and, 76–77, 82–83
  cellular mechanisms of, 79–82
    FFA role in, 75–77, 83–84
  discussion on, 82–84
  during fasting, 79, 83–84
  in growth hormone-deficient patients, 77–79, 83
  in healthy subjects, 75–77
  insulin sensitivity and, 76, 78–79, 83–84
  insulin signaling crosstalk in, 79–81
  lipolysis and, 76, 81–83
  overview of, 75, 84–85
  substrate metabolism and, 76, 79, 83–84
  therapeutic strategies for, 211, 239

Gonadal steroids, acromegaly and, 120
  female. See *Estrogens.*
  growth hormone action regulation by, **57–73**
    neurosecretory modulation, 57–60
    on body composition, 64–68
    on growth, 63–64
    on substrate metabolism, 64–67
    overview of, 57, 67, 69

peripheral biologic effects
modulation, 60–63
therapeutic indications for
adults, 205, 207
growth hormone secretion regulation
by, 47–48
male. See *Androgens; Testosterone.*

Growth. See *Human growth.*

Growth disorders, growth hormone
receptor mutations in, 5–8
non—growth hormone-deficient
growth hormone treatment of,
131–186
consensus statement
highlights on, 167–168
definitions for, 133–135
mechanisms of, 135–167
overview of, 131–133,
168–169
products available for,
131–132

Growth failure, 133

Growth hormone (GH), age-dependent
decline in, 188, 237–239. See also *Adult
onset growth hormone deficiency
(AOGHD).*
glucose/fat metabolism effects of,
75–87
acromegaly and, 76–77, 82–83
cellular mechanisms of, 79–82
FFA role in, 75–77, 83–84
discussion on, 82–84
during fasting, 79, 83–84
in growth hormone-deficient
patients, 77–79, 83
in healthy subjects, 75–77
insulin sensitivity and, 76, 78–79,
83–84
insulin signaling crosstalk in,
79–81
lipolysis and, 76, 81–83
overview of, 75, 84–85
substrate metabolism and, 76, 79,
83–84
therapeutic strategies for, 239
gonadal steroid regulation of action,
57–73
neurosecretory modulation,
57–60
on body composition, 64–68
on growth, 63–64
on substrate metabolism,
64–67
overview of, 57, 67, 69
peripheral biologic effects
modulation, 60–63
IGF-I mediation by, 90–92, 94

secretion regulation as, 37, 43–44,
113–114
synthesis regulation as, 112–113
tumorigenesis and tumor
progression with, 248–249
in puberty to adulthood transition,
187–201
deficiency diagnosis of, glucose
metabolism and, 77–79
re-establishing, 191–192
overview of, 187, 198
physiology of, 187–188
replacement therapy during,
191–198
discontinuation at final
height, 188–191
dosing of, 193
effect on body and fat mass,
193–194
effect on bone, 195–196
effect on glucose, 197
effect on lipids, 196
quality of life aspects,
197–198
sex steroids modulation of, 58.
See also *Gonadal steroids.*
measurements of, **101–108**
assay design for, 103–104
assay heterogeneity in, 101–102
growth hormone binding protein
impact on, 104
growth hormone molecules
heterogeneity in, 102
overview of, 101, 105
standard assay preparations and
units, 104–105
protein metabolism and, **89–100**
animal studies of, 90
during prolonged fasting, 91–92
in vitro studies, 90–91
mechanisms of, 93–95
obesity and, 92–93
overview of, 89–90, 95–96
postabsorptive, 90–91
secretion regulation factors of, **37–55**
aging-related, 39
ghrelin as, 42–43
gonadal steroids as, 47–48
growth hormone—releasing
hormone as, 37–40
insulin-like growth factor as, 37,
43–44
negative feedback mechanism,
43–45
nutritional, 48–49
overview of, 37, 49–50
schematic diagram of, 49
sleep as, 45–47
somatostatin as, 40–42

Growth hormone (GH) (*continued*)
  status assessment of, insulin-like
    growth factor in, **109–129**. See
    also *Insulin-like growth factor
    system markers.*

Growth hormone binding protein (GHBP),
  generation of, 1, 3
  growth hormone receptor mutations
    and, 5–6
  impact on growth hormone assays, 104
  in chronic renal insufficiency, 137

Growth hormone deficiency (GHD), adult
  onset, diagnosis of, 203–205
      IGF-I as marker of,
        116–117
      management of, **203–220**. See
        also *Growth hormone
        replacement therapy.*
      pathophysiology of, 78, 188–190,
        197
      quality of life and, 110–113
  childhood onset of, **187–201**
      diagnosis of, glucose metabolism
        and, 189–191
        IGF-I as marker of,
          116–117
        re-establishing, 191–192
      overview of, 187, 198
      physiology of, 187–188
      replacement therapy for, 191–198
        discontinuation at final
          height, 188–191
        dosing of, 193
  definition of, 134–135
  diagnosis of, 101–105
      assay design for, 103–104
      growth hormone assay
        heterogeneity, 101–102
      growth hormone binding protein
        impact on, 104
      growth hormone molecules
        heterogeneity, 102
      measurement overview of, 101,
        105
      standard assay preparations and
        units, 104–105
  features of, 234–235
  genetics of, **17–36**
      cellular analysis of, 31–32
      classification based on, 21–22
      familial isolated types in, 22–27
      isolated type II, as evolving
        pituitary deficit, 28–30
        clinical impact of, 29, 32
        gene mutations in, 23, 26–27
        in transgenic mice, 27–28
      overview of, 17–19
      splice site mutation in, 23, 31

      endocrine data on, 29–30
      transcription factors in, 19–21
      variable phenotypes in, 31
  glucose/fat metabolism effects of,
    77–79, 83
  IGF-I as diagnostic marker of,
    acid-labile subunit and, 115–116
    efficacy of, 115–116
    free concentration and, 119–120
    in adults, 116–117
  IGFBP-3 as diagnostic marker of, 117
  in the elderly, 234
      functional impairment with, 233,
        235–237, 240
      growth hormone for, 237–239
        potential risks of, 239–240
      growth hormone secretagogues
        for, 240
      growth hormone therapy for,
        233–245. See also *Growth
        hormone replacement
        therapy.*
  isolated, familial, 22–27
      genetic classification of, 21–22
  quality of life in, **221–232**
      acromegaly and, 222, 225–229
      adult patients and, 222–225
      as health versus disease indicator,
        221, 229
      questionnaires for, 221–222
      therapeutic implications of,
        210–211
  tests for, 191–192

Growth hormone excess. See also
    Acromegaly.
  diagnosis of, 101–105
      assay design, 103–104
      growth hormone assay
        heterogeneity, 101–102
      growth hormone binding protein
        impact on, 104
      growth hormone molecules
        heterogeneity, 102
      measurement overview, 101, 105
      standard assay preparations and
        units, 104–105

Growth hormone insensitivity syndrome, 5
  genetics of, 18
  growth hormone receptor mutations
    in, 6–8
  immunodeficiency and, 8
  in chronic renal insufficiency, 137

Growth hormone receptor (GHR), deficit
  of, in idiopathic short stature of SGA
  children, 150, 153–154
  growth hormone for non-GHD
    growth disorders and, 135–136
  in growth, **1–16**

mutation-related disorders of, 5–8
pharmacogenomics of, 9–13
polymorphisms of, 8–9
signaling cascades of, 3–5
signaling overview, 1–3
structure-function representation of, 2–3
in growth hormone deficiency, 18
signaling in muscle and fat, 79–81

Growth hormone receptor signaling, growth and, 3–5
overview of, 1–3

Growth hormone replacement therapy, cancer risk and, **247–263**
brain tumors in, 251–252
recurrence of, 255–256
clinical benefits versus, 247
colorectal tumors in, 251–253
de novo malignancies in, 250–253
epidemiologic observations of, 249–250
evidence for, 247
growth hormone—insulin-like growth factor axis in, 248–249
Hodgkin lymphoma in, 251–253
in vitro experiments on, 248
in vivo animal studies on, 248–249
leukemia/lymphoma in, 251–252
recurrence of, 253–255
management implications of, 257
renal tumors in, 251, 253
second malignant neoplasms in, 256–257
tumor recurrences in, 253–256
during puberty to adulthood transition, 191–198
deficiency diagnosis of, 77–79
re-establishing, 191–192
discontinuation at final height, 188–191
dosing of, 193
effect on body and fat mass, 193–194
effect on bone, 195–196
effect on glucose, 197
effect on lipids, 196
quality of life aspects, 197–198
for adults, **203–220**
baseline assessment for, 205–206
contraindications to, 215
dose titration for, 208–210
during pregnancy, 215–216
estrogen replacement therapy and, 205–206
glucose metabolism with, 211
goal of, 205
HPA axis evaluation for, 205
initial dose monitoring of, 206–208
initial dose selection for, 206
lipid metabolism with, 211
long-term monitoring and safety of, 211–215
overview of, 203, 216
quality of life and well-being with, 210–211
side effects of, 208–209
for non—growth hormone-deficient growth disorders, **131–186**
consensus statement highlights on, 167–168
definitions for, 133–135
mechanisms of, 135–167
overview of, 131–133, 168–169
products available for, 131–132
genomics of, 9–13
glucose/fat metabolism effects of, 75–85
IGF-I role in, growth response prediction, 117–118
monitoring as, 118
in the elderly, **233–245**
body composition changes and, 234–235
clinical indications for, 234
effects of, 237–239
functional indications for, 233, 235–237
muscle mass loss implications, 235–237
overview of, 233, 240–241
potential risks of, 239–240
secretagogues for, 240
quality of life benefits from, 223–225
for adults, 210–211
for children, 197–198
therapeutic implications of, 197–198, 210–211

Growth Hormone Research Society, on growth hormone for non-GHD growth disorders, 167

Growth hormone secretagogue receptor (GHSR), in idiopathic short stature, 160

Growth hormone secretagogues (GHSs), therapy potential in the elderly, 240

Growth hormone—releasing hormone (GH-RH), growth hormone secretion regulation by, 37–40
gonadal steroids modulation of, 57–58

Growth hormone (*continued*)
  in IGF-I deficiency versus growth hormone deficiency, 17–18, 204–205
Growth plate, deficit of, in idiopathic short stature of SGA children, 150, 156
  GH-IGF axis of, 135–136
    in non—GHD growth disorders, 149, 159–161
Growth velocity, 133

H

Haploinsufficiency, in short stature homeobox-containing *(SHOX)* gene deficiency, 146–147
Haplotypes, in growth hormone receptor polymorphisms, 8
Head trauma, growth hormone deficiency related to, 204
Health problems, patient perception of, with growth hormone deficiency, 221–223
Health-related quality of life (HRQoL), 221–222
  acromegaly and, 225–229
  growth hormone deficiency and, 222–225
Hearing deficits, growth hormone therapy risk for, 145
Height gain models, of growth hormone treatment, for non—GHD growth disorders, 168
Height growth, during puberty, growth hormone replacement therapy and, 188–191
    physiology of, 187–188
Height velocity, 133
  in growth failure, 133
Hemoglobin A_{1c}, growth hormone impact on, 78
HESX1 gene, in growth hormone deficiency, 20, 28
Heterogeneity, of growth hormone assays, 101–102
  of growth hormone molecules, 102
Histidine, in growth hormone receptor mutations, 6
Hodgkin lymphoma, growth hormone treatment related to, 251–253

Homodimerization, in growth hormone receptor mutations, 6
Hormone replacement therapy, growth hormone deficiency and, 58, 65–66
Hormone-sensitive lipase (HSL), in glucose/fat metabolism, growth hormone mediation of, 82–83
Human growth, disorders of, growth hormone receptor mutations in, 5–8
  non—growth hormone-deficient growth hormone treatment of, **131–186**
    consensus statement highlights on, 167–168
    definitions for, 133–135
    mechanisms of, 135–167
    overview of, 131–133, 168–169
    products available for, 131–132
  growth hormone effects on, gonadal steroids regulation of, 63–64
  growth hormone receptor in, **1–16**
    mutation-related disorders of, 5–8
    pharmacogenomics of, 9–13
    polymorphisms of, 8–9
    signaling cascades of, 3–5
    signaling overview, 1–3
    structure-function representation of, 2–3
  pharmacogenomics of, 9–13
Human growth hormone. See *Growth hormone (GH)*.
11β-Hydroxysteroid dehydrogenase, in growth hormone status assessment, 205
Hyperglycemia, growth hormone secretion regulation and, 48
Hypertension, benign intracranial, growth hormone therapy and, 166, 215
Hypoglycemia, fasting, in growth hormone-deficient children, 77–78
  insulin-induced, growth hormone release and, 60
Hypogonadotropic hypogonadism, 58
Hypopituitary men, growth hormone and testosterone replacement therapy for, 66–68

Hypothalamus, deficit of, in growth
hormone deficiency, 18,
203–206
growth hormone secretion regulation
role, 37–38, 47
semiquantification of, 39
Hypothalamus-pituitary-adrenal axis
(HPA), in growth hormone deficiency,
18, 205–206
therapeutic surveillance
indications for, 214–215

I

Idiopathic short stature (ISS), 158–167
diagnostic criteria for, 230
etiology of, 204, 231–233
genetics of, 159–161
growth hormone treatment of,
consensus statements on,
167–168
efficacy of, 162–165
mechanism of, 9–13, 135–136
overview of, 132, 158–159
safety of, 165–167
growth pattern in, 231
phenotypes of, in SGA children, 150,
152–156
prevalence of, 231
untreated adult height in, 231

IGF-I gene, 112, 150

IGF-II gene, 150

IGF-IR gene, 150

Immunoassays, of growth hormone, design
of, 103–104
growth hormone binding protein
impact on, 104
heterogeneity of, 101–102
standard assay preparations and
units, 104–105

In vitro studies, of growth hormone effects,
on protein metabolism, 90–91
of growth hormone—insulin-like
growth factor axis in
tumorigenesis, 248–249
of isolated growth hormone deficiency
type II, 27–28

In vivo studies, of growth
hormone—insulin-like growth factor
axis in tumorigenesis, 248–249
of isolated growth hormone deficiency
type II, 27–28
of protein metabolism, 90, 95

Indirect calorimetry, of glucose/fat
metabolism, 76–77

Insulin, growth hormone therapy effects on,
197
protein metabolism role of, 95

Insulin-like growth factor (IGF)-I,
age-dependent decline in, 236–237
growth hormone therapy effects
on, 197, 238
deficiency of, genetics of, 17–18
in idiopathic short stature of
SGA children, 150, 155
fasting effects on, 84
free concentration of, growth hormone
deficiency diagnosis and, 119–120
IGFBP-3 changes impact on,
118–119
growth hormone effects on, 90–92, 94
action modulation as, 60
in non—GHD growth disorders,
135–136, 139, 150, 160–161,
165, 168–169
secretion regulation as, 37, 43–44,
113–114
synthesis regulation as, 112–113
tumorigenesis and tumor
progression with, 248–249
growth hormone receptor regulation
by, 3–5
growth hormone secretagogues impact
on, 240
in growth hormone status assessment,
109–129. See also *Insulin-like
growth factor system markers.*
in growth hormone therapy
assessment, 204, 206–210,
214–216
in quality of life assessments, 225, 227,
229
in sarcopenia, 236
mitogenic and anti-apoptotic activity
of, 247
peak height attainment and, 188–189,
192–193
protein metabolism role of, 95
secretion regulation of, by growth
hormone, 37, 43–44, 113–114
IGFBPs changes and, 118–119
variables of, 109–112
synthesis of, growth hormone
regulation of, 112–113

Insulin-like growth factor binding protein
(IGFBPs), IGF-I regulation by,
111–112, 114–115
free concentration changes and,
118–119
in non—GHD growth disorders
response, to growth hormone
treatment, 136, 139, 168–169
in tumorigenesis, 248–250

Insulin-like growth factor binding protein-3 (IGFBP-3), acromegaly and, 120–121
   efficacy in growth hormone deficiency diagnosis, 115–117
   IGF-I concentration regulation by, 111–112, 114–115
      free concentration changes, 118–119
   in growth hormone therapy assessment, 214
   in tumorigenesis, 248–250

Insulin-like growth factor system markers, in growth hormone status assessment, **109–129**
      acid-labile subunit in, efficacy in growth hormone deficiency diagnosis, 115–116
         growth hormone therapy and, 207
         idiopathic short stature and, 161
         regulation of, 110–111, 114–115
      acromegaly and, 120–122
      free concentration of, growth hormone deficiency diagnosis and, 119–120
         IGFBP-3 changes impact on, 118–119
      growth hormone deficiency diagnosis, 204
         efficacy of, 115–116
         free IGF-I and, 119–120
         in adults, 116–117
      growth hormone therapy in, growth response prediction after, 117–118
         monitoring role, 118, 206–207
      IGFBP-3 in, efficacy in growth hormone deficiency diagnosis, 115–117
         free IGF-I concentration changes resulting from, 118–119
         regulation of, 114–115
      regulation of IGF-I concentration, by growth hormone, 113–114
         IGFBPs changes and, 118–119
         variables of, 109–112
      synthesis of IGF-I, growth hormone regulation of, 112–113

Insulin receptor substrate 1 (IRS1)—associated phosphoinositol-3 (PI3) kinase, 81, 248

Insulin resistance, growth hormone deficiency and, 234
   measures of, 197
   with growth hormone treatment, in children, 140, 145, 151, 157–158, 166–167

Insulin sensitivity, growth hormone therapy impact on, 76, 78–79, 83–84, 239

Insulin signaling, growth hormone crosstalk between, 79–81

Insulin tolerance test (ITT), for growth hormone deficiency, 192, 204–205

International reference preparations (IRP), for growth hormone, 104

International units, in growth hormone assays, 104–105

Intima-media thickness (IMT), growth hormone therapy effects on, 196–197

Intracranial hypertension, benign, growth hormone therapy and, 166, 215

IRP 22-kD preparation, of recombinant human growth hormone, 104

IRP 88/624 preparation, of recombinant human growth hormone, 104

Irradiation, cranial, growth hormone deficiency related to, 204–205

Isolated growth hormone deficiency, familial types of, 22–27
   type IA, 22–23
   type IB, 23–26
   type II, as evolving pituitary deficit, 28–30
      clinical impact of, 29, 32
      gene mutations in, 23, 26–27
      in transgenic mice, 27–28
   type III, 23, 27

Isolated growth hormone deficiency type II (IGHD II), as evolving pituitary deficit, 28–30
   diagnosis of, 29
   endocrine parameters of, 29–30
   general aspects of, 28
   gland size and, 30
   hormonal data missense mutations in, 30
   hormonal data splice-site mutation in, 29–30
   variable clinical course of, 29
   clinical impact of, 29, 32
   gene mutations in, 23, 26–27
      variable clinical course depending on, 29
   in transgenic mice, 27–28

J

Jacob-Creutzfelt disease, 131

Janus tyrosine kinase (Jak2), growth
    hormone receptor signaling and, 2–4
    cellular mechanisms of, 80
    mutation of, 6–7
    in growth hormone status assessment,
        113

K

K^+ ion channels, in sarcopenia, 236

Kabi International Growth Study (KIGS),
    166, 253

Kinases, growth hormone receptor signaling
    and, 1–4
    cellular mechanisms of, 79–81
    protein metabolism and, 90

Knock-in models, of growth hormone
    receptor signaling cascade, 4

Knock-out models, of growth hormone
    receptor signaling cascade, 3–5
    mutations of, 7
    of growth hormone therapy effects,
        239–240

Kyphosis, growth hormone therapy risk for,
    145–146

L

Langerhans histiocytosis, growth hormone
    treatment related to, 256

Laron mice, 3

Laron syndrome, 5–6

Lawson Wilkins Pediatric Endocrine
    Society, on growth hormone for
    non-GHD growth disorders, 167–168

Leukemia, growth hormone treatment
    related to, 251–252
    recurrences of, 253–255

LHX3 gene, in growth hormone deficiency,
    20

LHX4 gene, in growth hormone deficiency,
    20

Life events, impact on quality of life
    perception, 223

Life expectancy, growth hormone deficiency
    and, 233, 240

Ligand-binding properties, of growth
    hormone receptor polymorphisms, 9

Linear growth, failure of, 133
    growth hormone treatment of,
        135–136

Lipids. See also *Free fatty acids (FFAs)*.
    growth hormone secretion regulation
        and, 48–49
    growth hormone therapy effects on,
        196
        during fasting, 79
        short- versus long-term, 211
    oxidation of. See *Lipolysis*.
    protein metabolism role of, 93–94

Lipolysis, growth hormone therapy impact
    on, 76, 81–83

Longitudinal growth, during puberty,
    187–188

Lymphomas, growth hormone treatment
    related to, 251–252
    Hodgkin's, 251–253
    recurrences of, 253–255

M

MAP kinase, growth hormone receptor
    signaling and, 2–4
    mutation of, 7

Mass units, in growth hormone assays, 104

Membrane proximal domain, of growth
    hormone receptor, 2

Menopause, growth hormone concentration
    during, 58–59
    substrate metabolism and,
        65–66

Menstrual cycle, growth hormone
    concentration and, 58, 60

Metabolism. See *Substrate metabolism;
    specific substrate*.

Mice models, of growth hormone effects, on
    tumorigenesis, 248–249
    of growth hormone receptor signaling
        cascade, 3–5
        mutations in, 7
    of growth hormone secretion
        regulation, 42
    transgenic, of isolated growth
        hormone deficiency type II,
        27–28

Microdialysis technique, for glucose/fat
    metabolism detection, 81–82

Missense mutations, in isolated growth
    hormone deficiency type II,
    23, 30

Mitogenic activity, of growth hormone, 213, 247, 256
  of IGF-I, 247

Molecular mechanisms, of growth hormone deficiency, 31–32
  of growth hormone heterogeneity, 102
  of growth hormone therapy effect on glucose/fat metabolism, 79–82
    FFA role in, 75–77, 83–84

Mood, growth hormone deficiency impact on, 234

mRNA, growth hormone secretion regulation and, 37

mU/L, in growth hormone assays, 104

Muscle force, loss with aging, functional implications of, 236
    growth hormone therapy impact on, 238

Muscle mass, growth hormone and testosterone impact on, 238–239
  loss with aging, 234–235
    functional implications of, 235–237, 240
    sarcopenia impact on, 236–237

Muscle strength, growth hormone therapy effects on, 212–213
  loss with aging, growth hormone and testosterone impact on, 238
    sarcopenia impact on, 236–237

Muscle wasting, cancer-associated, 236, 239
  in the elderly, 234–235
    functional implications of, 235–237, 240

N

National Cooperative Growth Study (NCGS), 166, 252–253

Near-final height, 134

Negative feedback mechanism, of growth hormone secretion regulation, 43–45

Neoplasia, growth hormone therapy effects on, 166, 197–198
  malignant. See *Cancer(s)*.

Neuroimaging, in growth hormone therapy surveillance, 198–199

Neurosecretory modulation, of growth hormone, by gonadal steroids, 57–60

Nevi, benign cutaneous melanocytic, growth hormone therapy risk for, 146

Non—growth hormone-deficient growth disorders, growth hormone treatment of, **131–186**
  consensus statement highlights on, 167–168
  definitions for, 133–135
  mechanisms of, 135–167
  overview of, 131–133, 168–169
  products available for, 131–132

North American Pediatric Renal Transplant Cooperative Study (NAPRTCS), 137, 139

Notthingham Health Profile (NHP), 221–222, 227

NPR2 gene, in idiopathic short stature, 161

Nutrition, growth hormone secretion regulation and, 48–49
  idiopathic short stature related to, 161
  in growth hormone status assessment, 113

O

Obesity, growth hormone secretion regulation and, 49
  protein metabolism and, 92–93

Octreotide, growth hormone secretion regulation and, 37–39
    continuous infusion suppression of, 41–42

Oncogenic processes. See also *Cancer(s)*.
  growth hormone receptor polymorphisms in, 9

Oral glucose tolerance testing (OGTT), for growth hormone-deficient patients, 77–79

Otitis media, growth hormone therapy risk for, 145, 166

P

"Partial" growth hormone deficiency, 134

Pegvisomant, 77

Peripheral biologic effects, of growth hormone, gonadal steroids modulation of, 60–63

Pharmacogenomics, of human growth, 9–13

Phenotypes, in growth hormone deficiency, transcription factors of, 19–21
    variable, 31
  of idiopathic short stature, in SGA children, 150, 152–156

Phenylalanine, growth hormone effects on, 91–93

Phosphatidylinositol-3 (PI-3) kinase, growth hormone receptor and, 2

Phosphorylation, protein metabolism and, 90, 94

Physical activity, decreased, sarcopenia associated with, 236–237

PIT1 transcription factor, in growth hormone deficiency, 19–21

Pituitary gland, deficit of, in growth hormone deficiency, 18, 28–30, 203–205
    in idiopathic short stature of SGA children, 150, 152
    in males, 66–68
    glucose metabolism role, 75
    growth hormone derived from, 131
    in growth hormone deficiency, 18, 203–205
        isolated type II, 28–30
            diagnosis of, 29
            endocrine parameters of, 29–30
            general aspects of, 28
            gland size and, 30
            hormonal data missense mutations in, 23, 30
            hormonal data splice-site mutation in, 29–30
            variable clinical course of, 29
        therapeutic surveillance indications for, 214–215

Polymorphisms, gene, IGF-I secretion regulation and, 112
    in IGFBP-3 regulation, 115
    of growth hormone receptor, 8–9

Post-growth hormone receptor defects, of growth hormone receptor, 7–8

Postnatal growth, in SGA children, 149–150

POU1F1 gene, in growth hormone deficiency, 19–21

Prader-Willi syndrome (PWS), 132

Pregnancy, growth hormone therapy during, 215–216

Prenatal growth, in SGA children, 149

PRL deficiency, growth hormone deficiency and, 20–21

Progestogens, growth hormone action modulation by, 62–63

Proline, in growth hormone receptor mutations, 7

PROP1 gene, in growth hormone deficiency, 20–21

Protein metabolism, growth hormone effects on, **89–100**
    animal studies of, 90
    during prolonged fasting, 91–92
    gonadal steroids modulation of, 65–67
    in vitro studies, 90–91
    mechanisms of, 93–95
    obesity and, 92–93
    overview of, 89–90, 95–96
    postabsorptive, 90–91
    sarcopenia associated with, 236

Protein synthesis, growth hormone and testosterone impact on, 238–239

Psychiatric morbidity, patient perception of, with growth hormone deficiency, 221–223

Psychologic well being. See *Quality of life (QoL)*.

Psychological General Well Being Schedule (PGWS), acromegaly and, 225–227
    growth hormone deficiency and, 221–223
    reliability of, 228

Puberty to adulthood transition, growth hormone and, **187–201**
    deficiency diagnosis of, glucose metabolism and, 77–79
    re-establishing, 191–192
    overview of, 187, 198
    physiology of, 187–188
    replacement therapy during, 191–198
        discontinuation at final height, 188–191
        dosing of, 193
        effect on body and fat mass, 193–194
        effect on bone, 195–196
        effect on glucose, 197
        effect on lipids, 196
        quality of life aspects, 197–198
    sex steroids modulation of, 58. See also *Gonadal steroids*.

Q

Quality of life (QoL), in growth hormone deficiency, **221–232**
    acromegaly and, 222, 225–229

adult patients and, 203, 222–225
as health versus disease indicator, 221, 229
effects and measurement of, 234, 240–241
growth hormone therapy for, 203, 206, 210–211, 237–238. See also *Growth hormone replacement therapy.*
questionnaires for, 221–222
therapeutic implications of, 197–198, 210–211

Questions on Life Satisfaction-Hypopituitarism (QLS-H), 222–224

R

Radioimmunoassays (RIAs), of growth hormone, 103

Radiolabeled glucose infusion, for glucose/fat metabolism detection, 76

Randle cycle, of glucose-fatty acid metabolism, 76–77

Rat models, of growth hormone effects, on protein metabolism, 90
on tumorigenesis, 248
of growth hormone secretion regulation, 38, 40–41
gonadal steroids and, 47–48

Receptor shedding, 3

Recombinant human growth hormone (rhGH), for growth hormone deficiency, 131, 203. See also *Growth hormone replacement therapy.*
quality of life benefits from, 223
for non—growth hormone deficient growth disorders, 131–132, 150, 162
IRP 22-kD preparation of, 104
IRP 88/624 preparation of, 104

Reference preparations, for growth hormone assays, 104–105

Renal transplantation, growth hormone effects on, 139–140

Renal tumors, growth hormone treatment related to, 251, 253

Rhabdomyosarcoma, growth hormone treatment related to, 256

mRNA, growth hormone secretion regulation and, 37

S

Sarcopenia, in the elderly, 233, 236
etiology of, 236
functional impairment with, 236–237

Scoliosis, growth hormone therapy risk for, 145–146

Second malignant neoplasms (SMNs), growth hormone treatment related to, 256–257

Selective estrogen receptor modulators, of growth hormone action, 63

Sex drive, growth hormone therapy effects on, 224–226

Sex steroids. See *Gonadal steroids.*

Sexual dimorphism, in growth hormone secretion regulation, 40–41
gonadal steroids and, 47–48

SF-36 Questionnaire, on QoL, 221–222, 225, 227

"Short-feedback," in growth hormone secretion regulation, 43

Short stature, criteria for, 133
genetics of, 22
growth hormone effects on, 135–136
idiopathic, 204. See also *Idiopathic short stature (ISS).*
persistent, in SGA children. See *Small for gestational age (SGA).*

Short stature homeobox-containing *(SHOX)* gene deficiency, 146–148
etiology of growth disturbance in, 147
genetics of, 146
growth hormone treatment of, efficacy of, 147–148
mechanism of, 135–136
overview of, 132
safety of, 148
growth pattern in, 146–147
prevalence of, 146
Turner syndrome and, 141, 146, 148
untreated adult height in, 146–147

SHOX gene deficiency, short stature homeobox-containing, etiology of growth disturbance in, 147
genetics of, 146
growth hormone treatment of, efficacy of, 147–148
mechanism of, 135–136
overview of, 132
safety of, 148
growth pattern in, 146–147
prevalence of, 146

SHOX (continued)
    Turner syndrome and, 141, 146, 148
    untreated adult height in, 146–147
Signal transducers, of growth hormone receptor, 3
Signaling cascades, of growth hormone receptor, growth and, 3–5
    mutations of, 6–7
    overview of, 1–3
Skeletal muscle mass, loss with aging, 234–235
    functional implications of, 235–237, 240
    sarcopenia impact on, 236–237
    sex steroids impact on, 238
Skeletal muscle mass index (SMI), 237
Sleep, growth hormone secretion regulation and, 45–47
Slipped capital femoral epiphysis, growth hormone therapy risk for, 146, 166
Small for gestational age (SGA), definition of, 148–149
    persistent short stature and, etiology of, 149–150
    growth hormone treatment of, 148–158
        efficacy of, 150–151, 157
        mechanism of, 9–13, 135–136
        overview of, 132
        safety of, 151, 157–158
    growth pattern in, 149
    phenotypes of, 150, 152–156
    untreated adult height in, 149
SOCS domain, of growth hormone receptor signaling, 3–4
Sodium retention, growth hormone regulation of, gonadal steroids modulation of, 67
Somatomedin hypothesis, 3–5
Somatostatin, growth hormone secretion regulation by, 40–42
    gonadal steroids modulation of, 57–58
Somatotropic axis, growth hormone secretion regulation and, 47–48
    nutritional influences, 49
Somatotropin release—inhibiting factor (SRIF), growth hormone secretion regulation and, 37–39
    animal models of, 40–42

SOX2 gene, in growth hormone deficiency, 20
SOX3 gene, in growth hormone deficiency, 20
Splice site mutation, in growth hormone deficiency, 23, 31
    endocrine data on, 29–30
    in isolated growth hormone deficiency type II, 29–30
Stat 3 pathway, of growth hormone receptor signaling, mutation of, 7
Stat 5 pathway, in growth hormone status assessment, 113
    of growth hormone receptor signaling, 3–4
        cellular mechanisms of, 80
        mutation of, 7
Stat SH2 domains, of growth hormone receptor, 3
Strength. See Muscle strength.
Stress, growth hormone resistance related to, 113
Substrate metabolism. See also specific substrate.
    growth hormone deficiency impact on, 234, 239
    growth hormone effects on, gonadal steroids modulation of, 65–66
    growth hormone therapy impact on, 76, 79, 83–84
Suppressor of cytokine signaling-2 (SOCS2), in growth hormone status assessment, 113
    in tumorigenesis, 249

T

Target organs, in growth hormone deficiency, 18
Testosterone, for sarcopenia, 238–239
    growth hormone action regulation by, linear growth and, 64
        neurosecretory modulation, 58–59
        peripheral biologic effects modulation, 60–61
        substrate metabolism and, 66–67
    growth hormone secretion regulation by, 47–48
    IGF-I secretion regulation by, 112, 214
    IGFBP-3 regulation by, 114

Thyroid disease, in idiopathic short stature, 161, 166

Thyroid function, IGFBP-3 regulation by, 114–115
protein metabolism role of, 95

Thyroid stimulating hormone (TSH), deficiency of, growth hormone deficiency and, 19–21, 30–32

Thyroxine, acromegaly and, 120
IGF-I secretion regulation by, 112
protein metabolism role of, 95

Thyroxine replacement therapy, growth hormone and, 205

Transcription activators, of growth hormone receptor, 3

Transcription factors, in growth hormone deficiency, 18–21

Translation factors, of protein metabolism, 90

Triiodothyronine, protein metabolism role of, 95

"Trophic-feedback," in growth hormone secretion regulation, 43

Tumor progression, growth hormone—insulin-like growth factor axis in, 248–249

Tumorigenesis, growth hormone therapy associated with, 213–214
neuroimaging surveillance of, 214–215
growth hormone—insulin-like growth factor axis in, 248–249

Turner syndrome (TS), growth disturbance related to, 140–146
etiology of, 141
growth hormone treatment of,
efficacy of, 11–15
mechanism of, 135–136
overview of, 10, 132
safety of, 15–16
growth pattern in, 140–141
SHOX gene deficiency and, 141, 146, 148
untreated adult height in, 140–141

Type 2 diabetes, growth hormone therapy risk for, 140, 145, 151

Tyrosine phosphatases, growth hormone receptor regulation by, 3
cellular mechanisms of, 80–81

U

ug/L, in growth hormone assays, 104

Urea synthesis, growth hormone effects on, 90–92

V

Visual analog scale (EQ-VAS), for QoL, 222

VO_2max, growth hormone and testosterone impact on, 238–239

X

X-linked growth hormone deficiency, 22–23, 27

X-linked non—GHD growth disorders, 141, 146

Moving?

Make sure your subscription moves with you!

To notify us of your new address, find your **Clinics Account Number** (located on your mailing label above your name), and contact customer service at:

E-mail: elspcs@elsevier.com

800-654-2452 (subscribers in the U.S. & Canada)
407-345-4000 (subscribers outside of the U.S. & Canada)

Fax number: 407-363-9661

Elsevier Periodicals Customer Service
6277 Sea Harbor Drive
Orlando, FL 32887-4800

*To ensure uninterrupted delivery of your subscription, please notify us at least 4 weeks in advance of move.